T0255624

SMART ALGEBRA 1

If $ax^2 + bx + c = 0$, where a, b, and c are constants and a $\neq$ 0, then

$$x = \frac{-b \pm \sqrt{b^2 - 4ac}}{2a}$$

$$y = ax^2 + bx + c$$

$$\text{Vertex} : V\left(\frac{-b}{2a}, \frac{4ac - b^2}{4a}\right)$$

$$y = mx + b$$

$$m = \frac{y_2 - y_1}{x_2 - x_1}$$

$$y - y_1 = m(x - x_1)$$

$$y_2 - y_1 = m(x_2 - x_1)$$

TESFAYE LEMA BEDANE

PREFACE

This book, entitled Smart Algebra 1, is prepared as a reference to assist students who seek to understand the concept of Algebra 1. It covers 15 chapters. Each chapter begins with notes and brief explanations followed by helpful fully solved examples at the end of each chapter. There are practice problems to help students make sure of mastering each concept of Algebra 1 and improve their problem-solving skills. The last chapter is fully solved word problems, which explains how to perform mathematical operational word problems, like age, geometry and motion word problems. These will help students to develop logical thinking and mathematical reasoning. In addition to this, it helps the students the way how to practice different mathematical operations, such as addition, subtraction, multiplication and division and develop mathematical skills.

To realize the dreams of the author in a better way and test the fruits of this work, I highly recommend students to practice solving similar questions repeatedly. That is the only way to acquaint with basic mathematics and related disciplines. Besides, I strongly believe that this book will help interested students to understand the concept of Algebra 1 as it is presented in more simple, understandable and effective approach to solve the above-mentioned areas of study. As always, I welcome to hear your comments bearing in your mind that constructive comments are the most pleasurable, but critical comments are the most useful. With the help of Almighty, may we join in the near future having the next part of this work: Smart Algebra II!

Tesfaye Lema Bedane

ACKNOWLEDGEMENT

It gives me great pleasure to acknowledge and offer my warmest thanks to my caring, loving and supportive wife, Kidmalem Chernet, for her patience, encouragement and advice that made this book a reality. I would also like to extend my gratitude and appreciation to my friend, Abrham Terfa and my nephew, Solomon Heyi as they encouraged and inspired me through all the ways of preparing this book.

Also, I present a much appreciation to my best friend Engineer Berhanu Balcha, for his tremendous help in the reviewing and editing this book. I can't say enough for his wonderful intensive editing of this book. He has helped me more than I could have ever imagined.

Finally, I thank all of you who gave me constructive feedbacks for my first book; entitled, Basic Mathematics for Grade 9 Algebra and Geometry.

Tesfaye Lema Bedane

CONTENTS

Preface...v

Acknowledgement...vii

CHAPTER 1 FOUNDATION OF ALGEBRA...1

1.1.   Variable and Expression...1

1.2.   Writing Algebraic Expression in Words. 2

1.3.   Translating Words to Algebraic Expressions 3

        1.3.1.   Addition 3

        1.3.2.   Subtraction...3

        1.3.3.   Multiplication..4

        1.3.4.   Division...4

        Exercise 1.1-1.3.4...5

1.4.   Order of Operation and Evaluating Expression. 7

        1.4.1.   Order of Operations Step by Step...8

        1.4.2.   Order of Operation Rule in BODMAS..9

        Exercise 1.4 -1.4.2 11

1.5.   Real Numbers and Number Line. 11

        1.5.1.   Advance Note for Real Numbers...12

        1.5.2.   Different types of real numbers...12

1.6.   Real Number Line 13

        1.6.1.   Cartesian Coordinate System 13

        1.6.2.   Order of the Real Numbers..14

        1.6.3.   Graphing a Linear Inequality on a Number Line. 15

1.7.   Opposite Real Numbers 16

        Exercise: 1.5-1.7..17

1.8.   Properties of Real Numbers 19

        Exercise 1.8 21

1.9.   Adding and Subtracting Real Numbers 23

        1.9.1.   Addition on Number Line..23

        1.9.2.   Subtracting Real Numbers on the Real Number Line................24

        1.9.3.   Multiplying and Dividing Real Numbers 25

        1.9.4.   Division 26

        Exercise 1.9-1.9.4.1...27

1.10.   Distributive Property 29

1.11.   An Introduction to Equations..29

        1.11.1.   Classifying Equations..29

CHAPTER 2 SOLVING AND GRAPHING LINEAR EQUATIONS......................31
2.1. Definition of Linear Equation with One Variable..............................31
2.2. Solving Linear Equation with One Variable 31
2.3. Solving Equations Using Addition and Subtraction 31
   Exercise 2.1-2.3 33
2.4. Solving Equations Using Multiplication and Division.......................33
2.5. Solving an equation by dividing each side by the coefficient of an equation........34
   2.5.1. Solving an Equation by Multiplying Each Side of an Equation 34
   Exercise 2.4-2.5...36
2.6. Solving Two Step Equations 37
   2.6.1. Finding the Solution Set Using Distributive Property...................38
   Exercise 2.6-2.6.1 39
2.7. Finding Solution Set for Equations with Variables on Both Sides.........40
   Exercise 2.7 42
2.8. Graphing Linear Equations..42
   Exercise 2.8...47
2.9. Linear Equations 47
2.10. Slope (m), y-intercept and x-intercept of a Line..............................48
   Exercise 2.9-2.10 50
2.11. Y-intercept and X-intercept of a Line...51
   Exercise 2.11 53
2.12. Finding the Linear Equation of a Line 53
   2.12.1. Equation of a line from two given points (Slope-intercept form of a line)...............54
   2.12.2. Equation of a Line from a Point and Given Slope. (Point-Slope Form of a Line)............54
   Exercise 2.12-2.12.2 55
2.13. Graphing Linear Functions 56
   Exercise 2.13 62
2.14. Classification of Lines by Slope...62
2.15. Slope of Parallel and Perpendicular Lines 63
   2.15.1. Parallel Lines...63
   2.15.2. Perpendicular Lines..64
   Exercise 2.14-2.15.2...67

CHAPTER 3 RELATION AND FUNCTION...68
3.1. Relation 68
3.2. Function 69
3.3. Domain and Range of a Function 70
   3.3.1. Domain 70
   3.3.2. Range..70
3.4. Vertical Line Test for Functions..71
   Exercise 3.1-3.4 72

CHAPTER 4 SOLVING AND GRAPHING LINEAR AND ABSOLUTE INEQUALITIES 75

4.1.   Inequality 75

    4.1.1.   Property of Trichotomy 75

    4.1.2.   Transitive Property of Inequality 75

4.2.   Solving Linear Inequalities 76

    Exercise 4.1-4.2 79

4.3.   Graphing Linear Inequality on Number Line 79

4.4.   Compound Inequality 82

    Exercise 4.3-4.4 86

4.5.   Solving Absolute Value Equations 86

    Exercise 4.5 91

4.6.   Inequalities with Absolute Value 91

    Exercise 4.6 96

4.7.   Graphs of Absolute Value Function on Coordinate Plane 96

4.8.   Behaviors of Graphs of Parent Absolute Value Function: $f(x) = |x|$. 98

    Exercise 4.7-4.8 98

4.9.   The Graph of Linear Inequality in Two Variables 99

    4.9.1.   For Horizontal Line(s) $y \geq b$ or $y \leq b$ 108

    4.9.2.   For Vertical Line(s) $x \geq a$ or $x \leq a$: 109

    Exercise 4.9 - 4.9.2 112

CHAPTER 5 SIMULTANEOUS EQUATIONS 113

5.1.   Definition of Simultaneous Equation 113

5.2.   How to solve systems of linear equations 113

    5.2.1.   Solving Systems of Linear Equations by Elimination Method. 113

    Exercise 5.1-5.2.1 118

    5.2.2.   Solving Systems of Linear Equations Using Substitution Method 118

    5.2.3.   Solving Systems of Linear Equations by Graphing Method 125

5.3.   Types of Systems of Linear Equations 131

    Exercise 5.2.2-5.3 132

CHAPTER 6 EXPONENTS AND EXPONENTIAL FUNCTIONS 133

6.1.   Definition of Exponent and Exponential Functions 133

6.2.   Properties of Exponents 134

    6.2.1.   Multiplication Properties of Exponents 134

    Exercise 6.2.1 135

    6.2.2.   Division Properties of Exponents 136

    Exercise 6.2.2 144

6.3.   Graphs of Exponential Functions 148

6.4.   Basic Properties of Graphs of Exponential Functions 150

    Exercise 6.3-6.4 151

6.5.   Scientific Notation 151

6.6.   Converting Scientific Notation to Decimal Notation..................................153

         Exercise 6.5-6.6 153

CHAPTER 7   SETS...155

7.1.   Introduction to Sets 155

7.2.   Description of Sets 156

       7.2.1.   (1) Verbal Method 156

       7.2.2.   (2) Listing Method (The Roster Method) 156

       7.2.3.   (3) Set Building Method 156

7.3.   Finite and Infinite Set 157

       7.3.1.   Finite Set 157

       7.3.2.   Infinite Set 157

7.4.   Operation on Sets 157

       7.4.1.   Union of Sets. 157

       7.4.2.   Intersection of Sets 158

       7.4.3.   Difference of two sets...158

       7.4.4.   Cartesian Product (Cross Product) of a Set. 159

7.5.   Properties of Cartesian Product 160

       7.5.1.   Distributive Property over Set of Intersection.........................160

       7.5.2.   Distributive Property over Set of Union. 160

7.6.   Cartesian Product of three Sets 160

7.7.   Symmetric Difference between Two Sets 160

7.8.   Venn Diagram 161

7.9.   Universal Set...162

7.10.   Complement of a Set 164

       7.10.1. Some Properties of Complement Sets 165

7.11.   SUBSET..165

7.12.   Proper Subset 166

7.13.   Equal Sets and Equivalent Sets..167

       7.13.1. Equal sets...167

       7.13.2. Equivalent Sets 167

7.14.   Disjoint Sets 168

7.15.   Cardinality of a Set 168

7.16.   Ordered Pairs 169

7.17.   Power Set..170

         Exercise 7.1-7.17 179

CHAPTER 8   POLYNOMIAL FUNCTIONS...184

8.1.   Definition of a Polynomial Function..184

8.2.   Operation on Polynomial Functions...187

       8.2.1.   Commutative Law for Addition 187

       8.2.2.   Commutative Law for Multiplication 187

       8.2.3.   Associative Property of Addition. 187

       8.2.4.   Associative Property of Multiplication......................................187

8.2.5. Distributive Property 187

8.3. Addition of Polynomial Functions 188

8.4. Subtraction of Polynomial Functions... 189

8.5. Multiplication of Polynomial Functions 190

8.6. Division of Polynomial Functions.. 191

    8.6.1. The Division Algorithm 191

8.7. The Remainder Theorem 194

8.8. The Factor Theorem... 195

    8.8.1. The Remainder and The factor Theorem for the divisor (ax+b)............. 196

8.9. Special Formulas.. 201

8.10. Zero of Polynomial Function 203

    Exercise 8.1 - 8.10 206

CHAPTER 9 REVIEW OF ALGORITHM FOR SIMPLIFYING AND SOLVING EQUATIONS 209

9.1. Rational Expressions..209

9.2. Reduction of a Rational Expression 210

9.3. Complex Rational Expressions 214

9.4. Evaluating Rational Expression 217

9.5. Solving Rational Equations..222

    Exercise 9.1 - 9.5 229

CHAPTER 10 EVALUATING RADICAL EXPRESSION 232

10.1. Radical Expression... 232

10.2. Simplifying Radical Terms 235

10.3. Rational Exponents 239

    Exercise 10.1-10.3.. 252

CHAPTER 11 QUADRATIC EQUATIONS AND FUNCTIONS............................256

11.1. Quadratic Equation 256

11.2. 11.2 Solving Quadratic Equations... 256

11.3. Properties of Quadratic Equations... 256

11.4. How to Solve Quadratic Equations by Factoring Method. 256

    Exercise 11.1-11.4 259

11.5. Solving Quadratic Equations by Factoring 259

    Exercise 11.5 262

11.6. Solving Quadratic Equation by the Quadratic Formula 262

11.7. Discriminant of Quadratic Equation.. 265

    Exercise 11.6-11.7.. 266

11.8. Solving Quadratic Equation by Square Root Property................. 266

    Exercise 11.8 269

11.9. Solving a Quadratic Equation by Completing the Square Method..................... 269

    11.9.1. Completing the Square Method..269

    Exercise 11.9 - 11.9.1 274

11.10.   Sum and Product of Roots (or zero) of Quadratic Equation: 275
          11.10.1. Sum of the Roots 275
          11.10.2. Product of the Roots 276
          Exercise 11.10-11.10.2 284
11.11.   Graphing Quadratic Functions 284
          Exercise 11.11 293
11.12.   Important Features of a Parabola 293
          Exercise 11.12 298
11.13.   Maximum and Minimum Values of a Quadratic Function 298
          Exercise 11.13 301
11.14.   Application of Quadratic Functions 301
          Exercise 11.14 309
11.15.   Quadratic Inequality 310
11.16.   Steps to Solve Quadratic Inequality Using Sign Chart Method. 310
          Exercise 11.15-11.16 315
11.17.   Finding Solution Set of Quadratic Inequalities Using Graphs 316
          Exercise 11.17 326

CHAPTER 12 RATIO, PROPORTION, RATES AND PERCENTAGE 327
12.1.   Ratio 327
12.2.   Proportion 336
          12.2.1. Direct and Inverse Proportionality. 336
          12.2.2. Properties of Proportion 336
12.3.   Rates 338
          12.3.1. Calculating the Unit Rate 339
12.4.   Percentage 341
          12.4.1. Notation of Percent 341
          Exercise 12.1 - 12.1.4 347

CHAPTER 13 GRAPH OF SQUARE ROOT FUNCTION 351
13.1.   Square Root Function 351
13.2.   Graphing Square Root Function 351
13.3.   Transformation of the Square Root Function. 352
13.4.   Shift 352
          13.4.1. Horizontal Shift 352
          13.4.2. Horizontal Shift to the Left 353
          13.4.3. Horizontal Shift to the Right 356
13.5.   Stretching and Shrinking 359
13.6.   Vertical Stretch and Shrink Graphs 359
13.7.   Horizontal Stretching and Shrinking Graph of Square Root Functions. 363
13.8.   Reflection 366
          13.8.1. Reflection About the x-Axis 367
          13.8.2. Reflection About the y-axis. 368
          Exercise 13.1-13.8.2 371

13.9.   Domain and Range of Square Root Function 372
      13.9.1. Domain of Square Root Function.. 372
      13.9.2. Range of Square Root Function... 372
      Exercise 13.9-13.9.2... 373

CHAPTER 14 STATISTICS AND PROBABILITY.. 375
14.1.   Statistics 375
      14.1.1. Data 375
      14.1.2. Raw Data 375
      14.1.3. Qualitative Data 375
      14.1.4. Quantitative Data... 375
      14.1.5. Variable 375
14.2.   Frequency Distribution Table... 375
14.3.   Measures of Location and Dispersion for Grouped Data. 376
14.4.   Summation Notation 376
14.5.   Mean... 377
      14.5.1. Properties of Mean 380
14.6.   The Median (Md) 381
14.7.   The Mode 386
14.8.   Measure of Dispersion 387
14.9.   Range 387
14.10.   Standard Deviation... 388
14.11.   Variance 389
      14.11.1. 389
      Exercise 14.1-14.11.1 396
14.12.   Probability.. 398
14.13.   Probability of an Event (E).. 399
14.14.   Types of Events 402
      14.14.1. Simple Event (or Single Event) 402
      14.14.2. Compound Event 403
14.15.   Occurrence or Non-occurrence of an Event... 403
14.16.   Complement of an Event 403
      14.16.1. Complement Rule 403
14.17.   Algebra of Events... 404
      14.17.1. Exhaustive Events... 404
      14.17.2. Mutually Exclusive Events 404
14.18.   Exhaustive and Mutually Exclusive Events. 405
      14.18.1. Exhaustive Event 405
      14.18.2. Mutually Exclusive Events 405
14.19.   Independent Events 405
14.20.   Dependent Events 406
      Exercise 14.12-14.20... 406

CHAPTER 15 SOLVED WORD PROBLEMS 407

    15.1.   Solved Algebra Word Problems 407

           15.1.1. Addition 407

           15.1.2. Subtraction 407

           15.1.3. Multiplication 407

           15.1.4. Division 408

    15.2.   Solved Age Word Problems 419

    15.3.   Solved Geometry Word Problems 426

    15.4.   Solved Motion Word Problems 435

           Exercise 15.1-15.4 443

ANSWER TO EXERCISES 445

CHAPTER 1

FOUNDATION OF ALGEBRA

1.1. Variable and Expression

Before we start this lesson, we need to know some mathematical terms and how to translate algebraic expression into words. In order to do this, we have to recognize what words mean and what operations are.

Note: - It is helpful to know the following vocabularies to make sure that this chapter has to be clearly understood and properly well done.

Vocabularies

1. Variable: A letter or symbol that stands for a number.
   Examples: x, y, a, b

2. Equation: A mathematical sentence that have an equal sign.
   Examples: $2x + 1 = 0$, $x^2 + 3x + 2 = 0$

3. Quantity: The concept that something has a magnitude and can be represented in algebraic expressions by a constant or a variable.
   Example: How much oranges are in a barrel?

4. Constant: A fixed and well-defined number or mathematical object.
   Example: $y = 2x + 5$, Here 5 is a constant.

5. Solution Set: A set of values that make the open sentence true.
   Example: $x + 5 = 11$, In this open sentence, when the value of x is 6, this equation becomes true. Hence, $x = 6$ is the solution set of this equation.

Note: Sometimes, the solution set of a given equation is also called the true set of that equation.

6. Numerical Expression: An expression that contains numbers and one or more operations.
   Examples: $10 + 6$, $75 + 4 - 2$

7. Open Sentence: - A mathematical statement that contains one or more variables that might be true or false depending upon the value that is substituted for that variable.

Examples:
a) A rectangle has n sides.

Here, n is a variable. This open sentence can be either true or false depending up on the value of n.

Thus, if n = 4, the open sentence is true, and

If n ≠ 4, the open sentence is false.
b) In equation 2y = 3x – 4, here x and y are variables, this open sentence can be either true or false depending up on the values of x and y.

8. Closed sentence: A mathematical statement or an equation that is always true or always false.
Examples:
1) 3 is an odd number.
2) A triangle has three sides.
3) –1 is less than 0.

In the above three examples, all of them are closed and always true sentences.
Examples:
4) 8 is an odd number.
5) A rectangle has five sides.
6) –2 is greater than 6.

In contrast, the above three examples are all closed and always false sentences.

9. Algebraic Expression: - An expression that contains variables, constants and (or) operations.
10. Operation: It is a duty of performing + (Addition), – (Subtraction), × (Times or Multiplication), and ÷ (Division).

Note: As a student, you need to know or understand the above mathematical symbols and how they could be operated.

1.2. Writing Algebraic Expression in Words.

Here are some examples that shows the way how to write each algebraic expression to words.
Examples:
a) x + 5
   •   Sum of x and 5
   •   x increased by 5.
   •   x plus 5.
b) y + 9
   •   Sum of y and 9.
   •   y increased by 9
   •   y plus 9
c) x – 6
   •   Difference of x and 6.

- x decreased by 6.
- x minus 6.

d) y − 8
- Difference of y and 8.
- y decreased by 8.
- y minus 8.

e) 5d
- Product of 5 and d.
- 5 times d.
- d increased 5 times.

f) m ÷ 7
- quotient of m and 7.
- m divided by 7.

1.3. Translating Words to Algebraic Expressions

Here below are some examples that shows the way how to translate words to algebraic expression.
Examples:

1.3.1. Addition

- The sum of eight and the number, 8 + x
- Six more than the number, n + 6
- The number plus 7, n + 7
- The number increased by 10, n + 10

1.3.2. Subtraction

- The difference of four and a number, 4 − n
- Nine less than a number; x − 9
- Ten minus a number, 10 − y
- A number decreased by two, m − 2

Note: - When we translate words to algebraic expression, the order is very important for subtraction and division but not for addition and multiplication.

Examples:
- Six less than a number,
  Representing these words in:
  n − 6 is correct expression, but 6 − n is not. Likewise,
- For expression, Nine less than a number,
  n − 9 is correct representation, whereas 9 − n is not.

Note: - The same is true for division, in case of addition and multiplication the order does not matter.

Examples:
- The sum of 3 and a number.
  This can be translated either as (n + 3) or (3 + n).
- The sum of 5 and two times a number can be translated either as (5 + 2x) or (2x + 5).

1.3.3. Multiplication

- The product of eight and a number can be written as: 8n or 8(n) or 8×n or 8*n
- Five times a number, can be put as: 5m, or 5(m), or 5× m or 5*m, Similarly,
- A number multiplied by seven can be: 7y, or 7(y), or $7 \times y$ or 7*y

1.3.4. Division

- Ten divided by a number can be expressed as: $10 \div x \text{ or } \frac{10}{x}$
- The quotient of a number and 5 can be: $y \div 5 \text{ or } \frac{y}{5}$

Note: - In the operation of division, the divisor must be different from zero.

Thus, the expression $\frac{x}{y}$ has a meaning whenever the value of $y \neq 0$. In other words, to have a meaningful expression, a number cannot be divided by zero.
- We can use two or more operations at once.

Examples:
Translate the phrase into algebraic expression.

a) Nine more than six times a number.
   6x + 9

b) Two less than eight times a number n.
   8n – 2

c) Seven less than the sum of four and a number y.
   (y + 4) – 7

d) Eight less than the quotient of ten and a number m.
   $\frac{10}{m} - 8$

e) The quotient of nine and the difference of 15 and a number y.
   $\frac{9}{15-y}$

f) The quotient of six and the sum of four and a number x.
   $\frac{6}{4+x}$

Exercise 1.1-1.3.4

(i) Fill in each blank space below with the most appropriate word or phrase.
   a) _____ is a letter or symbol that stands for a number.
   b) _____ is a mathematical sentence that has an equal sign.
   c) _____ is a concept that something has a magnitude and can be represented in algebraic expression by a constant or a variable
   d) _____ is a fixed and well-defined number or mathematical object.
   e) _____ is an expression that contains number and one or more operations.
   f) _____ is an expression that contains variables, constants and (or) operations.
   g) The mathematical symbols + (Addition), − (Subtraction), × (Multiplication), and ÷ (Division) are called _____.
   h) _____ is a set of values of a variable that makes the open sentence true.
   i) _____ is a mathematical statement that contain one or more variables that may be true or false depending up on the value that is substituted for that variable.
   j) A statement or an equation that is always true (or always false) is called _____.

(ii) Translate each of the following algebraic expression into words.
   a) $x + 6$
   b) $m - 4$
   c) $8y$
   d) $\dfrac{3n}{2}$
   e) $\dfrac{12}{x}$

(iii) Translate each of the words into algebraic expression.
   a) The sum of a number and 6.
   b) Five times a number x.
   c) The difference of y and 10.
   d) The quotient of 4m and 13.
   e) 4 less than six times a number x.
   f) 3 more than seven times a number x.
   g) The quotient of ten and the difference of eight and five times a number x.
   h) Two less than the sum of four and a number m.
   i) 5 less than 6 times a number n.

(iv) Multiple Choice
   Choose the best answer for each of the following questions.
   1) Which of the following is an open sentence?
      a) A rectangle has four sides.
      b) 2 is an even number.
      c) A triangle has n sides.
      d) 0 is greater than −5.
      e) none

5

2) Which of the following is a closed sentence?
    a) 1 is an odd number.
    b) $y = 3x + 4$
    c) A square has n sides.
    d) $y = x$
    e) none

3) When we translate the word "Two less than the sum of four times x and seven" into algebraic expression is _____
    a) $(4x + 7) - 2$
    b) $(7 + 4x) - 2$
    c) $(4x - 2) + 7$
    d) $(4x - 7) + 2$
    e) a and b

4) _____ is a mathematical sentence that has an equal sign.
    a) constant
    b) variable
    c) equation
    d) quantity
    e) none

5) The set of values of a variable that makes the open sentence true is called _____.
    a) Numerical operation
    b) Variables
    c) Solution Set
    d) True Set
    e) c and d

6) Which of the following is true when we translate the algebraic expression $\left(\frac{10}{8-x}\right)$ into words?
    a) The quotient of ten and the sum of 8 and x.
    b) The quotient of ten and the difference of x and 8.
    c) The quotient of ten and the difference of 8 and x.
    d) The quotient of $(8 - x)$ and 10.
    e) The product of 10 and $(x - 8)$.

7) Which of the following is true when the algebraic expression $(m - 5) + 2$, is translated to words?
    a) Two more than the difference of the number m and 5.
    b) Two more than the difference of the number 5 and m.
    c) Five less than the sum of the number m and 2.
    d) Five more than the difference of the number m and 2.
    e) a and b.

8) Which of the following is true when we write the phrase "Seven less than the quotient of the number x and 5" in to algebraic expression?

a) $\dfrac{x}{7} - 5$

b) $\dfrac{x}{5} - 7$

c) $\dfrac{x-5}{7}$

d) $\dfrac{x-7}{5}$

e) $\dfrac{x-5}{x-7}$

9) Which one is the equivalent expression of $(y + 3) - 6$.
   a) Six less than the sum of y and 3.
   b) Six less than the sum of 3 and y.
   c) Six more than the difference of y and 3.
   d) Six more than the difference of 3 and y.
   e) a and b

10) Which one is the equivalent expression of "5 more than two times a number z"?
   a) $(2z - 5)$
   b) $(2z + 5)$
   c) $(5 + 2z)$
   d) $(5z + 2)$
   e) b and c

1.4. Order of Operation and Evaluating Expression.

In mathematics, the order of operation is the rule which is used to solve, to simplify and/ or to find the possible solution for given mathematical expression(s) or equations. The order of operation is the rule that someone has to follow so as to make/do different mathematical operations $(+, -, x, \div)$. In addition to these mathematical operations, parentheses and exponents are also included in this rule to guide us for grouping symbols and solving the operations based on their sequences.

Note: 1) Parenthesis is the symbol ( ) used to group the different operation or things together.

Example: $(5 \times 4) + 6$
Here, 5×4, are inside parenthesis and they are together in one group.

Examples: - Evaluate each of the following operation.
   a) $2x (6 + 4)$
   b) $3x (9 - 4)$

Solutions:

    a)   2x (6 + 4)

First, we have to evaluate what is inside parenthesis (6 + 4) = 10 and multiplying by 2, which is $2 \times 10 = 20$

    b)   3 x (9 – 4)

First, we have to evaluate what is inside parenthesis (9 – 4) = 5 and multiplying by 3, which is $3 \times 5 = 15$

2) Exponent: tells us that how many times to use the number in multiplication and it is written in the form of a^n

Where a is called base and n is called exponent or power.

Example: 5^3, Here, 5 is the base and 3 is exponent, it means that 5 is multiplied three times,
$$5 \times 5 \times 5 = 125$$

Examples

Evaluate each of the following questions.

    a) 3×4^2

    b) 2×3^4

Solutions

    a) 3×4^2

First, we have to evaluate the exponential notation 4^2. Here 4^2 means = $4 \times 4 = 16$

Then, multiplying this value by 3 gives us $3 \times 4^2 = 3 \times 16 = 48$.

    b) $2 \times 3^4 = 2x (3 \times 3 \times 3 \times 3)$

                $= 2 \times 81$

                $= 162$

In general, in mathematics, the rules of operation or rules of order are as follows: -

1.4.1. Order of Operations Step by Step

Step 1) Evaluate all of the parentheses or brackets that are found in the given equation before you do anything.

Example:

    $3 \times (4 + 5) = 3 \times 9 = 27$

Step 2) Evaluate exponents before adding, subtracting, multiplying or dividing numbers/ expressions.

Example: $4 \times 5^3 = 4x (5 \times 5 \times 5) = 4 \times 125 = 500$

Step 3) Evaluate all multiplication and division from left to right.

Example: $12 \times 6 \div 2$

             $72 \div 2$

              36

Step 4) Evaluate all addition and subtraction, by working from left to right.

$$5 + 9 - 4$$
$$14 - 4$$
$$10$$

Some people remembered the order of operations by using the acronym "PEMDAS", which is common in United States of America. This acronym is taken from the first letter of the following mathematical operations and sign.

P = Parentheses
E = Exponent
M = Multiplication
D = Division
A = Addition
S = Subtraction

Here, in the case of multiplication and division always evaluate in order from left to right whichever comes first. The same is true for addition and subtraction.

In other countries some people remembered the order of operation by acronym BODMAS. Where each letter again stands for the first letter of each mathematical operations and sign.

B = Bracket   M = Multiplication
O = Of or Order   A = Addition
D = Division   S = Subtraction

1.4.2. Order of Operation Rule in BODMAS

If a given expression has ( ), { }, [ ], then we have to solve the bracket first, then of or order (numbers involving square roots or exponents) follows. Division, multiplication, addition, and subtraction are performed as their order by starting evaluating from the left to right.

Note: - Whenever evaluating order of operation, make sure to follow the correct order, otherwise your final answer will be wrong.

Examples

    Evaluate each of the following expression:

1) $18 \div 6 + 3(5 + 7) - 4 + 9 \times 7$

Solution: $18 \div 6 + 3(5 + 7) - 4 + 9 \times 7$

             $18 \div 6 + 3(12) - 4 + 9 \times 7$ .... Step 1) Evaluate parenthesis

             $3 + 3(12) - 4 + 9 \times 7$ .... Step 2) Evaluate Division

             $3 + 36 - 4 + 63$ .... Step 3) Evaluate Multiplication

             $39 - 4 + 63$ .... Step 4) Evaluate Addition

             $39 + 63 - 4$ .... Step 5) Evaluate Addition

             $102 - 4$ .... Step 6) Last Subtraction

             98

2) $5^2 + 4(8 - 2)^2 \div 12$

Solution: $5^2 + 4(8 - 2)^2 \div 12$

             $5^2 + 4(6)^2 \div 12$ .... Step 1) Evaluate Parenthesis

             $5 * 5 + 4(6 * 6) \div 12$ .... Step 2) Evaluate Exponent

             $25 + 144 \div 12$

             $25 + 144 \div 12$ .... Step 3) Evaluate Division

             $25 + 12 = 37$ .... Step 4) Evaluate Addition

3) $4 + 8(6 + 3) \div 18 - 2$

Solution: $4 + 8(6 + 3) \div 18 - 2$ .... Step 1) Evaluate Parenthesis

             $4 + 8(9) \div 18 - 2$ .... Step 2) Evaluate Multiplication

             $4 + 72 \div 18 - 2$ .... Step 3) Evaluate Division

             $4 + 4 - 2$ .... Step 4) Evaluate Addition

             $8 - 2$ .... Step 5) Evaluate Subtraction

             6

4) $(5 \times 6) + (7 \times 3) - 9 + 33 \div 11 - 4$

Solution: $(5 \times 6) + (7 \times 3) - 9 + 33 \div 11 - 4$

             $30 + 21 - 9 + 33 \div 11 - 4$

             $30 + 21 + 3 - 9 - 4$

             $54 - 9 - 4$

             $51 - 13$

             41

5) $160 - [40 + 80 - (28 - 6)]$

Solution: $160 - [40 + 80 - (28 - 6)]$

             $160 - [40 + 80 - 22]$

             $160 - [40 + 58]$

             $160 - 98$

             62

6) Evaluate $5(6 \times 2^3) \div 10 + 4 - 7$

Solution: $5(6 \times 2^3) \div 10 + 4 - 7$ ....Evaluate Bracket and Exponent

$5(6 \times 2 * 2 * 2) \div 10 + 4 - 7$

$5(48) \div 10 + 4 - 7$

$5 \times 48 \div 10 + 4 - 7$ ....Evaluate $5 \times 48 = 240$

$240 \div 10 + 4 - 7$ ....Evaluate $240 \div 10 = 24$

$24 + 4 - 7$ ....Evaluate from left to right

$28 - 7$

21

Exercise 1.4 -1.4.2

Evaluate each of the following expression using "PEMDAS" or "BODMAS" rule.

1) $9 + 16 \div 8 \times 6 - 1$

2) $12 \div 3 + 15 \div 5 \times 9$

3) $4 + 8 [6 + (5 - 2)^2]$

4) $39 - 28 \div 4 \times 5 + 2$

5) $24 \div 8 - 12 + 3 \times 6$

6) $8 \times 6 \div 16 + 36 \div 3 - 7$

7) $(24 - 8 \div 4) + 3 \times 7$

8) $(3 \times 4^2 \div 12) - (11 - 3^2)$

9) $(84 \div 3 - 4 \times 2)^2 - (8 - 27 \div 9)^2$

10) $6 + [3(8 + 2 - 4)] \div 9 \times 3$

1.5. Real Numbers and Number Line.

A real number is a number that describes quantities in daily life business, industry, agriculture, and in any scientific world. The symbol for real number is **R**, or a blackboard bold $\mathbb{R}$.

- A real number is any number that can be written as a decimal.

We use real numbers to describe quantities like age, weight, area, density, volume and distance etc. Some of real number examples are: -

$$2, 6.3, -8, \frac{2}{3}, \frac{2}{5}, \pi, \sqrt{6}$$

Since a real number is any number that can be written as a decimal, we can write the above real numbers as a decimal respectively as: -
2.0, 6.3, –8.0, $0.\overline{6}$, 3.1415..., 2.4494...

The bar above a block of digit (i.e., $\overline{6}$) shows that the block repeats itself indefinitely. These examples show that any real number can be expressed as a decimal notation.

1.5.1. Advance Note for Real Numbers

Definition: - A set is a well-defined collection of distinct objects. The object that makes up a set is called an element or a member.
Note: - In mathematics, when we say 'well-defined', we mean that there is a rule which can be used to identify the members of the set. In contrast, some sets whose members may be difficult to identify, for example, the collection of all students in a class who are at least 250 years old; does not define a set because "at least 250 years old" is not well-defined. Thus, to be a set, collection of objects must be well-defined.
- A set is denoted by placing its members between a pair of curly brackets i.e. { }.
- A set is usually represented by upper case letter, such as A, B, C, or D. An element or a member of a set is represented by lower case letter, such as a, b, x, y or z.
- A set with no element in it is called Empty Set.

Examples: -
1) A = {1, 2, 3, a, b}, Here, we read this set as set A. It has five members or elements. 1, 2, 3, a, and b are elements of set A.
2) B = { }, Here, set B is an empty set and set B has no member or element. The number of set B is Zero.

Note: - An empty set is also called null set or void set or vacuous set.

1.5.2. Different types of real numbers.

- Natural number (N): - These are sets of the numbers that we use for counting.
  N = {1, 2, 3, 4, ...}
  Examples of natural numbers: 1, 2, 11, 13
- Whole number (W): - These are sets of positive natural numbers and zero.
  W = {0, 1, 2, 3, ...}
  Examples of whole numbers: 0, 3, 4, 5
- Integer (Z): - These are the set of negative natural numbers and the whole numbers.
  Z = {..., –3, –2, –1, 0, 1, 2, 3,...}
  Examples of integer numbers: –2, –5, 0, 4, 7

- Rational number (Q): The set of rational number is the set of all numbers that can be expressed as a quotient of two integers, where the numerator can be any number whereas its denominator should be any numbers excluding zero.

$$Q = \left\{ \frac{p}{q} : p \text{ and } q \text{ are integers and } q \neq 0 \right\}$$

Examples: 5, 6, $\frac{2}{3}$, $\frac{-4}{5}$, 0.3, 0.$\overline{4}$, $\frac{-1}{3}$, 0

Note: - Rational numbers can be expressed as terminating decimals. It is a number that contains repeating decimals or else finite number of digits after the decimal point. (In other words, they are numbers which contains digits that repeats themselves endlessly).

Examples: -
Terminating Decimals
0.5, 13.42, 2.65, 0.8, 0.3
Repeating Decimals
0.$\overline{3}$, 0.32$\overline{6}$, 74.2$\overline{53}$, 0.$\overline{9}$
Irrational Numbers (P): - These are real numbers that cannot be expressed as the ratio of two integers.

Note: - Irrational numbers are non-terminating and non-repeating decimals.

Examples: π, 5.349..., $\sqrt{2}$, $\sqrt{7}$, 1.24...

1.6. Real Number Line

A real number line is a representation of real numbers as a point on a horizontal line. Each point on the number line corresponds to exactly one real number. The real number corresponding to a point is called the coordinate of a point.

1.6.1. Cartesian Coordinate System

Crossing of two number lines perpendicularly (i.e., at right angle) on the same plane makes a coordinate plane. The horizontal number line is called x-axis whereas the vertical one is y-axis. The point where these lines intersect (cross) each other is called the origin. On the horizontal line (x-axis), those points that are found at the right side of the origin (point 0) are positive numbers while those at the left side are negative numbers. Similarly, those points that are found on the vertical line (y-axis) above the origin (point 0) are positive numbers whereas those below the origin are negative numbers.

The real number corresponding to a point is called the coordinate of that point. Coordinate system is a means to represent (identify) any number on the plane specifically (uniquely). To do this, naming of a point on the x-axis is mentioned first and then the y-axis follows.

On real number line, the unit distance of a point chosen to the right of zero is labeled as 1. In other words, this distance that extends from zero to 1 is called the unit distance. Numbers to the right of the origin are positive numbers. But those that are found to the left of the origin are negative numbers.

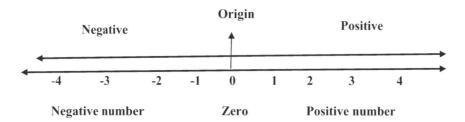

Real numbers can be indicated on a number line by placing a solid dot at the exact lactation for each number.

Example: Graph the integers
−4, −2, 0, 1 and 5 on the number line.

Note: - There is one-to-one correspondence between the real numbers and points on the number line. For every point on the line, there is a unique real number; which means, for every real number there is a unique point on the line.

1.6.2. Order of the Real Numbers

On the number line, if the plot of a real number 'a' is to the right of the plot of a real number 'b', we say that the real number 'a' is greater than the real number 'b'. That is,
a > b, (read as the number a is greater than the number b).
Similarly, we can say that b < a, (read as the number b is less than that of a). Let us see some examples below.

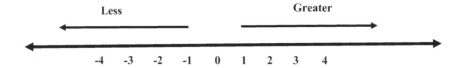

On the number line above, observe that −2 is to the right of −4, this means that −2 is greater than −4.

Similarly; −4 is to the left of −2 as seen on the number line, this means, −4 is less than −2.

Thus, −4<−2 and −2>−4 are the different saying for the same fact.

Note: • Zero is greater than any negative and less than any positive number.

- The symbols < and > are called inequality symbols. There are other inequality symbols such as ≤ (read as less than or equal to) and ≥ (read as greater than or equal to).
  Examples: when we write:
  x is non-negative numbers; x ≥ 0.
  x is non-positive numbers; x ≤ 0.

1.6.3. Graphing a Linear Inequality on a Number Line.

When graphing a linear-inequality on a number line, we use a closed or solid circle (●) for "greater than or equal to" and an open circle (○) for "less than or greater than".

Example 1) Indicate "x is greater than or equal to 3" on the real number line.

Answer:

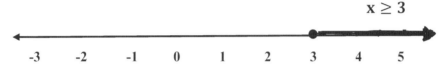

Example 2) Indicate "x is less than or equal to –2" on the real number line.

Answer:

Example 3) Indicate "x is less than 2" on the real number line.

Answer:

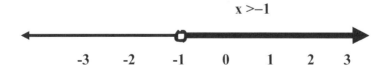

Example 4) Indicate "x is greater than –1" on the real number.

Answer:

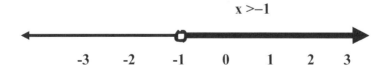

Example 5) Indicate {x:x ≠ 0} on the real number line.

Answer:

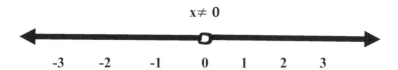

Note that we use the signs "=" reads as "equal to" and "≠" reads as "not equal to" in mathematics in early grade levels. We can also label numbers with these "equal to" and "not equal to" signs.

Example 6) Indicate "x is equal to 4" on the number line.

Answer:

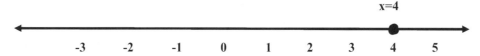

Example 7) Indicate "x is not equal to 3" on the number line.

Answer:

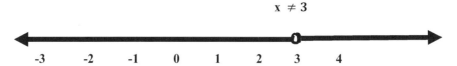

1.7. Opposite Real Numbers

Opposite real numbers are real numbers that have the same distance from the origin on the number line but their graph lie on the opposite sides of the origin and the numbers have opposite signs.

Example:

In the above real number line, point –3 is three units away from zero (origin) to the left side and point 3 is also three units away from the zero (origin) but to the right side. Thus, they are the same distance but opposite in direction. Hence, we say –3 is the opposite of 3. Also, –5 is the opposite of 5.

Note: - The opposite of zero is zero itself as there is no negative zero number.

16

Exercise: 1.5-1.7

Choose the best answer for each of the following questions.

1. Which of the following is not rational number?
    a) 5
    b) $\dfrac{56}{7}$
    c) π
    d) $2.3\overline{42}$

2. Which of the following is not a member of integers?
    a) $\dfrac{5}{9}$
    b) 5
    c) 0
    d) −1

3. Which of the following is correct when we graph x < 5 on the real number line?

    a)
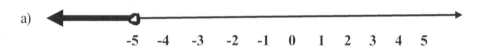

    b)

    c)

    d)

4. Which of the following is correct when we graph **x > 3** on the real number line?

    a)

    b)

    c)
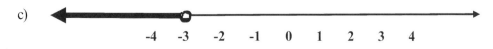

d)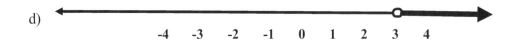

-4   -3   -2   -1   0   1   2   3   4

5. Which of the following is true when we graph y ≥ 0 on the real number line?

    a)

    0

    b)

    0

    c)

    0

    d)

    0

6. Which of the following corresponds to the set of positive integers less than 8?
    a)   {0, 1, 2, 3, 4, 5, 6, 7}
    b)   {1, 2, 3, 4, 5, 6, 7, 8}
    c)   {0, 1, 2, 3, 4, 5, 6, 7, 8}
    d)   {1, 2, 3, 4, 5, 6, 7}

7. Which of the following is the set of whole numbers?
    a)   W = {..., –3, –2, –1, 0, 1, 2, 3,...}
    b)   W = {0, 1, 2, 3,...}
    c)   W = {1, 2, 3,...}
    d)   W = {..., –3, –2, –1, 0}

8. Which of the following is <u>not</u> an example of a set?
    a)   The set of all students in a class whose age is less than 25.
    b)   The set of all students in a class who obtained greater than 50 percent marks in mathematics exam.
    c)   The set of all students in a class who have six legs and nine eyes.
    d)   The set of all students who love physics.

9. Which of the following number is not a member of natural numbers?
    a)   1
    b)   3
    c)   0
    d)   5

10. Which of the following is not rational number?
    a) $5.\overline{3}$
    b) 3.121212....
    c) 0.54132...
    d) $\dfrac{6}{5}$

1.8. Properties of Real Numbers

In this lesson, we are going to learn different basic properties of the real numbers. Learning these properties of real numbers will help us to solve different mathematical equations, simplify algebraic and numerical expressions and applying them in studying mathematics and other related disciplines.

There are different properties of real numbers: These are discussed below:
Let a, b, and c be real numbers.
1) Commutative Property of Addition.
    (a + b) = (b + a)

    This means changing the order when adding numbers does not affect the result.
    Example: (2 + 5) = (5 + 2)
    $\qquad\qquad\quad$ 7 = 7

2) Commutative Property of Multiplication.
    (a • b) = (b • a)

    This means, changing the order when multiplying numbers does not affect the result.
    Example: (6 × 8) = (8 × 6)
    $\qquad\qquad\quad$ 48 = 48

    Note: - Subtraction and division do not have commutative property.

3) Associative Property of Addition.
    a + (b + c) = (a + b) + c

    This means, changing the group when adding numbers does not affect the result.
    Example: 2+(3+6) = (2 + 3) + 6
    $\qquad\qquad\quad$ 2+9 = 5+6
    $\qquad\qquad\quad\;$ 11 = 11

4) Associative Property of Multiplication.
    a • (b • c) = (a • b) • c

    This means, changing the group when multiplying does not affect the result.
    Example: 3× (4×5) = (3× 4) × 5

$$3 \times 20 = 12 \times 5$$
$$60 = 60$$

Note: Subtraction and division do not have associative property. To illustrate these facts, let us see some examples:

$$6 - (3 - 2) \neq (6 - 3) - 2$$
$$6 - 1 \neq 3 - 2$$
$$5 \neq 1$$

Similarly, $16 \div (8 \div 4) \neq (16 \div 8) \div 4$
$$16 \div 2 \neq 2 \div 4$$
$$8 \neq 0.5$$

5) Distributive Property of Multiplication over addition.
$$a \cdot (b + c) = a \cdot b + a \cdot c$$

This means, multiplication can be distributed over addition and does not affect the result.

Example: $5 \times (3 + 2) = 5 \times 3 + 5 \times 2$
$$5 \times 5 = 15 + 10$$
$$25 = 25$$

6) Identity Property of Addition
$$a + 0 = a$$

This means, when someone adds any real number to zero, the result is the real number itself, not zero.

Example: $9 + 0 = 9$

7) Identity Property of Multiplication.
$$a \cdot 1 = a$$

This means, when any real number is multiplied by 1, the result is that real number itself.

Example: $5 \times 1 = 5$

8) Zero Property of Multiplication.
$$a \times 0 = 0$$

Example: $5 \times 0 = 0$

This means, when any real number is multiplied by 0, the result is 0.

9) Additive Inverse Property.
$$a + (-a) = 0$$
$$-a + (a) = 0$$

This means, the sum of any real number and its opposite (additive inverse) is zero.

Example: $4 + (-4) = 0$
$$(-4) + 4 = 0$$

10) Multiplicative Inverse Property

$$a \times \left(\frac{1}{a}\right) = 1, \text{ where } a \neq 0$$

$$\left(\frac{1}{a}\right) \times a = 1, \text{ where } a \neq 0$$

This means, the product of a non-zero real number and its multiplicative inverse is 1.

Example: $6 \times \left(\frac{1}{6}\right) = 1$

$$\left(\frac{1}{6}\right) \times 6 = 1$$

11) Closure Property of Addition.
If a and b are real numbers, then a + b is a real number.

Example: Since 2 and 13 are real numbers, then their sum (2 + 13) = 15 is also the real number.

12) Closure Property of Multiplication.
If a and b are real numbers, then a • b is a real number.
Example: Since –5 and 9 are real numbers, then –5 × 9 = –45. Here, the result –45 is also the real number.

Exercise 1.8

Choose the best answer for each of the following questions

1) Which of the following property of real number belongs to 5 + 6 = 6 + 5?
   a) Commutative Property of Multiplication.
   b) Commutative Property of Addition.
   c) Associative Property of Multiplication.
   d) Associative Property of Addition.

2) Which of the following property of real number belongs to 5 + (4 + 6) = (5 + 4) + 6 ?
   a) Associate Property of Multiplication.
   b) Distributive Property of Addition.
   c) Commutative Property of Addition.
   d) Associative Property of Addition.

3) The additive inverse of –5 is
   a) –5
   b) 0

c) 5

d) $\dfrac{-1}{5}$

4) Which of the following property of real number belongs to $19 \times 20 = 20 \times 19$?
   a) Commutative Property of Multiplication.
   b) Multiplicative Inverse.
   c) Associative Property of Multiplication.
   d) Distributive Property of Multiplication.

5) Which of the following property of real number belongs to $(4 \times 3) \times 6 = 4 \times (3 \times 6)$?
   a) Associative Property of Addition.
   b) Distributive Property of Multiplication.
   c) Associative Property of Multiplication.
   d) Commutative Property of Multiplication.

6) Multiplicative inverse of 12 is
   a) −12
   b) 12
   c) $\dfrac{1}{12}$
   d) $\dfrac{-1}{12}$

7) When $6 \times (3 + 7)$ is evaluated, the result is _____
   a) 16
   b) 25
   c) 24
   d) 60

8) When $3 \times (7 \times 4)$ is evaluated, the result is
   a) 84
   b) 48
   c) 33
   d) 40

9) If a, b and c are real numbers, which of the following property of real number belongs to $a \cdot (b + c) = ab + ac$?
   a) Associative Property of Multiplication over Addition.
   b) Commutative Property of Multiplication over Addition.
   c) Distributive Property of Multiplication over Addition.
   d) Distributive Property of Addition over Multiplication.

10) The operation subtraction and division are not closed under Associative and Commutative Properties. This statement is
   a) True
   b) False

1.9. Adding and Subtracting Real Numbers

In mathematics, all operations of arithmetic are defined in terms of the operation addition by considering the direction on the real number line. In addition to this, we have more rules and properties that explain how to perform these basic operations.

1.9.1. Addition on Number Line

As we have learned before, the positive numbers represent the numbers that are to the right side of the origin (zero) while negative numbers represent the numbers that are to the left side of the origin (zero).
Examples: - Adding real numbers on number line.

1) Adding 3 and 4 on the real number line.
    Solution: - Since both 3 and 4 are positive numbers, they are located on the right side from the origin (zero). Hence,

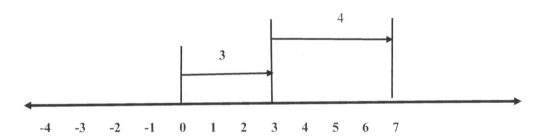

Move forward 3 units to the right from the origin (zero) and starting from 3 move forward 4 units to the right again. The total distance will be 7. That is 3 + 4 = 7
Adding –5 and (–3) on the real number line.
Solution: Since –5 and –3 are negative numbers; they are located on the left side of the origin (zero). Hence,

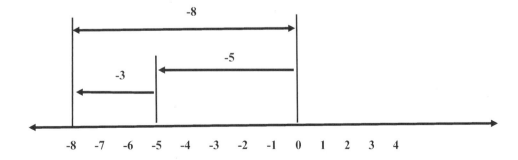

Move backward 5 units to the left from the origin (zero) and starting from –5 move 3 units to the left again. Your final point is –8. Thus, your final destination is 8 units to the left. In other words, –5 + (–3) = –8.

In conclusion, when adding two or more real numbers with the same sign, the final result will have the same sign as those numbers added.

Examples: 9 + 7 = 16 (positive)

–8 + (–4) = –12 (negative)

1.9.2. Subtracting Real Numbers on the Real Number Line.

Subtracting numbers on the number line is the same as changing the operation from subtraction to addition and moving backward from the origin. That means, if we have 6–4, then we can rewrite as 6 + (–4).

Examples

1) Subtract: 5 – 3 on the real number line.
Since 5 – 3 can be rewritten as 5 + (–3), it is easy to add these numbers like what we did before:

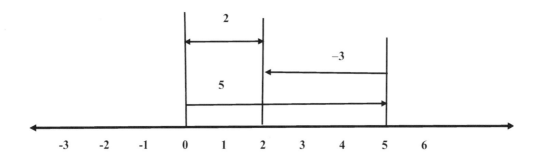

First, start from the origin (zero) and move forward 5 units to the right. Then, from 5 move backward 3 units i.e., to the left. Your final distance will be 2 units to the right side of the origin (zero): meaning 5 + (–3) = 2.

2) Subtract: (–4) – 2 on the real number line.
Since (–4) – 2, can be rewritten as –4 + (–2), it is easy to add these two negative numbers on real number line. Thus,

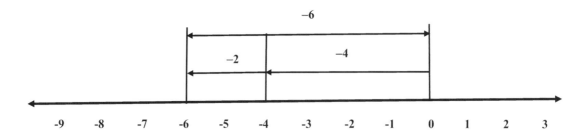

24

Start from the origin (zero) and move to the left side 4 units and then 2 units to the left again. Your final distance will be 6 units to the left side from the origin (zero).
Meaning: $-4 + (-2) = -6$.

Note: - Subtracting a number is the same as adding its opposite (Additive Inverse). In order to subtract numbers, rewrite subtraction as adding its inverse and as you learn before, use the rules for addition of real numbers.

Examples: - $-3 - 4 = -3 + (-4) = -7$
$6 - 5 = 6 + (-5) = 1$

1.9.3. Multiplying and Dividing Real Numbers

Multiplication is one of the four mathematical operations, often denoted by the cross-symbol x or by the mid-line dot operator (•). The result of multiplication operation is called a product.

If m and n are real numbers, then m × n, (read as 'm' times 'n')
i.e., m×n = n+n+...+n, ('n' is added 'm' times)

Example: 4 multiplied by 5, written as 4 × 5 and read as "4 times 5"
$5 + 5 + 5 + 5 = 20$

Here, 4 and 5 are called factors, and 20 is called the product.

As you learned previously, multiplication has Commutative Property.
The designation of multiplier and multiplicand does not affect the result of the multiplication.
Example: $2 × 6 = 6 + 6$
$= 12$
$6 × 2 = 2 + 2 + 2 + 2 + 2 + 2$
$= 12$

1.9.3.1 Multiplicative Inverse

If a is a number and different from zero, then $\frac{1}{a}$ is called the inverse of a or reciprocal of a.

Example: $\frac{1}{5}$ is an inverse of 5.
$\frac{8}{9}$ is an inverse of $\frac{9}{8}$.

1.9.3.2 Rules of Multiplying Integers

Positive × Positive = Positive
$(+) × (+) = (+)$
Example: $6 × 7 = 42$

Negative × Negative = Positive

$$(-) \times (-) = (+)$$

Example: $(-4) \times (-8) = 32$

Positive × Negative = Negative

$$(+) \times (-) = (-)$$

Example: $20 \times (-3) = -60$

Negative × Positive = Negative

$$(-) \times (+) = (-)$$

Example: $-40 \times 5 = -200$

Note: Any number multiplied by zero is zero.

1.9.4. Division: The mathematical operation division is denoted by the symbol ÷ or / (fraction bar). It is basically the inverse of the operation multiplication.

If a, b and c are integers where $b \neq 0$ and

$\dfrac{a}{b} = c$, then

a is called Dividend

b is called Divisor

c is called Quotient.

If any leftover amount as a whole number is there, it is called the remainder. Thus, the above formula can be written as: $a = bc + r$, where 'r' is a remainder.

Example 1) $30 \div 5 = 6$

Here, 30 is Dividend;

5 is Divisor; and

6 is Quotient.

Example 2) $27 \div 4$

Here, the quotient is 6 and the remainder is 3. This can be written as: $27 = 4 \times 6 + 3$

1.9.4.1 Rules of Division

- Division by zero is undefined. This means, we cannot divide any number by zero to have meaningful number. Example: $30 \div 0$ is undefined.
- When we divide a number by a fraction, we need to multiply by its reciprocal where the numerator and denominator are different from zero.

Examples   (a) $\dfrac{5}{4} \div 3 = \dfrac{5}{4} \times \dfrac{1}{3} = \dfrac{5}{12}$

           (b) $\dfrac{2}{3} \div \dfrac{4}{7} = \dfrac{2}{3} \times \dfrac{7}{4} = \dfrac{14}{12} = \dfrac{7}{6}$

(c) $\dfrac{\frac{5}{3}}{8} = (5)\left(\dfrac{8}{3}\right) = \dfrac{40}{3}$

- When we divide 0 by another number other than zero, the result is always 0.

Examples: a) $\dfrac{0}{4} = 0$

b) $\dfrac{0}{8} = 0$

- If the dividend and the divisor have the same signs, then the quotient is always positive.

Examples a) $30 \div 3 = 10$
b) $-20 \div -4 = 5$

- If the dividend and the divisor have different signs, then the quotient is always negative.

Examples a) $-25 \div 5 = -5$
b) $24 \div -4 = -6$

Exercise 1.9-1.9.4.1

Choose the correct answer for each of the following questions.

1) $5 \div \dfrac{3}{4} =$ _____

a) $\dfrac{3}{20}$

b) $\dfrac{20}{3}$

c) $\dfrac{15}{4}$

d) $\dfrac{4}{15}$

2) $-3 \times 0 =$ _____
a) 0
b) -30
c) -3
d) Undefined

3) Which of the following is a multiplicative inverse of $\dfrac{2}{9}$?
a) 18
b) $\dfrac{9}{2}$

c) 9

d) 2

4) $-3 + 2 =$ _____

   a) 5

   b) -5

   c) -1

   d) 1

5) $-3 + (-9) =$ _____

   a) -12

   b) -7

   c) 12

   d) -27

6) $(-12) + 6 =$ _____

   a) 6

   b) -18

   c) -6

   d) 18

7) $-5 + (-9) =$ _____

   a) -4

   b) -14

   c) 14

   d) -45

8) $(-4) + 4 =$ _____

   a) -8

   b) 0

   c) 8

   d) -16

9) $(-6) + 8 =$ _____

   a) -14

   b) -2

   c) 2

   d) 14

10) $(-13) + 18 =$ _____

   a) 5

   b) 31

   c) -5

   d) -31

1.10. Distributive Property

The distributive property is the property where the product of an expression and a sum is equal to the sum of the product of the expressions. Thus, if a, b and c are numbers, then

$$a (b + c) = ab + ac.$$

Note: - The rule also holds true for the difference. That is,

$$a (b - c) = ab - ac.$$

Examples: a) $2(3 + 4) = 2 \times 3 + 2 \times 4$
$$= 6 + 8$$
$$= 14$$

    b) $(2 + 5)4 = 4 \times 2 + 4 \times 5$
$$= 8 + 20$$
$$= 28$$

    c) $6(2 - 1) = 6 \times 2 - 6 \times 1$
$$= 12 - 6$$
$$= 6$$

    d) $(8 - 6)5 = 5 \times 8 - 5 \times 6$
$$= 40 - 30$$
$$= 10$$

1.11. An Introduction to Equations

An equation is a statement where two expressions are equal.

Examples: - a) $3 + 1 = 2 + 2$, Here, the expression $3 + 1$ is equal to the expression $2 + 2$. This is because both of them have the same result 4. Therefore, it is an example of equation.

(b) $9 + 4 = 6 + 7$, both the right and left side expressions have the same result 13. So, it is an example of equation.

In algebraic equation, the left-hand side is equal to the right-hand side. For example, $2x + 3 = 9$ is an equation. In this equation, 2x, 3 and 9 are called Terms and $2x + 3$ is called an Expression. The way of finding the values of the variable in the given equation is called Solving an Equation.

Note: $5x + 3$ is not an equation because it doesn't have equal sign. It is an expression.

1.11.1. Classifying Equations

Equations can be classified as identities and conditional equations. An identity equation is true for all values of the variable and a conditional equation is only true for some particular values of the variables.

Example 1) xy = yx is an example of an identity equation because when x = 2 and y = 4, the left side and the right side are equal 2 × 4 = 4 × 2. Thus, the solution set is defined by {2, 4}. If x= 6 and y = 7, both the left and the right expressions are equal; i.e., xy = yx implies

6 × 7 = 7 × 6. Hence, the solution set is defined by {6, 7}. In replacing values of variables of the given equations, an identity equation is always true for all the numbers that are presented in the sets. In the case of conditional equations, for example 5x = 15, the only solution that makes this equation true is x = 3. If we substitute any number for x other than 3, the equation will not be equal. Thus, conditional equation is only true for specific value of the variable.

As we learned before, equations can be open, closed, true or false. Thus, an open equation is an equation that contains one or more variables and may be true or false depending on the values of these variables.

Example: x + 5 = 9, Here, this equation may be true or false depending on the value of x. If x =2, then 2 + 5 = 9 is false. If x = 4, then 4 + 5 = 9, which is true. Thus, x + 5 = 9 is an example of an open equation.

- A closed sentence is always true or always false.

Examples: 1)   8 is an even number.
                     It is an example of a closed sentence. (Always true)
              2)   Seven is an even number.
                     It is an example of closed sentence. (Always false).

There are different types of equations. We will see them more in detail in the next chapters. Some of the lists of mathematical equations involved in algebra are: -

- Linear Equations
- Quadratic Equations
- Radical Equations
- Exponential Equations
- Rational Equations
- Cubic Equations
- Simultaneous Equations.

CHAPTER 2

Solving and Graphing Linear Equations

2.1. Definition of Linear Equation with One Variable

Linear equation in one variable is an equation where the variable has an exponent of 1. Thus, it is written in the form of $ax + b = 0$, where a and b are real numbers and $a \neq 0$. a is called coefficient of x and x is a variable.

Note: - A linear equation of the form $ax + b = 0$ has only one solution set.

Examples

1) $2x + 5 = 0$, this is an example of linear equation
2) $5x - 4 = 0$, this is an example of linear equation.

2.2. Solving Linear Equation with One Variable

In mathematics, finding the solution set of (solving) an equation means to find the value of the variable that makes the equation true. As I mentioned before, the value that makes the given equation true is called solution set of the equation.

2.3. Solving Equations Using Addition and Subtraction

Note: - In solving linear equation, addition and subtraction are inverse operations.

To solve such equations using addition, subtract the same number from both sides of the equation. This helps to isolate the variable on one side and the numbers or other terms on the other side.

Thus, $x + a = m$

$\Rightarrow x + a - a = m - a$

$\therefore x = m - a$

Here, $m - a$ is the solution set of the equation $x + a = m$.

Examples

1) Solve for x: $x + 5 = 6$

Solution: $x + 5 = 6$

$\qquad x + 5 - 5 = 6 - 5$ ....(Subtract 5 from each side of the equation)

$\qquad x + 0 = 1$

$\qquad x = 1$

2) Solve for x:$2x + 4 = 12$

$\qquad$ Solution: $\quad 2x + 4 = 12$

$\qquad\qquad\qquad 2x + 4 - 4 = 12 - 4$ ....(Subtract 4 from each side of the equation)

$\qquad\qquad\qquad 2x + 0 = 8$

$\qquad\qquad\qquad 2x = 8$

$\qquad\qquad\qquad \dfrac{2x}{2} = \dfrac{8}{2}$ ....(Divide both sides by the coefficient of x; i.e., 2)

$\qquad\qquad\qquad x = 4$

To solve equations using subtraction, add the same number to each side of the equation. This helps to isolate the variable on one side and the other terms (numbers) on the other side.

Thus, $\quad x - a = m$

$\qquad\qquad x - a + a = m + a$ .... (Add 'a' on both sides)

$\qquad\qquad x + 0 = m + a$

$\qquad\qquad\quad x = m + a$

Here, m + a is the solution set of the equation x – a = m

Examples

1) Solve for x: $x - 6 = 4$

Solution: $\qquad x - 6 = 4$

$\qquad\qquad\qquad x - 6 + 6 = 4 + 6$ ....(Add 6 on both sides of the equation)

$\qquad\qquad\qquad x + 0 = 10$

$\qquad\qquad\qquad x = 10$

2) Solve for x: $3x - 7 = 8$

Solution: $\qquad 3x - 7 = 8$

$\qquad\qquad\qquad 3x - 7 + 7 = 8 + 7$ ....(Add 7 on both sides of the equation)

$\qquad\qquad\qquad 3x + 0 = 15$

$\qquad\qquad\qquad 3x = 15$

$\qquad\qquad\qquad \dfrac{3x}{3} = \dfrac{15}{3}$ ....(Divide both sides by the coefficient of x; i.e., 3)

$\qquad\qquad\qquad x = 5$

Exercise 2.1-2.3

Choose the best answer for each of the following questions.

1) Which of the following is an example of linear equation?
   a) $x + 5 = 2$
   b) $x + 4 = 1$
   c) $2x + 4 = 0$
   d) $3x + 5 = 6$
   e) All

2) The solution set of the linear equation $x - 1 = 8$ is _____
   a) 4
   b) 3
   c) 9
   d) $\dfrac{8}{3}$

3) The solution set of the linear equation $x + 6 = 16$ is _____
   a) 3
   b) 10
   c) 5
   d) 4

4) The value of x in the linear equation $2x - c = m$ is _____
   a) $m + c - 2$
   b) $m + c + 2$
   c) $\dfrac{m + c}{2}$
   d) $m + c$

5) The value of y in the linear equation $y + 3 = 6$ is _____
   a) -3
   b) 3
   c) 6
   d) 7

2.4. Solving Equations Using Multiplication and Division

Like solving equations by addition and subtraction, we can solve linear equations by using multiplication and division. But the way we solve is different. To solve such equations, we are going to multiply or divide each side by the same non-zero numbers. As you learned in the previous lesson, multiplication and division are inverse operations.

2.5. Solving an equation by dividing each side by the coefficient of an equation

Examples

1) Solve for x: $6x = 5$

Solution: $6x = 5$, Here, to isolate x, divide each side of an equation by its coefficient 6.

$$\frac{6x}{6} = \frac{5}{6}$$

$$x = \frac{5}{6}$$

2) Solve for x: $-8x = 3$

Solution: $-8x = 3$ Here, to isolate x, divide each side of an equation by its coefficient -8.

$$\frac{-8x}{-8} = \frac{3}{-8}$$

$$x = \frac{-3}{8}$$

3) Solve for x: $7x = 5$

Solution: $7x = 5$ Here, to isolate x, divide each side of an equation by its coefficient 7.

$$\frac{7x}{7} = \frac{5}{7}$$

$$x = \frac{5}{7}$$

2.5.1. Solving an Equation by Multiplying Each Side of an Equation

Examples:

1) Solve for $x : \frac{x}{8} = 9$

Solution: $\frac{x}{8} = 9$

$$8\left(\frac{x}{8}\right) = 9(8)$$ .... (To isolate x, multiply each side of an equation by 8; i.e.,

the reciprocal of the coefficient of x: $\frac{1}{8}$)

$$x = 72$$

2) Solve for y: $\frac{y}{6} = -4$

Solution: $\frac{y}{6} = -4$

$6\left(\dfrac{y}{6}\right) = -4(6)$ (To isolate y on the left side, multiply each side of an equation by 6; i.e., the reciprocal of the coefficient of y: $\dfrac{1}{6}$

$y = -24$

Note: As you learned before, the reciprocal of $\dfrac{a}{b}$, where a and b are different from zero is $\dfrac{b}{a}$. Understanding this helps to solve an equation having a fractional coefficient.

Examples:

1) Solve for x: $\dfrac{3x}{5} = 6$

Solution: $\dfrac{3x}{5} = 6$

        $\left(\dfrac{5}{3}\right)\left(\dfrac{3x}{5}\right) = 6\left(\dfrac{5}{3}\right)$ ....(Multiply both sides of the equation by $\dfrac{5}{3}$, i.e., the reciprocal of the x coefficient: 3/5.)

        $x = \dfrac{30}{3}$

        $x = 10$

2) Solve for x: $\dfrac{5x}{7} = 3$

Solution: $\dfrac{5x}{7} = 3$

        $\dfrac{7}{5}\left(\dfrac{5x}{7}\right) = 3\left(\dfrac{7}{5}\right)$ ....(Multiply both sides of the equation by $\dfrac{7}{5}$, i.e., the reciprocal of the x coefficient: $\dfrac{5}{7}$.

        $x = \dfrac{21}{5}$

3) Solve for x: $\dfrac{1}{9}x = 5$

Solution: $\dfrac{1}{9}x = 5$

        $9\left(\dfrac{1}{9}x\right) = 5(9)$

        $x = 45$ ....(Likewise, multiply each side by 9)

4) Solve for x: $\dfrac{-3x}{4} = 7$

Solution: $\dfrac{-3x}{4} = 7$

        $\left(\dfrac{-4}{3}\right)\left(\dfrac{-3x}{4}\right) = 7\left(\dfrac{-4}{3}\right)$ ....(Multiply each side by $\dfrac{-4}{3}$)

        $x = \dfrac{-28}{3}$

Exercise 2.4-2.5

Choose the best answer for each of the following questions.

1) The solution set of $\frac{2}{3}x = 1$ is

   a) $\frac{-3}{2}$

   b) $\frac{2}{3}$

   c) $\frac{3}{2}$

   d) 6

2) The reciprocal of $\frac{9}{4}$

   a) 36

   b) $\frac{4}{9}$

   c) $\frac{-9}{4}$

   d) $\frac{1}{36}$

3) The solution set of $\frac{-5x}{4} = \frac{1}{8}$ is

   a) $\frac{-1}{5}$

   b) $\frac{-1}{10}$

   c) $\frac{4}{5}$

   d) $\frac{-1}{40}$

4) The solution set of $\frac{3}{4}x = \frac{1}{4}$ is

   a) $\frac{1}{3}$

   b) 3

   c) $\frac{16}{3}$

   d) $\frac{3}{16}$

5) The solution set of $\frac{4}{9}y = \frac{-1}{3}$ is

   a) $\frac{3}{4}$

   b) $\frac{4}{3}$

c) $\dfrac{-3}{4}$

d) $\dfrac{-9}{4}$

2.6. Solving Two Step Equations

As previously explained, solving an equation means finding the value of the given variable that makes the equation true. When we say two step equations, it means that it takes two steps to find the solution set of the given equation. The first step is to subtract from or add to a constant number to both sides, and the second step is to divide both sides by the coefficient of the given variable in the equation.

Examples

1. Solve for x: $2x + 5 = 9$

   Solution: $2x + 5 = 9$

   $2x + 5 - 5 = 9 - 5$..........................(Subtract 5 from each side)

   $2x = 4$

   $\dfrac{2x}{2} = \dfrac{4}{2}$ ....(Divide both sides by 2)

   $x = 2$

2. Solve for x: $3x - 6 = 12$

   Solution: $3x - 6 = 12$

   $3x - 6 + 6 = 12 + 6$........................(Add 6 to each side)

   $3x = 18$

   $\dfrac{3x}{3} = \dfrac{18}{3}$ ....(Divide each side by 3)

   $x = 6$

3. Solve for x: $6x - 4 + 2x = 20$

   Solution: $6x + 2x - 4 + 4 = 20 + 4$...............(Collect like terms and then add 4 on each side)

   $8x = 24$

   $\dfrac{8x}{8} = \dfrac{24}{8}$ ....(Divide each side by 8)

   $x = 3$

4) Solve for x: $-5x + 6 - 4x = 28$

   Solution: $-5x + 6 - 4x = 28$

   $-5x - 4x + 6 - 6 = 28 - 6$...............(Collect like terms and then subtract 6 from each side)

   $-9x = 22$

   $\dfrac{9x}{-9} = \dfrac{22}{-9}$ ....(Divide each side by –9)

   $x = \dfrac{-22}{9}$

Note: - To check the solution set of an equation whether it is correct or not, substitute the value of the variable in the original equation.

In the above, example 3, the value of x is 3, thus

$$6x - 4 + 2x = 20$$

Check: $6x - 4 + 2x = 20$

$$6(3) - 4 + 2(3) = 20$$
$$18 - 4 + 6 = 20$$
$$24 - 4 = 20$$
$$20 = 20\checkmark$$

Therefore, the answer is correct.

2.6.1. Finding the Solution Set Using Distributive Property

Examples: -

1) Solve for x: $3x + 4(2x - 3) = 10$

    Solution: $3x + 4(2x - 3) = 10$

                $3x + 8x - 12 = 10$ ....(Distributive property)

                $11x - 12 = 10$.................................(Add like terms)

                $11x - 12 + 12 = 10 + 12$.................(Add 12 to each side)

                $11x = 22$

                $\dfrac{11x}{11x} = \dfrac{22}{11}$ ....(Divide both sides by 11)

                $x = 2$

2) Solve for x: $2x - 3(x - 6) = 9$

    Solution: $2x - 3(x - 6) = 9$

                $2x - 3x + 18 = 9$ ....(Distributive property)

                $-x + 18 = 9$.................................(Add like terms)

                $-x + 18 - 18 = 9 - 18$.....................(Subtract 18 from each side)

                $-x = -9$

                $\dfrac{-x}{-1} = \dfrac{-9}{-1}$ ....(Divide both sides by –1)

                $x = 9$

3) Solve for y: $\dfrac{3}{4}(y - 6) = 6$

    Solution: $\dfrac{3}{4}(y - 6) = 6$

                $(4)\left(\dfrac{3}{4}\right)(y - 6) = 6 \times 4$ ....(Multiply each side by 4)

                $3(y - 6) = 24$

                $3y - 18 = 24$ ....(Distributive property)

                $3y - 18 + 18 = 24 + 18$ ....(Add 18 on each side)

$$3y = 42$$

$$\frac{3y}{3} = \frac{42}{3} \quad \dots\dots\dots\dots\dots\dots\dots\dots\dots\text{(Divide each side by 3)}$$

$$y = 14$$

4) Solve for x: $\dfrac{2}{5}(2x-1) = \dfrac{3}{4}$

   Solution: $\dfrac{2}{5}(2x-1) = \dfrac{3}{4}$

Here, for the left and the right side of this equation, the denominators are 5 and 4 respectively. These numbers have the LCM (Least Common Multiple) of 20. Therefore, we multiply each side of this equation by 20. Thus,

$$\frac{2}{5}(2x-1) = \frac{3}{4}$$

$$(20)\left(\frac{2}{5}\right)(2x-1) = \frac{3}{4}(20) \quad \dots\dots\dots\dots\text{(Multiply each side by LCM)}$$

$$4(2)\,(2x-1) = 3 \times 5$$

$$8(2x-1) = 15 \quad \dots\dots\dots\dots\dots\dots\dots\text{(Distributive property)}$$

$$16x - 8 = 15$$

$$16x - 8 + 8 = 15 + 8 \quad \dots\dots\dots\dots\dots\text{(Add 8 on each side)}$$

$$16x = 23$$

$$\frac{16x}{16} = \frac{23}{16} \quad \dots\dots\dots\dots\dots\dots\dots\dots\text{(Divide each side by 16)}$$

$$x = \frac{23}{16}$$

Exercise 2.6-2.6.1

Solve each of the following equations.

1) $3x - 1 = 0$

2) $5x - 9 + 3x = 12$

3) $3m + 6 = 9$

4) $\dfrac{13}{2}x + 4 = 7$

5) $2x + 5(3x - 1) = 9$

6) $3x - 8 + x = 5$

7) $2 - \dfrac{3}{4}x = \dfrac{1}{2}$

8) $\dfrac{2}{3}(x - 1) = \dfrac{3}{14}$

9) $5x - 7 = 0$

10) $x - \dfrac{1}{4} = 3x - 1$

2.7. Finding Solution Set for Equations with Variables on Both Sides

Some equations have unknown quantity (variable) on both sides. To solve such equations, we must add to or subtract from both sides of the equation, the variable that contains the greater coefficient.

Note: There is no strict rule to follow, rather we can add to or subtract from both sides of the equation the variable that contains the smaller coefficient. Thus, it is easier to add or subtract the variable with smaller coefficient from the larger coefficient.

Examples: -

1. Solve for x: $3x + 2 = x - 4$

   Solution: $3x + 2 = x - 4$

   $3x + 2 = x - 4$

   $3x - x + 2 = x - x - 4$ ....(Subtract the variable with smaller coefficient from both sides)

   $2x + 2 = -4$

   $2x + 2 - 2 = -4 - 2$ ....(Subtract 2 from each side)

   $2x = -6$

   $\dfrac{2x}{2} \quad \dfrac{6}{2}$ ....(Divide both sides by 2)

   $x = -3$

2) Solve for x: $8x - 3 = 9 - 4x$

   Solution: $8x - 3 = 9 - 4x$

   $8x - 3 + 4x = 9 - 4x + 4x$ ....(Add the variable with smaller coefficient to both sides)

   $12x - 3 = 9$

   $12x - 3 + 3 = 9 + 3$ ....(Add 3 on both sides)

   $12x = 12$

   $\dfrac{12x}{12} = \dfrac{12}{12}$ ....(Divide both sides by 12)

   $x = 1$

3) Solve for x: $5x - 1 = 7x + 2$

   Solution: $5x - 1 = 7x + 2$

$$5x - 5x - 1 = 7x - 5x + 2 \quad \text{.................(Subtract smaller coefficient variable from both sides)}$$

$$-1 = 2x + 2$$

$$-1 - 2 = 2x + 2 - 2 \quad \text{........................(Subtract 2 from both sides)}$$

$$-3 = 2x$$

$$\frac{-3}{2} = \frac{2x}{2} \quad \text{.........................(Divide both sides by 2)}$$

$$x = \frac{-3}{2}$$

4) $\frac{3}{2}x + 4 = \frac{5}{4}x - 8$

Solution: $\frac{3}{2}x + 4 = \frac{5}{4}x - 8$

Here; the denominators 2 and 4 have LCM of 4; so multiply both sides by 4.

$$(4)\left(\frac{3}{2}x + 4\right) = (4)\left(\frac{5}{4}x - 8\right)$$

$$6x + 16 = 5x - 32$$

$$6x - 5x + 16 = 5x - 5x - 32 \quad \text{............(Subtract 5x from both sides)}$$

$$x + 16 = -32$$

$$x + 16 - 16 = -32 - 16 \quad \text{...................(Subtract 16 from both sides)}$$

$$x = -48$$

5) Solve for x: $-5x + 3 = -8x - 6$

Solution: $-5x + 3 = -8x - 6$

$$-5x + 8x + 3 = -8x + 8x - 6 \quad \text{...........(Add 8x on both sides)}$$

$$3x + 3 = -6$$

$$3x + 3 - 3 = -6 - 3 \quad \text{........................Subtract 3 from both sides.}$$

$$3x = -9$$

$$\frac{3x}{3} = \frac{-9}{3} \quad \text{.........................(Divide both sides by 3)}$$

$$x = -3$$

6) Solve for x: $3(2 - 4x) + 5 = -2(x - 1)$

Solution: $3(2 - 4x) + 5 = -2(x - 1)$

$$6 - 12x + 5 = -2x + 2 \quad \text{....................(Distributive property)}$$

$$11 - 12x = -2x + 2$$

$$11 - 12x + 12x = -2x + 12x + 2 \quad \text{.....(Add 12x on both sides)}$$

$$11 = 10x + 2$$

$$11 - 2 = 10x + 2 - 2 \quad \text{......................(Subtract 2 from both sides)}$$

$$9 = 10x$$

$$\frac{9}{10} = \frac{10x}{10} \quad \text{.........................(Divide both sides by 10)}$$

$$x = \frac{9}{10}$$

Exercise 2.7

Solve each of the following questions

1) $2(3x - 4) = -3x + 9$

2) $4x - 1 = 7(6 - x) + 2$

3) $-5x + 3 = 7x - 4$

4) $\frac{1}{3}(x - 4) = \frac{3}{2}(8 - x)$

5) $-3(8 - 4x) = 5(2x + 4)$

6) $\frac{1}{4}\left(\frac{1}{3}x - 1\right) = \frac{3}{5}(x - 1)$

7) $3x - 4 = \frac{2}{3}(2x - 4)$

8) $(x - 9) - (3x - 3) = x + 9$

9) $\frac{1}{2}x + 3x = 6$

10) $x - 6 = \frac{1}{2}x - 1$

2.8. Graphing Linear Equations

The graph of the linear equation is the set of points in the given coordinate that all are members of the solutions to the equation. In order to graph for linear equation, we need to know about a coordinate plane system and how to plot points in a coordinate plane system.

Definition: - A coordinate plane system is a two-dimensional plane formed by two real number lines that intersect at a right angle. Each point in a coordinate plane system corresponds to an ordered pair of real numbers.

The first number (the first entry) in an ordered pair is called x-coordinate (abscissa)and the second number (second entry) is called y-coordinate(ordinate). The ordered pairs are always written in the form, (x, y). For example, the ordered pair (4, –3) has an x coordinate 4 and a y-coordinate –3.

Below are some more descriptions of the coordinate plane system.

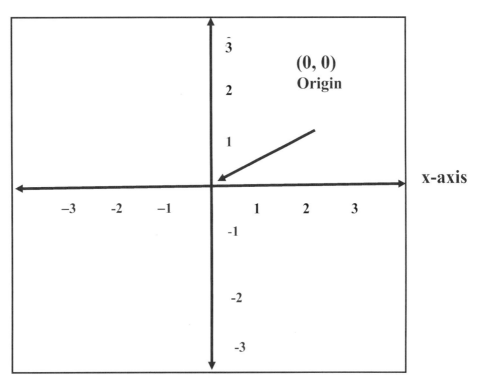

Fig. 2.1

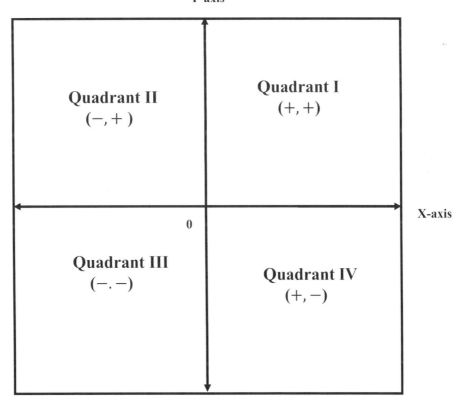

Fig. 2.2

NOTE: •   The horizontal line is called the x-axis or abscissa.
 •   The vertical line is called the y-axis or ordinate.
 •   The point of intersection of the x-axis and the y-axis is called the Origin.
 •   The x-axis and the y-axis divide the coordinate plane into four sections called Quadrants.
 •   Quadrants are named using the Roman numerals I, II, III, and IV, starting with the top right and moving counter clockwise.
 •   In the first quadrant, both x and y values are positive; (+, +).
 •   In the second quadrant, x value is negative and y-value is positive;(–, +).
 •   In the third quadrant, both the x and y values are negative; (–, –).
 •   In the fourth quadrant, x-value is positive and y-value is negative; (+, –).
 •   The point in the plane that corresponds to an ordered pair (x, y) is called the coordinate of the point.
 •   The coordinate plane extends infinitely in all directions. To indicate this, we usually put arrows at the ends of the axes.
 •   As mentioned before, the point at which the two axes intersect is called the origin. This origin has the ordered pair (0, 0).
   Recall that the value of x on the y-axis is always 0, the same is true for the value of y on the x-axis as it is always 0.

Plotting the Point on the Coordinate System

-   The way of locating points on the rectangular coordinate system is called plotting the points.
-   To plot the location of an ordered pair (x, y) on the coordinate plane system: -
-   Start from the origin, (0, 0) and move x unit to the right if x is positive; and x unit to the left, if x is negative.
-   From new x location found move y unit up if y is positive; and y-units down if y is negative.

Examples: -

Plot the ordered pairs given below in the coordinate plane system.
a)   A(2, 4)
b)   B(–3, 1)
c)   C(0, 0)
d)   D(–4, –5)
e)   E(3, 0)
f)   F(0, –2)
g)   G(0, 4)
h)   H(–5, 0)
i)   I(3, –4)

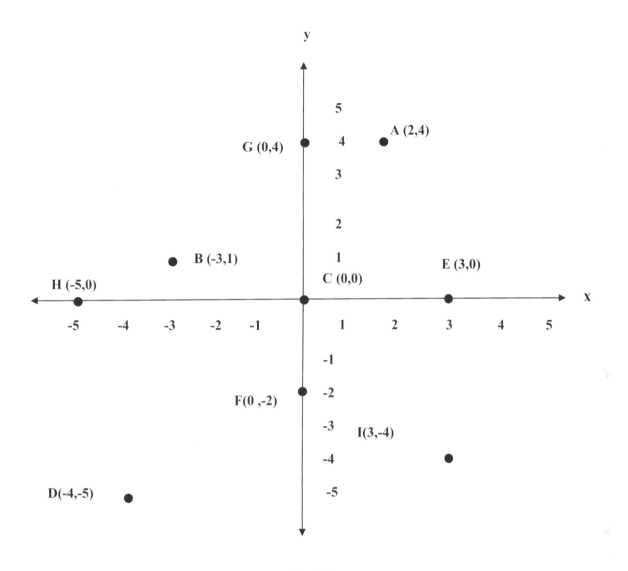

Fig. 2.3

A) To plot the point A(2, 4), start at the origin. Move 2 units to the right and 4 units up. Then, place a dot at the tip.

B) To plot the point B(–3, 1), start at the origin. Move 3 units to the left and 1 unit up. Then, place a dot at the tip.

C) Here the point C(0, 0) is the intersection of x and y axes and it is called the origin.

D) To plot the point D(–4, –5), start at the origin. Move 4 units to the left and 5 units down. Then, place a dot at the tip.

E) To plot the point E(3, 0), start at the origin. Move 3 units to the right and at this location place a dot since the y value on the x-axis is always 0.

F) To plot the point F(0, –2), start at the origin, as the x value on the y-axis is always 0, move 2 units down on the y-axis and place a dot.

G) To plot the point G(0, 4), start at the origin, as the x value on y-axis is always 0, move 4 units up and place a dot.

H) To plot the point H(–5, 0), start at the origin. Move 5 units to the left and at this point place a dot.

I) To plot the point, I(3, –4), start at the origin. Move 3 units to the right and 4 units down. Then place a dot at the tip.

Note: To plot the points (0, y) and (0, –y), we don't move any x-units to the right or to the left direction from the origin along the x-axis. This is because of the fact that the value of x is always zero on the y axis.

- To plot the points (x, 0) and (–x, 0), we do not move any y-units up or down from the origin along the y-axis. This is because of the fact that the value of y is always zero on the x axis.

Examples:

In which quadrant does each of the following points located?

a) (2, 4)
b) (–3, 1)
c) (0, 0)
d) (–4, –5)
e) (3, 0)
f) (0, –2)
g) (0, 4)
h) (–5, 0)
i) (3, –4)

Solutions

a) (2, 4) lies in the first quadrant.
b) (–3, 1) lies in the second quadrant.
c) (0, 0) is located at the origin.
d) (–4, –5) lies in the third quadrant.
e) (3, 0) lies at the positive x-axis.
f) (0, –2) lies at the negative y-axis.
g) (0, 4) lies at the positive y-axis.
h) (–5, 0) lies at the negative x-axis.
i) (3, –4) lies in the fourth quadrant.

Exercise 2.8

1) In which quadrant does each of the following points located?
   a) (2, 3)
   b) (–4, 5)
   c) (–5, –3)
   d) (6, –2)
   e) (0, 6)
   f) (4, 0)
   g) (0, 0)

2) Find the coordinates of the point for each of the following requirements.
   a) The point is located four units below the x axis and two units to the right of the y-axis.
   b) The point is on the x-axis and eight units to the right of the y-axis.
   c) The point is located five units to the left of the y-axis and three units above the x-axis.
   d) The point is on the y-axis and six units above the x-axis.
   e) The point is on the y-axis and three units below the x-axis.
   f) The point is located two units above the x-axis and three units to the left of y-axis.
   g) The point is located eight units to the left of the origin and three units below the x-axis.
   h) Both points lie at the origin.

3) Plot and label the following ordered pairs in a coordinate plane system.
   A) M(2, 5), N(–3, –2), O(4, 0)
   B) R(–1, 4), S(0, 0), T(5, 0.5)
   C) G(–2, –5), H(–1, 5), I(0.5, 3)

2.9. Linear Equations

Linear equation is an algebraic equation involving two variables and its graph is a straight line in the coordinate plane system. The general form of linear equation in two variables is

AX + BY + C = 0, where both A and B cannot be zero at the same time.

Example: $3x + 5y + 6 = 0$ is a linear equation in general form where A = 3, B = 5 and C = 6

As we learned before, the general linear equation AX + BY + C = 0, can be written as

$$y = \frac{-A}{B}x - \frac{C}{B}, \text{ where, } \frac{-A}{B} \text{ is the slope of a line and } \frac{-C}{B} \text{ is the y-intercept of a line.}$$

Example: $4x + 3y + 5 = 0$ is an example of a linear equation in general form and it can be rewritten as $y = \frac{-4}{3}x - \frac{5}{3}$, where $\frac{-4}{3}$ is the slope of a line and $\frac{-5}{3}$ is the y-intercept of a line.

2.10. Slope (m), y-intercept and x-intercept of a Line.

A linear equation can be written as $y = mx + b$, where m is the slope of a line and b is the y-intercept of a line.

Definition: The slope of a line is defined as the ratio of the vertical change between two y-values, the rise, to the horizontal change between two x-values, the run. The slope of a line is usually represented by lower case 'm'.

$$\text{Slope} = \frac{\text{rise}}{\text{run}} = \frac{\text{Change in } y}{\text{Change in } x} = \frac{y_2 - y_1}{x_2 - x_1}$$

$$m = \frac{y_2 - y_1}{x_2 - x_1}$$

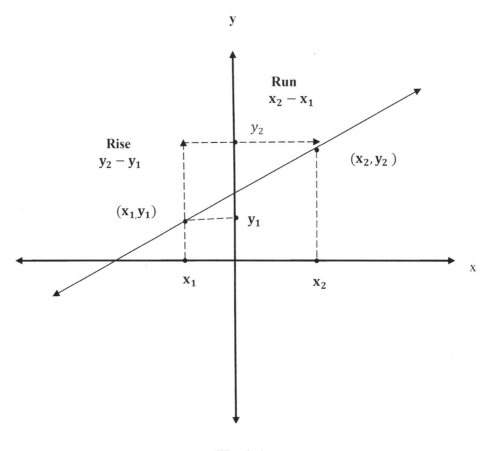

Fig. 2.4

As we move along the line from the point (x_1, y_1) to the point (x_2, y_2), the y-value changes from y_1 to y_2, the rise, an amount which is equal to $y_2 - y_1$. As y varies from y_1 to y_2, the value of x changes from x_1 to x_2, the run; an amount equal to $x_2 - x_1$. The ratio of the change in y to the change in x is called the slope of the line. If we have two points on a straight line, (x_1, y_1) and (x_2, y_2), where $x_1 \neq x_2$, then the slope of the line through these two points is given by:

$$\text{Slope} = \frac{y_2 - y_1}{x_2 - x_1}$$

$$m = \frac{y_2 - y_1}{x_2 - x_1}$$

Examples: -

a) Find the slope of the line passing through the points (–5, 2) and (4, 3),
   Solution:
   Given the points: (–5, 2) and (4, 3)
   $$(x_1, y_1) \qquad (x_2, y_2)$$

$$m = \frac{y_2 - y_1}{x_2 - y_1}$$

$$= \frac{3 - 2}{4 - (-5)}$$

$$= \frac{1}{4 + 5}$$

$$= \frac{1}{9}$$

Thus, the slope of the line is $\frac{1}{9}$.

Note: There is no strict rule to determine which point is (x_1, y_1) and the other (x_2, y_2) for the given two points. You can assume either of the given points (x_1, y_1) or (x_2, y_2).

b) Find the slope of the line passing through the points (–3, 4) and (5, 8).
   Solution:

$$m = \frac{y_2 - y_1}{x_2 - x_1}$$

$$= \frac{8 - 4}{5 - (-3)}$$

$$= \frac{4}{5 + 3}$$

$$= \frac{4}{8}$$

$$= \frac{1}{2}$$

Thus, the slope of the line is $\frac{1}{2}$.

c) Find the slope of the line passing through the points (1, 3) and (4, 3).
   Solution:

$$m = \frac{y_2 - y_1}{x_2 - x_1}$$

$$= \frac{3 - 3}{4 - 1}$$

$$= \frac{0}{3}$$

$$= 0$$

Thus, the slope of the line is 0.

d) Find the slope of the line passing through the points (5, 2) and (5, 3).
   Solution:

$$m = \frac{y_2 - y_1}{x_2 - x_1}$$

$$m = \frac{3 - 2}{5 - 5}$$

$$m = \frac{1}{0}$$

Since $x_2 - x_1 = 0$ and $\frac{y_2 - y_1}{x_2 - x_1} = \frac{1}{0}$, the line does not have slope.

That means, the slope is undefined.

Exercise 2.9-2.10

1. Find the slope of the line passing through the points:
   a) (5, 4) and (4, 3)
   b) (6, 3) and (7, 8)
   c) (4, 7) and (0, 6)
   d) (−9, 4) and (3, −4)
   e) (8, 5) and (8, 4)
   f) (6, 2) and (7, 2)

2. Find the slope of the line passing through the points:
   a) (a, b), (4, 3), a ≠ 4
   b) (a, 2b), (3, 7), a ≠ 3
   c) (2x − 1, 5), (3x + 4, 2y), x≠−5

2.11. Y-intercept and X-intercept of a Line

As we learned before, the equation of a straight line is written as $y = mx + b$, where m is the slope of a line and b is its y-intercept. The y-intercept of a line is the value of y at the point where the line crosses the y-axis and the x-intercept is the point where the line crosses the x-axis.

Note: - For the equation of a line given as $y = mx + b$, its slope is m, y-intercept is b and the x-intercept is $\frac{-b}{m}$. Writing such equation of a line in this form is called slope-intercept form of a line. To find the y-intercept of the equation of a line $y = mx + b$, we plug 0 for x and get $y = b$, represented as (0, b). This implies as b is the y-intercept of the line. Thus, in the equation $y = mx + b$, it is necessary to plug 0 for x so as to get its y-intercept.

$y = m(0) + b$

$y = b$

b is the y-intercept of a line.

To find the x-intercept of a line, we plug 0 for y and get $x = \frac{-b}{m}$. Thus, $\frac{-b}{m}$ is the x-intercept of the given line. Thus, in the equation $y = mx + b$, plugging 0 for y, yields:

$0 = mx + b$

$0 - b = mx + b - b$

$-b = mx$

$x = \frac{-b}{m}$

$\frac{-b}{m}$ is the x-intercept of a line.

Examples: Find the slope, y-intercept and x-intercept of the line passing through the points:
  a) $y = 3x + 5$
  b) $y = -4x - 9$
  c) $-2y = 8x - 6$
  d) $y = 3x$
  e) $y = x$
  f) $y = 0$
  g) $x = 0$

Solution
  a) $y = 3x + 5$
  $y = 3x + 5$
  Slope = 3, y-intercept = 5 and to find x-intercept, plug 0 for y.
  $y = 3x + 5$
  $0 - 5 = 3x + 5 - 5$
  $-5 = 3x$
  $-5/3 = 3x/3$
  $x = \frac{-5}{3}$

Thus, x-intercept is $\dfrac{-5}{3}$ and it occurs at $\left(\dfrac{-5}{3}, 0\right)$.

b) $y = -4x - 9$

Solution: $y = -4x - 9$

Slope $= -4$, y-intercept $= -9$. And to find x-intercept, plug 0 for y.

$y = -4x - 9$

$0 = -4x - 9$

$9/-4 = -4x/-4$

$x = \dfrac{-9}{4}$

x-intercept is $\dfrac{-9}{4}$ and it occurs at $\left(\dfrac{-9}{4}, 0\right)$

c) $-2y = 8x - 6$

Solution: –

$2y = 8x - 6$

$\dfrac{-2y}{-2} = \dfrac{8x}{-2} - \dfrac{6}{-2}$ ....(Divide both sides by –2)

$y = -4x + 3$

Slope $= -4$, y-intercept $= 3$ and to find x-intercept, plug 0 for y.

$0 = -4x + 3$

$0 - 3 = -4x + 3 \,\text{-}3$

$-3 = -4x$

$-3/-4 = -4x/-4$

$x = \dfrac{3}{4}$

x-intercept is $\dfrac{3}{4}$ and it occurs at $\left(\dfrac{3}{4}, 0\right)$.

d) $y = 3x$

Solution: $y = 3x$

Slope $= 3$, y-intercept $= 0$ and to find x-intercept, plug 0 for y.

$y = 3x$

$0 = 3x$

$\dfrac{0}{3} = \dfrac{3x}{3}$

$x = \dfrac{0}{3}$

$x = 0$

x-intercept is 0 and it occurs at (0, 0).

e) $y = x$

Solution: $y = x$

$y = 1x$

Slope = 1, y-intercept = 0 and to find x-intercept, plug 0 for y.

$$y = x$$
$$0 = x$$

x-intercept is 0 and it occurs at (0, 0).

f) $y = 0$

Solution: $y = 0$

Here, the value of m and b are 0 and for any value of x, the value of y is 0. It can be rewritten as $y = 0x + 0$

The slope (m) = 0, y-intercept is 0 and the x-intercept is also 0, it occurs at (0, 0).

Note: - The graph of $y = 0$ is the x-axis.

g) $x = 0$

Solution: $x = 0$

In this equation, for any value of y, the value of x is always 0. It shows the run or change in x is 0.

Thus, $m = \dfrac{y_2 - y_1}{x_2 - x_1}$, $x_2 - x_1 = 0$, the difference of any two real numbers $(y_2 - y_1)$ divided by 0 is undefined. So, the line $x = 0$ does not have slope. The y-intercept is 0 and it occurs at (0, 0). Furthermore; the x-intercept is 0 and it occurs at (0, 0).

Note: - The graph of $x = 0$ is the y-axis.

Exercise 2.11

Find the slope, y-intercept and x-intercept of the following linear equations.

a) $3y = -6x + 9$

b) $y = \dfrac{1}{3}x - 8$

c) $y = 2x$

d) $3x = y - 7$

e) $6y = 4x$

f) $5x + 7y - 8 = 0$

g) $-6x + y = 12$

2.12. Finding the Linear Equation of a Line

There are different ways finding the equation of the line. These are: -

2.12.1. Equation of a line from two given points (Slope-intercept form of a line)

Examples:
Find the equation of a line passing through the points:
a) (2, 3) and (4, –5).
b) (4, –2) and (3, 5).

Solutions
a) (2, 3) and (4, –5) , where $x_1 = 2$, $x_2 = 4$, $y_1 = 3$ and $y_2 = -5$. Use the following general formula to answer the above equations and the like.

$$\frac{y - y_1}{x - x_1} = \frac{y_2 - y_1}{x_2 - x_1}$$

$$\frac{y - 3}{x - 2} = \frac{-5 - 3}{4 - 2}$$

$$\frac{y - 3}{x - 2} = \frac{-8}{2}$$

$$\frac{y - 3}{x - 2} = -4$$

$y - 3 = -4(x - 2)$...................................(Cross multiplication)

$y - 3 = -4x + 8$

$y - 3 + 3 = -4x + 8 + 3$

$y = -4x + 11$

b) (4, –2) and (3, 5) where $x_1 = 4$, $x_2 = 3$, $y_1 = -2$ and $y_2 = 5$.

$$\frac{y - y_1}{x - x_1} = \frac{y_2 - y_1}{x_2 - x_1}$$

$$\frac{y - (-2)}{x - 4} = \frac{5 - (-2)}{3 - 4}$$

$$\frac{y + 2}{x - 4} = \frac{7}{-1}$$

$$\frac{y + 2}{x - 4} = -7$$

$(y+2) = -7(x-4)$...................................(Cross multiplication)

$y + 2 = -7x + 28)$

$y + 2 - 2 = -7x + 28 - 2$

$y = -7x + 26$

2.12.2. Equation of a Line from a Point and Given Slope. (Point-Slope Form of a Line)

Examples
1) Find the equation of a line with a slope 3 and passing through a point (2, 5).
    Solution:
    Given slope = 3, and point (2, 5), where $x_1 = 2$ and $y_1 = 5$.

$$\frac{y - y_1}{x - x_1} = m$$

$$\frac{y - 5}{x - 2} = 3$$

$y - 5 = 3(x - 2)$..(Cross multiplication)

$y - 5 = 3x - 6$

$y - 5 + 5 = 3x - 6 + 5$......................................(Add 5 on both sides)

$y = 3x - 1$

2) Find the equation of a line with a slope 4 and passing through a point (6, –3).

Solution:

Given slope = 4, and point (6, –3), where $x_1 = 6$ and $y_1 = -3$.

$$\frac{y - y_1}{x - x_1} = m$$

$$\frac{y - (-3)}{x - 6} = 4$$

$$\frac{y + 3}{x - 6} = 4$$

$y + 3 = 4(x - 6)$...(Cross multiplication)

$y + 3 = 4x - 24$...(Distributive property)

$y + 3 - 3 = 4x - 24 - 3$.....................................(Subtract 3 from both sides)

$y = 4x - 27$

3) Find the equation of a line with a slope –2 and passing through a point (0, 0).

Solution:

Given slope = –2 and point (0, 0), where $x_1 = 0$ and $y_1 = 0$.

$$\frac{y - y_1}{x - x_1} = m$$

$$\frac{y - 0}{x - 0} = -2$$

$y - 0 = -2(x - 0)$.................................(Cross multiplication)

$y - 0 = -2x + 0$

$y = -2x$

Exercise 2.12-2.12.2

1) Find the equation of the line passing through the points: -
   a) (–5, 4) and (–1, 3)
   b) (–2, 3) and (5, 2)
   c) (–1, 4) and (1, 0)
   d) (–3, 5) and (6, –2)

e) (0, –3) and (4, 3)
f) (1, 1) and (2, 5)
g) (2, 2) and (4, 4)

2) Find the equation of a line having the given slope m and passing through the given point.
   a) m = 4, (–3, –2)
   b) m = 3, (–3, –6)
   c) m = 1, (–2, –8)
   d) m = 2, (6, –4)
   e) m = –4, (–5, –3)

2.13. Graphing Linear Functions

The graph of a linear function is a straight line. The domain (value of x) of a linear function is the set of all real numbers and the range (value of y) is also the set of real numbers. We can draw the graph of a linear function in different methods. Let us see how to draw the graph of a linear function by plotting a point first and then drawing a line that connects these points.

Examples:

1) Draw the graph of each of the following linear equations.
   a) $y = 2x + 1$
   b) $y = -x - 2$
   c) $y = x$
   d) $y = x - 1$
   e) $y = 3$
   f) $y = -x$
   g) $x = -2$

Solutions:
   a) $y = 2x + 1$
      1) Begin by making a table of values
         (Choose convenient values of x)
      2) Plug the values of x into the original equation and find their corresponding values of y.
      3) Complete the table.
      4) Plot the points on the graph and draw a line by connecting such points.

$y = 2x + 1$	x	–2	–1	0	1	2
	y	–3	–1	1	3	5

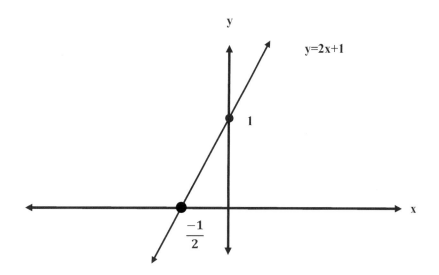

Fig. 2.5

b) $y = -x - 2$

$y = -x - 2$	x	–3	–2	–1	0	1	2
	y	1	0	–1	–2	–3	–4

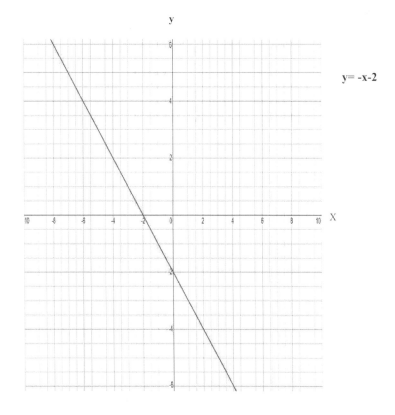

Fig. 2.6

c) y = x

y = x	x	−3	−2	−1	0	1	2	3
	y	−3	−2	−1	0	1	2	3

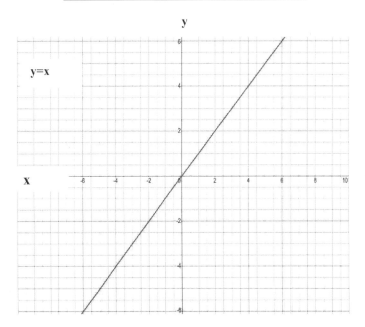

Fig. 2.7

d) y = x − 1

y = x − 1	x	−2	−1	0	1	2	3
	y	−3	−2	−1	0	1	2

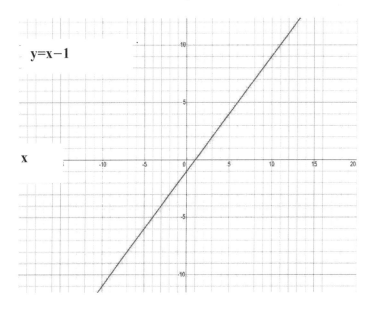

Fig. 2.8

58

e) y = 3

y = 3	x	–3	–2	–1	0	1	2	3
	y	3	3	3	3	3	3	3

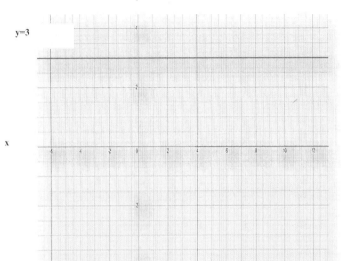

Fig. 2.9

f) y = –x

y = –x	x	–3	–2	–1	0	1	2	3
	y	3	2	1	0	–1	–2	–3

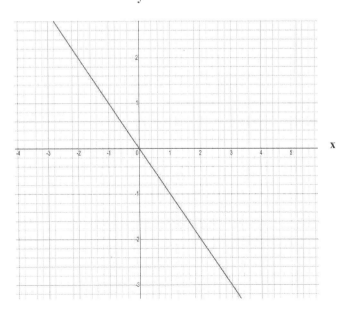

Fig. 2.10

g) x = –2

x = –2	x	–2	–2	–2	–2	–2
	y	–2	–1	0	1	2

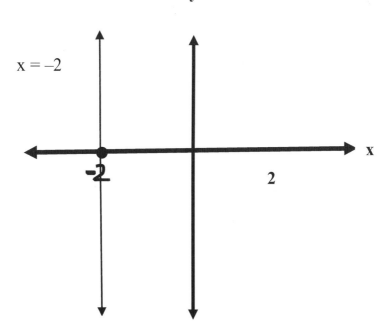

Fig. 2.11

2) Find the slope of the line passing through the points.
   a) (4, –2) and (0, 4)
   b) (–1, 3) and (5, 6)
   c) (0, 4) and (3, 4)
   d) (5, –2) and (5, 6)

Solutions: -
   a) (4, –2) and (0, 4)

   Assume $x_1 = 4$, $x_2 = 0$ and $y_1 = $ -2 and $y_2 = 4$.

   Hence, slope $m = \dfrac{y_2 - y_1}{x_2 - x_1} = \dfrac{4-(-2)}{0-4} = \dfrac{4+2}{-4} = \dfrac{-6}{4} = \dfrac{-3}{2}$

   b) (–1, 3) and (5, 6)

   Assume $x_1 = $ -1, $x_2 = 5$ and $y_1 = 3$ and $y_2 = 6$.

   $m = \dfrac{y_2 - y_1}{x_2 - x_1} = \dfrac{6-3}{5-(-1)} = \dfrac{3}{5+1} = \dfrac{3}{6} = \dfrac{1}{2}$

   c) (0, 4) and (3, 4)

   Assume $x_1 = 0$, $x_2 = 3$ and $y_1 = 4$ and $y_2 = 4$.

$$m = \frac{y_2 - y_1}{x_2 - x_1} = \frac{4-4}{3-0} = \frac{0}{3} = 0$$

d) (5, –2) and (5, 6)

Assume $x_1 = 5$, $x_2 = 5$ and $y_1 = -2$ and $y_2 = 6$

$$m = \frac{y_2 - y_1}{x_2 - x_1} = \frac{6-(-2)}{5-5} = \frac{6+2}{0} = \frac{8}{0}$$

Note: - Division by zero is undefined.

The expression $\frac{8}{0}$ does not have meaning. Thus, the line has no slope.

3) Find the value of x so that the line passing through the points (–3, 1) and (x, 5) has a slope 6.

Solution: Let $(x_1, y_1) = (-3, 1)$ and $(x_2, y_2) = (x, 5)$

$$\mathbf{m} = \frac{\mathbf{y_2 - y_1}}{\mathbf{x_2 - x_1}}$$

$$6 = \frac{5-1}{x-(-3)}$$

$$6 = \frac{4}{x+3}$$

$6(x+3) = 4$

$6x+18 = 4$

$6x+18-18 = 4-18$ ....(Subtracting 18 from both sides)

$6x = -14$

$6x = -14$

$$\frac{6x}{6} = \frac{-14}{6}$$

$$x = \frac{-7}{3}$$

$$x = \frac{-7}{3}$$

Thus, the value of x is $\frac{-7}{3}$

4) Find the value of y such that the line passes through the points (4,y) and (6, 8) has a slope $\frac{2}{5}$

Solution: Let $(x_1, y_1) = (4, y)$ and $(x_2, y_2) = (6, 8)$.

$$m = \frac{y_2 - y_1}{x_2 - x_1}$$

$$\frac{2}{5} = \frac{8-y}{6-4}$$

$$\frac{2}{5} = \frac{8-y}{2}$$

$$\frac{8-y}{2} = \frac{2}{5}$$

$5(8 - y) = 2 \times 2$ .... Cross multiplication.

$40 - 5y = 4$

$40 - 40 - 5y = 4 - 40$ .... Subtract 40 from both sides.

$-5y = -36$

$\frac{-5y}{-5} = \frac{-36}{-5} = 36/5$ .... Divide both sides by –5.

$$y = \frac{36}{5}$$

Exercise 2.13

1) Draw the graph of each of the following linear equations.
   a) $y = -2x + 1$
   b) $y = \frac{1}{2}x$
   c) $x + y = 1$
   d) $y = 2x$

2) Find the value of x so that the line that passes through the points (5, 3) and (x, –3) has a slope of 2

3) Find the value of y so that the line passes through the points (4, y) and (–8, 2) has a slope of $\frac{-1}{3}$

2.14. Classification of Lines by Slope

As we defined earlier, slope is the ratio of rise over run or vertical increase over horizontal increase. We can determine the line whether it has positive, negative, zero or has no slope just by looking at the rise and run.

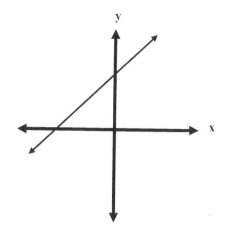

Fig.2.12
m>0
Positive Slope
Lines rises from left to right

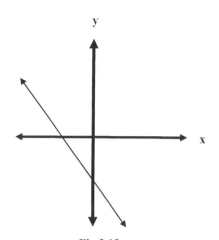

Fig.2.13
m<0
Negative slope
Lines falls from left to right

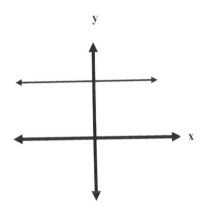

Fig.2.14
m=0
Zero Slope
Slope is constant
(Horizontal line)
Parallel to the x-axis

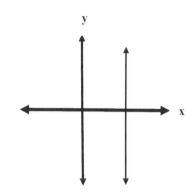

Fig.2.15
'm' is undefined
Has no slope
(Vertical line)
Parallel to the y-axis

2.15. Slope of Parallel and Perpendicular Lines

2.15.1. Parallel Lines

- Parallel lines are lines in a plane that do not intersect at any point. Parallel lines have the same slope.

In the figure below, the lines $\overleftrightarrow{XY}$ and $\overleftrightarrow{MR}$ are parallel.

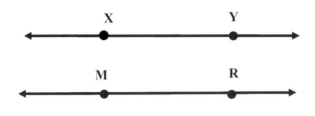

Fig. 2.16

Thus; $\overleftrightarrow{XY}$ // $\overleftrightarrow{MR}$

Note: We use the symbol // for showing parallel lines.

2.15.2.Perpendicular Lines

Perpendicular lines are lines that intersect at right angle, i.e., **90⁰**. If two lines are perpendicular to each other, then the product of their slopes is equal to –1.

In the figure below, the lines AB and CD are perpendicular to each other.

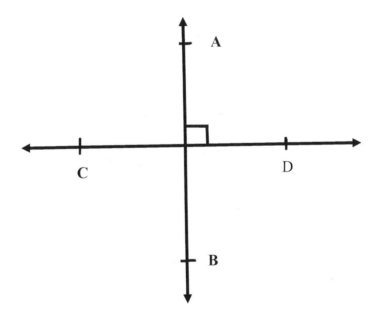

Fig. 2.17

Thus, $\overleftrightarrow{AB}$ ⊥ $\overleftrightarrow{CD}$

Note: We use the symbol ⊥ for showing perpendicular lines.
Examples of two parallel lines:

As you see from the diagram below, the parallel lines L_1 and L_2: -
   • Have the same slope, m = 2
   • Never going to intersect each other.

$L_1: = y = 2x$
$L_2: = y = 2x - 1$

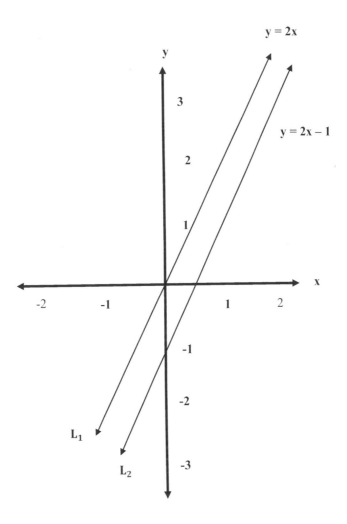

Fig. 2.18

Since L_1 is parallel to L_2, both of them have the same slope 2.

Example of two perpendicular lines.

As you see from the figure below:

- The slope of the two perpendicular lines are negative reciprocals. (Thus, the slopes 4 and $\frac{-1}{4}$ are negative reciprocals)
- These two lines are perpendicular and intersect at 90 degrees.

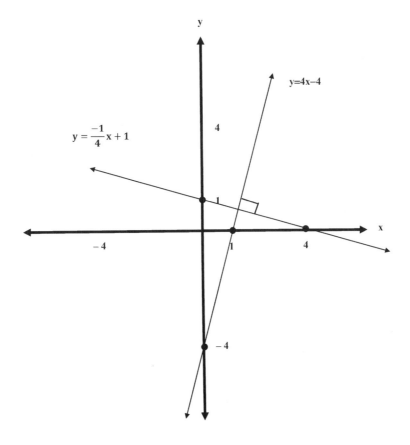

Fig. 2.19

$L_1 := y = 4x - 4$
$L_2 := y = \dfrac{-1}{4}x + 1$

As we can see, the slope of L_1 is 4 whereas that of L_2 is -1/4. In other words, their slopes are negative reciprocal.

Note: - The product of the slopes of $L_1 = 4$ and $L_2 = \dfrac{-1}{4}$ is:

$$m_1 m_2 = 4 \times \left(\dfrac{-1}{4}\right) = -1$$

Therefore, L_1 and L_2 are perpendicular lines.

Exercise 2.14-2.15.2

Decide whether the graphs of the following pairs of linear equations are parallel, perpendicular or neither nor.

a) $y = 3x$ and $y = \dfrac{-1}{3}x + 6$

b) $y = \dfrac{-5}{4}x + 5$ and $y = 5x + 5y - 10$

c) $y = -x$ and $y = x + 5$

d) $3x - y = 6$ and $y = 6x - 4$

e) $y + 3x = 6x + 5$ and $4y = -8x + 12$

f) $y = \dfrac{-1}{8}x + 8$ and $y = 8x$

g) $y = 3x + 1$ and $y = 5x - 1$

CHAPTER 3

RELATION AND FUNCTION

3.1. Relation: A relation is a set of ordered pairs.

Example: -
R = {(1, 4), (2, 5), (4, 6)}
In addition to set notation, we can represent the relation through tables, mapping diagram or by plotting on x-y axis.

X	Y
1	4
2	5
4	6

Relation in table

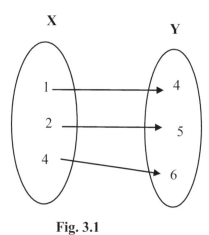

Fig. 3.1

Relation in mapping diagram

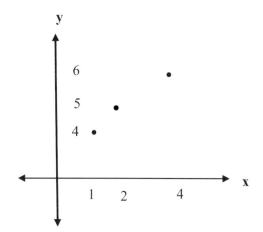

Relation in Graph

Fig.3.2

3.2. Function: A function is a relation in which for every element of the first set, there corresponds exactly one element of the second set.

Example 1)

Example 2)

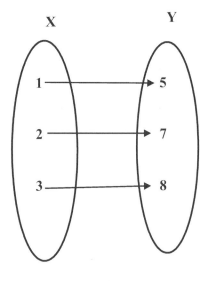

Fig.3.3

Function

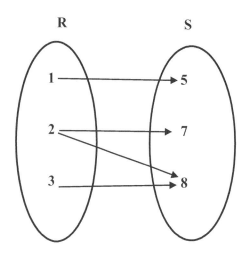

Fig.3.4

Not Function

In the above examples, example **1**, a relation from **X to Y** is an example of a function because every element of **X**, corresponds exactly one element of **Y**. Whereas, in example **2**, a relation from **R to S** is **not** a function because the first entry; value of **R** = 2 associates for **two** different second entries; values of **S**, 7 and **8**. S = 7 and S = 8. i.e., **(2,7)** and **(2,8)**.

3) More Examples:

  A = {(1, 2), (3, 4), (0, 5)}
  B = {(2, 3), (4, 7), (2, 5)}
  C = {(0, 5), (4, 8), (6, 7)}
  D = {(1, 6), (2, 4), (1, 8)}

In the above examples, all are examples of relations. But sets A and C are examples of functions because for every value of x there is only one corresponding y value; but sets B and D are not examples of functions because in set B, the first entry 2, corresponds to two different second entries, 3 and 5. And in set D, the first entry 1 corresponds to two different second entries 6 and 8. Thus, sets B and D are not examples of functions.

Note: Every function can be always a relation. But its inverse is not always holds true. That means, every relation cannot always be a function.

3.3. Domain and Range of a Function

3.3.1. Domain: The domain of a function is the set of all its possible inputs. In other words, it is the set of all possible values of the first entry or values of x.

Examples

1) A = {(1, 5), (2, 4), (3, 7)}

Here, in this example, the domains are values of the first entries. i.e., Domain of A = {1, 2, 3}

2)

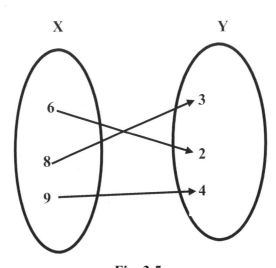

Fig. 3.5

In the above example 2, the set {6, 8, 9} is the domain of the function.

3.3.2. Range: The range of a function is the set of all possible values of its output. That means, it is the set of all possible values of second entries or values of y.

In the above example 1, set A has the range of {5, 4, 7}. Similarly, in example 2, the range of the function is {3, 2, 4}.

Example 3)

A = {(1, 5), (2, 7), (3, 9)}

Domain of A = {1, 2, 3}

Range of A = {5, 7, 9}

Example4)

P={(a,6), (t,7), (0,4), (8,8)}

Domain of P={a,t,0,8}

Range of P={6,7,4,8}

3.4. Vertical Line Test for Functions

When the graph of a function is given, the vertical line test is a method that helps to easily determine if that curve is a graph of a function or not. A function can only have one output, or y-value, for each unique input, or x-value. In the vertical line test, the graph represents a function if and only if at least one of the possible drawn vertical line intersects the given graph at most once. This means every x-value has related to a unique y-value. In contrast, if the vertical line that we draw arbitrarily intersects the curve more than once for any value of x, then the graph is not representing a function. This is because every x- value has related to two or more y-values.

Examples

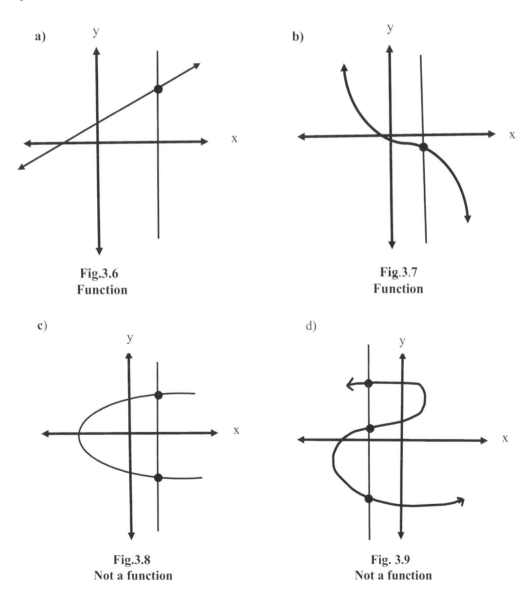

a)

Fig.3.6
Function

b)

Fig.3.7
Function

c)

Fig.3.8
Not a function

d)

Fig. 3.9
Not a function

Note: - In the above graphs 'a' and 'b', the vertical lines drawn cross the graph only once and due to this, these graphs are examples of functions. Whereas, the graphs of 'c' and 'd' are

71

crossed twice and three times respectively by the vertical test lines drawn. Thus, 'c' and 'd' are **not** functions.

Exercise 3.1-3.4

Answer each of the following questions

1) Which of the following graph has a positive slope?

a)

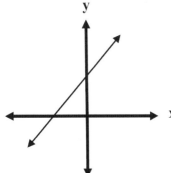

b)

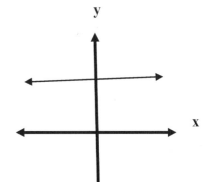

c)

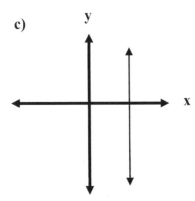

d)

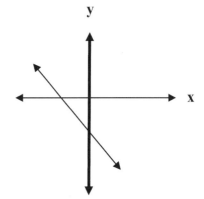

2) The graph of a horizontal line has
   a) Zero slope
   b) Negative slope
   c) Positive slope
   d) No slope

3) The graph of a vertical line has
   a) Positive slope
   b) Negative slope
   c) Zero slope
   d) No slope

4) _____ is the set of ordered pairs.
   a) Function
   b) Relation
   c) Domain
   d) Range

5) Which of the following relation is an example of a function?
   a) S = {(1, 5), (2, 6), (4, 5), (1, 8)}
   b) M = {(3, 7), (2, 9), (4, 5), (6, 4)}
   c) R = {(0, 2), (3, 5), (0, 3), (5, 7)}
   d) W = {(2, 5), (8, 7), (0, 0), (8, 9)}

6) If A = {(1, 5), (6, 3), (2, 8)} is a function, then the domain of set A is _____.
   a) {1, 6, 8}
   b) {5, 3, 8}
   c) {1, 2, 8}
   d) {1, 6, 2}

7) The range of set A in the above question number 6 is _____.
   a) {1, 6, 2}
   b) {5, 3, 8}
   c) {1, 6, 2, 5, 3, 8}
   d) {1, 5, 8}

8) Which of the following graph is a function?

a)
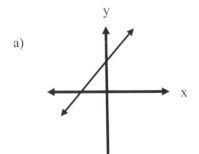

b)
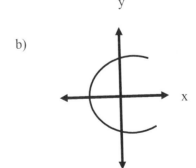

c)

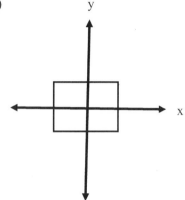

d)

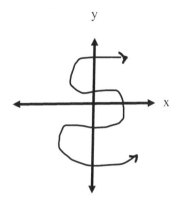

9) Which of the following mapping diagram is **not** a function?

a)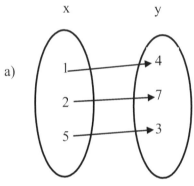

b)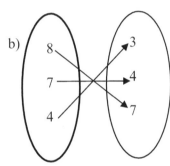

c)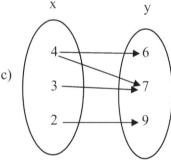

d)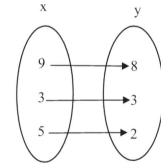

10) The graph represents a function if all possibly drawn vertical test lines intersect the graph at
_____.

a) least once
b) least twice
c) most once
d) no point

CHAPTER 4

Solving and Graphing Linear and Absolute Inequalities

4.1. Inequality: In mathematics, inequality tells us the relationship between two expressions which makes a non-equal comparison between two numbers or other mathematical expressions. It is used to compare two numbers on the number line by their size.

The followings are the most common inequalities. As they are involved in linear functions, they are called Linear Inequalities.

Symbols	Read as
<	Less than
>	Greater than
≤	Less than or equal to
≥	Greater than or equal to
≠	Not equal to

Examples: -

Take two numbers 6 and 9, we can express the relationship of these numbers in different ways using different inequality symbols. Thus,

$6 < 9$,   "6 is less than 9."

$9 > 6$,   "9 is greater than 6."

$6 \neq 9$,   "6 is different from 9" or

            "6 is not equal to 9"

Before we learn how to evaluate linear inequalities, we need to know some properties of inequalities.

4.1.1. Property of Trichotomy

For any two real numbers x and y, exactly one of the following holds true:

$x < y$, $x = y$ or $x > y$

4.1.2. Transitive Property of Inequality

If $x < y$ and $y < z$, then $x < z$

If $x > y$ and $y > z$, then $x > z$

These properties also hold true for "Less than or equal to" and "Greater than or equal to" linear inequalities; that is,

If $x \leq y$ and $y \leq z$, then $x \leq z$

If $x \geq y$ and $y \geq z$, then $x \geq z$

4.2. Solving Linear Inequalities

Rules of solving linear inequality are almost the same as solving linear equation. Same number can be added or subtracted without changing the inequality sign, just like linear equation. However, in the case of multiplying, dividing or square rooting numbers or terms, we have to check the given inequality sign. For instance, $8 > 6$. If we multiply both sides by negative one, we have $-8 < -6$. In general, when we multiply or divide the inequality term(s)/ equation(s) by negative number, the inequality sign will be reversed.

Examples

Solve for x.

    a) $x + 3 > 5$

    b) $x + 4 \geq 6$

    c) $x + 2 \leq 2x + 3$

    d) $2x + 3 \leq 9$

    e) $4x - 4 \geq 6x - 4$

    f) $-x + 3 > 2$

    g) $2x - 1 > 3$

Solutions

a) $x + 3 > 5$

    $x + 3 - 3 > 5 - 3$..(Subtract 3 from both sides)

    $x > 2$

Hence, the solution set is $x > 2$. We can write its solution set in different forms, Thus,

In set builder notation, Answer: $\{x: x > 2\}$

In interval notation, **Answer: $x \in (2, \infty)$**

Using number line,

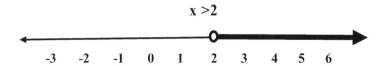

In other words, the solution set is the set of all real numbers that are greater than 2.

b) $x + 4 \geq 6$

    $x + 4 - 4 \geq 6 - 4$..(Subtract 4 from both sides)

    $x \geq 2$

In set builder notation, Answer: {x: x ≥ 2}

In interval notation, Answer: x ∈ [2,∞)

Using number line,

x ≥ 2

c) x + 2 ≤ 2x + 3

x − 2x + 2 − 2 ≤ 2x − 2x + 3 − 2.......................(Collect like terms)

−x ≤ 1

(−1) (−x) ≤ 1(−1)..(Multiply both sides by −1)

x ≥ −1 ....(When multiplying both sides by negative number, the inequality sign is reversed)

In set builder notation, Answer: {x: x ≥ −1}

In interval notation, Answer: x ∈ [−1,∞)

Using number line,

x ≥ −1

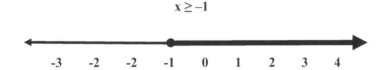

d) 2x + 3 ≤ 9

2x + 3 − 3 ≤ 9 − 3..(Subtract 3 from both sides)

2x ≤ 6

$\dfrac{2x}{2} \leq \dfrac{6}{2}$ ....(Divide both sides by 2)

x ≤ 3

- In set builder notation, Answer: {x: x ≤ 3}

  In interval notation, Answer: x ∈ (−∞, 3]

  Using number line,

x ≤ 3

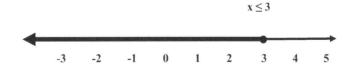

e) 4x − 4 ≥ 6x − 4

4x − 6x − 4 + 4 ≥ 6x − 6x − 4 + 4 ....(Collect like terms)

−2x ≥ 0

(−1) (−2x) ≥ 0 (−1)...(Multiply both sides by −1)

2x ≤ 0 ....(Inequality sign reversed)

$$\frac{2x}{2} \leq \frac{0}{2}$$ ....(Divide both sides by 2)
$$x \leq 0$$

In set builder notation, Answer: {x: x ≤ 0}
In interval notation, Answer: x ∈ (−∞, 0]
Using number line,

x ≤ 0

f)   −x + 3 > 2
    −x + 3 − 3 > 2 − 3...(Subtract 3 from both sides)
    −x > −1
    (−1) (−x) > (−1) (−1)...(Multiply both sides by −1)

    x < 1 ....(Inequality sign reversed)
    In set builder notation, Answer: {x: x < 1}
    In interval notation, Answer: x ∈ (−∞, 1)
    Using number line,

x < 1

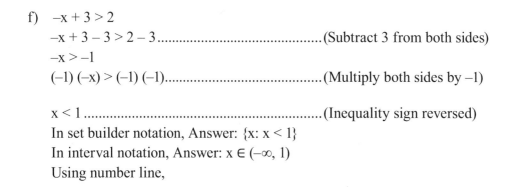

g)   2x − 1 > 3
    2x − 1 + 1 > 3 + 1...(Add +1 on both sides)
    2x > 4
    $$\frac{2x}{2} > \frac{4}{2}$$ ....(Divide both sides by 2)
    x > 2

In set builder notation, Answer: {x: x > 2}
In interval notation, Answer: x ∈ (2, ∞)
Using number line,

x > 2

Note: - The interval notation that we wrote in the above ways of writing solution in brackets and parentheses should be written in the same orders as they appear on the number line.

- The square brackets **[,]** indicates the boundary is included in the solution.
- The parenthesis **(,)** indicate the boundary is not included.
- An infinity symbol ∞ used to indicate the interval does not have an end point. Since ∞ is not a number, it should not be used with a square bracket.

Depending on the solution sets, the intervals can be written in the form of, **(,), [,], (,], [,)**

Exercise 4.1-4.2

A) Solve each of the following inequalities
   1) $2x - 1 \geq 3$
   2) $3x < x + 8$
   3) $8x + 4 \leq 5x - 14$
   4) $2x + 5 \leq 6x + 2$
   5) $-x + 4 \geq 3x + 6$

B) Solve each of the following and give your answer in interval notation.
   1) $2x - 1 \geq 3x - 4$
   2) $-5x + 2 \leq 2x - 12$
   3) $-x < 3x - 9$
   4) $3x - 5 < 6x + 3$
   5) $x - 4 \geq 2x + 5$

C) Solve each of the following and give your answer with number line.
   1) $-x + 2 > 5$
   2) $2x + 3 \geq x + 6$
   3) $3x - 1 \leq 2x + 4$
   4) $8x + 3 < 6x - 5$
   5) $2x + 3 \leq x + 3$
   6) $-x \leq 3$
   7) $-x > 2$
   8) $-2x \leq 4$

4.3. Graphing Linear Inequality on Number Line

The graph of a linear inequality in one variable is the set of points on the number line that represent all solutions of the inequality. Use an open circle (○) for < and > and a closed (a solid) circle (●) for ≤ and ≥.

Examples
Graph each of the following on the number line.

1) $-x \le 2$
2) All real numbers greater than 3.
3) All integers between -5 and 5.
4) All real numbers less than or equal to 0.
5) $x > 4$.
6) $x \le 3$
7) $2x < 4$
8) $-3x < 9$
9) $-x \ge -5$
10) $-2 < x$

Solutions

1) $-x \le 2$

$(-1)(-x) \le 2(-1)$..(Multiply both sides by -1)

$x \ge -2$..(inequality sign reversed)

$x \ge -2$

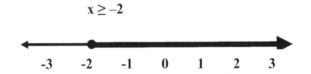

2) All real numbers greater than 3.
    All real numbers greater than 3 is the same as: $x > 3$.

$x > 3$

3) All integers between -5 and 5.
    The set of all integers between -5 and 5 is: $\{-4, -3, -2, -1, 0, 1, 2, 3, 4\}$

All integers between -5 and 5

4) All real numbers less than or equal to 0.
    $x \le 0$

x ≤ 0

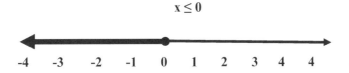

5) x > 4

x > 4

6) x ≤ 3

x ≤ 3

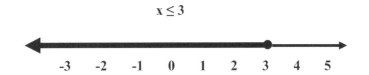

7) 2x < 4

$$\frac{2x}{2} < \frac{4}{2}$$ ....(Divide both sides by 2)

x < 2

x < 2

8) –3x < 9

$$\frac{-3x}{-3} < \frac{9}{-3}$$ ....(Divide both sides by –3)

x > –3 ....(The inequality sign reversed)

x > –3

9) –x ≥ –5

(–1)(–x) ≥ (–5)(–1) ....(Multiply both sides by –1)

x ≤ 5

$x \le 5$

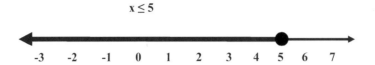

10) $-2 < x$

   $x > -2$

$x > -2$

4.4. Compound Inequality

A compound inequality consists of two inequalities joined either by conjunction "and" or "or". When two inequalities are joined by "and" the solution of the compound inequalities is the intersection of the solution of the compound inequalities. Whereas when two inequalities are joined by "or" the solution set is the union of the two individual inequalities solutions.

Examples: -
1) All real numbers that are greater than –2 and less than 5.

   Here, we have two statements
   (i)   All real numbers that are greater than –2
   (ii) All real numbers that are less than 5.

   These two statements are joined by "and"; thus, the solution set for this compound inequalities are the intersection of the solution of the compound inequalities. Thus,
   (i) $x > -2$

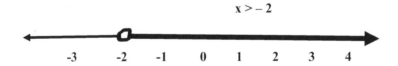

$x > -2$

(ii) $x < 5$

$x < 5$

Thus, the intersection of $x > -2$ and $x < 5$ is $-2 < x < 5$.

$$-2 < x < 5$$

2) All real numbers that are greater than 0 and less than or equal to 4.

   Here, we have two statements

   (i) All real numbers greater than 0.

   $x > 0$

   (ii) All real numbers less than or equal to 4.

   $x \leq 4$

   Thus,

   $x > 0$

$$x > 0$$

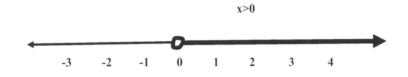

$x \leq 4$

$$x \leq 4$$

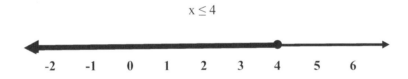

Since the compound statements are joined by conjunction "and", the solution set is the intersection of the two compound inequalities. That is, the solution set is $0 < x \leq 4$.

$$0 < x \leq 4$$

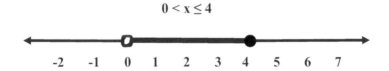

3) All real numbers that are less than –2 or greater than 3.

   Here, we have two statements that are connected by "or." Thus, the solution set of these compound inequalities is the union of the solution of the two statements.

   To brief it more,

   (i) All real numbers that are less than –2.

   $x < –2$

   (ii) All real numbers that are greater than 3.

   $x > 3$

   Both (i) and (ii) can be shown on number lines below: -

(i) x < –2

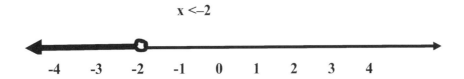

(ii) x > 3

Hence, the solution set is x < –2 or x > 3; that is, x < –2 ∪ x > 3

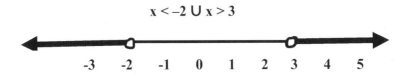

4) All real numbers that are less than 1 or greater than or equal to 2.
    We have two statements:
    (i)   All real numbers less than 1.
          x < 1

    (ii) All real numbers greater than or equal to 2.
          x ≥ 2
    Both (i) and (ii) can be shown on number lines as

    (i)   x < 1

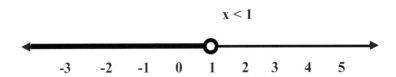

(ii) x ≥ 2

Thus, the solution set for this compound statement is the union of the solution set of (i) and (ii). That is,

$x < 1 \cup x \geq 2$

x < 1 or x ≥ 2 OR x < 1 ∪ x ≥ 2

5) Find the solution set and graph on the number line for $-3 < x + 1$ and $x + 1 \leq 2$

Solution: We can rewrite the equations as $-3 < x + 1 \leq 2$

$-3 - 1 < x + 1 - 1 \leq 2 - 1$ ....Subtract 1 from each term.

$-4 < x \leq 1$

Thus, the solution set is $-4 < x \leq 1$

$-4 < x \leq 1$

Alternate Method

$-3 < x + 1$

$-3 - 1 < x + 1 - 1$ ....Subtract 1 from both sides

$-4 < x$

$x > -4$

$x + 1 \leq 2$

$x + 1 - 1 \leq 2 - 1$ ....Subtract 1 from both sides

$x \leq 1$

$-3 < x + 1$ and $x + 1 \leq 2$

The connective word "and" tells us "Intersection."

$x > -4 \cap x \leq 1$

$-4 < x \leq 1$

Thus, the solution set is $-4 < x \leq 1$.

$-4 < x \leq 1$

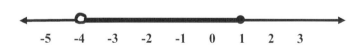

6) Find the solution set for $x \geq 1$ and $x < 1$ and graph your answer on number line.

Solution: For $x \geq 1$

x≥1

For x < 1

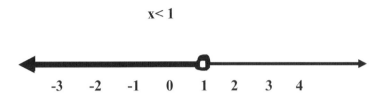

As we see the graphs of x ≥ 1 and x<1 above, they don't have any element in common. Thus, the solution set is empty set. i.e., **x ≥ 1 and x<1 = {   }**

N.B.: - In finding the solution set for compound sentences, the conjunction 'and' means intersection (∩) and the conjunction 'or' means union (∪).

Examples: x<6 and x ≥ 4 means x<6 ∩ x ≥ 4

x<6 or x ≥ 4 means x<6 ∪ x ≥ 4

Exercise 4.3-4.4

Find the solution set of the following compound inequalities and graph their solution sets on the number line.

1) 2x < 8 and x >–1
2) x < 3 or –x< 2
3) x + 3 ≥ 2 and x ≤ 3
4) x ≤ 0 or x ≥ 1
5) x >–1 or x ≤ –4
6) 3 < 2x + 1 and 2x +1 ≤ 5
7) 2 ≤ 2x – 1 < 1
8) x >–4 and 2x < 6
9) –x < 3 or x < 2
10) 3x < 9 and x >–2

4.5. Solving Absolute Value Equations

The absolute value of a real number x, denoted by |x| is the distance from zero on the number line and it is a non-negative number. It can be defined by:

$$|x| = \begin{cases} x, & \text{if } x \geq 0 \\ -x, & \text{if } x < 0 \end{cases}$$

Examples:

    a) $|5| = 5$
    b) $|0| = 0$
    c) $|6| = 6$
    d) $|-8| = -(-8) = 8$
    e) $|-4| = -(-4) = 4$

Absolute equation is an equation that contains an absolute value expression; just like linear equation, solving absolute value equation is easy. The only difference is that to solve absolute value equation, you can use the fact that the expression inside the absolute value symbol can be either positive or negative.

Examples
Solve each of the following expressions
    1) $|x + 6| = 4$
    2) $|2x - 5| = 6$
    3) $|x + 3| = -2$
    4) $|2x - 1| = 7$
    5) $|6x - 3| + 2 = 4$
    6) $|3x - 2| - 4 = 6$
    7) $|-2x - 3| = -4$
    8) $|x - 6| = 0$
    9) $|2x - 3| = 7$
    10) $|x| = 8$

Solutions
1) $|x + 6| = 4$

To solve such equation, we have two cases, the first one is $|x + 6|$ is positive and the second case is $|x + 6|$ is negative, that is, when $|x + 6| > 0$ and $|x + 6| < 0$. Thus,
Case 1
    $|x + 6|$ is positive
    $x + 6 = 4$
    $x + 6 - 6 = 4 - 6$..Subtract 6 from both sides.
    $x = -2$

Case 2
    $|x + 6|$ is negative
    $x + 6 = -4$
    $x + 6 - 6 = -4 - 6$..(Subtract 6 from both sides)
    $x = -10$

Thus, the solution set is $\{-2, -10\}$

2) $|2x - 5| = 6$

Case 1

$|2x - 5| = 6$

$2x - 5 = 6$

$2x - 5 + 5 = 6 + 5$

$2x = 11$

$\dfrac{2x}{2} = \dfrac{11}{2}$

$\dfrac{11}{2}$

Case 2

$|2x - 5| = 6$

$2x - 5 = -6$

$2x - 5 + 5 = -6 + 5$

$2x = -1$

$\dfrac{2x}{2} = \dfrac{-1}{2}$

$x = \dfrac{-1}{2}$

Thus, the solution set is $\left\{ \dfrac{-1}{2}, \dfrac{11}{2} \right\}$

3) $|x + 3| = -2$

Solution: The absolute value of a number is always greater than or equal to zero. That is, the absolute value of a number can never be negative. So, there is no solution set for this equation. Thus

The solution set is $\{\ \ \}$.

(The solution set is empty-set)

4) $|2x - 1| = 7$

Case 1

$|2x - 1| = 7$

$2x - 1 = 7$

$2x - 1 + 1 = 7 + 1$

$2x = 8$

$\dfrac{2x}{2} \quad \dfrac{8}{2}$

$x = 4$

Case 2

$|2x - 1| = 7$

$2x - 1 = -7$

$2x - 1 + 1 = -7 + 1$

$2x = -6$

$$\frac{2x}{2} = \frac{-6}{2}$$

$x = -3$

Thus, the solution set is $\{-3, 4\}$

5) $|6x - 3| + 2 = 4$

Case 1

$|6x - 3| + 2 = 4$

$6x - 3 + 2 = 4$

$6x - 1 = 4$

$6x - 1 + 1 = 4 + 1$

$6x = 5$

$$\frac{6x}{6} = \frac{5}{6}$$

$$x = \frac{5}{6}$$

Case 2

$|6x - 3| + 2 = 4$

$|6x - 3| + 2 - 2 = 4 - 2$

$|6x - 3| = 2$

$6x - 3 = -2$

$6x - 3 + 3 = -2 + 3$

$6x = 1$

$$\frac{6x}{6} = \frac{1}{6}$$

$$x = \frac{1}{6}$$

Thus, the solution set is $\left\{\dfrac{1}{6}, \dfrac{5}{6}\right\}$

6) $|3x - 2| - 4 = 6$

Case 1

$|3x - 2| - 4 = 6$

$|3x - 2| - 4 + 4 = 6 + 4$

$|3x - 2| = 10$

$3x - 2 = 10$

$3x - 2 + 2 = 10 + 2$

$3x = 12$

$$\frac{3x}{3} = \frac{12}{3}$$

$x = 4$

Case 2

$|3x - 2| - 4 = 6$

$|3x - 2| - 4 + 4 = 6 + 4$

$|3x - 2| = 10$

$3x - 2 = -10$

$3x - 2 + 2 = -10 + 2$

$3x = -8$

$\dfrac{3x}{3} = \dfrac{-8}{3}$

$x = \dfrac{-8}{3}$

Thus, the solution set is $\left\{\dfrac{-8}{3}, 4\right\}$

7) $|-2x - 3| = -4$

The absolute value of a number can never be negative. Such expression does not have solution set. Thus

The solution set is $\{\quad\}$.

8) $|x - 6| = 0$

In this case we have only one case.

$|x - 6| = 0$

$x - 6 = 0$

$x - 6 + 6 = 0 + 6$

$x = 6$

Thus, the solution set is $\{6\}$

9) $|2x - 3| = 7$

Case 1

$|2x - 3| = 7$

$2x - 3 = 7$

$2x - 3 + 3 = 7 + 3$

$2x = 10$

$\dfrac{2x}{2} = \dfrac{10}{2}$

$x = 5$

Case 2

$|2x - 3| = 7$

$2x - 3 = -7$

$2x - 3 + 3 = -7 + 3$

$2x = -4$

$\dfrac{2x}{2} = \dfrac{-4}{2}$

$x = -2$

Thus, the solution set is $\{-2, 5\}$

10) $|x| = 8$

Case 1

$|x| = 8$

$x = 8$

Case 2

$|x| = 8$

$x = -8$

Thus, the solution set is $\{-8, 8\}$

Exercise 4.5

Solve for x

1) $|x - 4| = 3$
2) $|2x - 3| - 5 = 6$
3) $|3x + 2| + 3 = -2$
4) $|5x - 1| - 3 = -9$
5) $|3x + 2| = 0$
6) $|3x - \dfrac{1}{2}| = 7$
7) $|8x - 4| = 2$
8) $|2x - 1| + 6 = 6$
9) $|x - 10| = 7$
10) $|2x - 1| = -4$

4.6. Inequalities with Absolute Value

As we have discussed before, the absolute value of a real number x, denoted by $|x|$ describes the distance of x from reference point zero on a real number line. We are going to use this description to solve an inequality with absolute value equations.

Consider $|x| < 4$.

This means that the distance of the point x from 0 is less than 4. The interval shows that the value of x is less than 4 units from the reference point 0. Thus, x is between -4 and 4. This means, x is greater than -4 and less than 4. We write this as $\{x: -4 < x < 4\}$. The number line below shows: -

$\{x: -4 < x < 4\}$, That is $|x| < 4$.

$$-4 < x < 4$$

Note: The absolute value expresses only the distance, not the direction of the number on a number line, it is always expressed as a positive number or zero.

- Find the solution set of $|x| > 4$ and indicate the solution set on the number line.

Solution:

$|x| > 4$

In this case, "greater than" inequality symbol, that is $|x| > 4$, is the distance of the point x from 0 is greater than 4. That means, x can be less than -4 or greater than 4 or $x < -4$ or $x > 4$. The figure below shows $|x| > 4$.

x<-4 or x>4

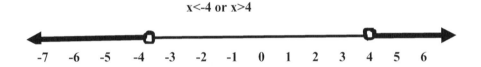

From the above description for solving an absolute inequality, we can generalize that: -

If x is an algebraic expression and "a" is a positive number, then

(i) The solutions of $|x| < a$ are all real numbers that satisfy: $-a < x < a$

(ii) The solutions of $|x| > a$ are all real numbers that satisfy: $x < -a$ or $x > a$

These rules also hold true if the inequalities < and > are replaced by ≤ and ≥ respectively. That is,

(iii) The solutions of $|x| \leq a$ are all numbers that satisfy $-a \leq x \leq a$.

(iv) The solutions of $|x| \geq a$ are all numbers that satisfy $x \leq -a$ or $x \geq a$.

Examples

A) Find the solution set of each of the following inequalities and indicate them on the number lines.

1) $|x| < 3$
2) $|x| > 2$
3) $|x| \leq 5$
4) $|x| \geq 1$

Solutions: -

To answer the questions 1-4, apply the above rules (i)-(iv).

1) $|x| < 3$

−3< x < 3 ………. This is the solution set

−3< x < 3

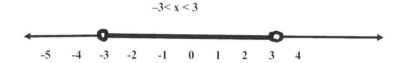

2) $|x| > 2$

x <−2 or x > 2……… This is the solution set.

x <−2 or x > 2

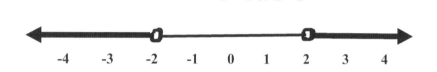

3) $|x| \leq 5$

−5 ≤ x ≤ 5…………. This is the solution set.

−5 ≤ x ≤ 5

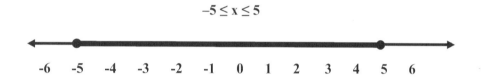

4) $|x| \geq 1$

x ≤ −1 or x ≥ 1……… This is the solution set.

x ≤ −1 or x ≥ 1

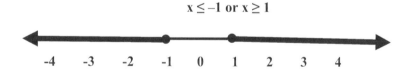

B) Solve and graph the solution set on the number line.
   1) $|x − 3| < 2$
   2) $|x − 1| > 3$
   3) $|x − 2| \leq 3$
   4) $|x − 4| \geq 1$
   5) $−3|x − 1| + 6 \geq −12$
   6) $−2|x − 3| \leq −8$
   7) $|x − 1| \geq 0$

Solutions: -

   1) $|x − 3| < 2$
   We have seen that the solution set of $|x| < a$ is -a< x< a, Thus,

-2<x-3<2

-2+3< x-3+3<2+3 ………. (Add 3 to the left, middle and right sides of the inequalities.)

1<x<5

Thus, the solution set is {x: 1<x<5}

1<x<5

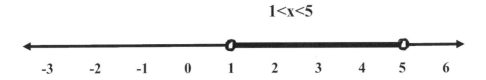

2) $|x - 1| > 3$

We are already familiar that the solutions set of $|x| > a$ are all real numbers that satisfy x < -a or x > a. Thus,

$|x - 1| > 3$

x-1 < -3 or x-1>3

x-1+1 < -3 + 1 or x-1+1 > 3+1 ……………………. (Add 1 on both sides of the inequalities.)

x <–2 or x > 4

Thus, the solution set is {x: x <–2 or x > 4}

x <–2 or x > 4

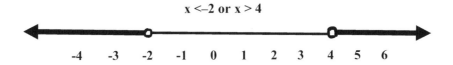

3) $|x - 2| \leq 3$
    $-3 \leq x - 2 \leq 3$
    $-3 + 2 \leq x - 2 + 2 < 3 + 2$ ……………………(Add 2 all sides of the inequalities)
    $-1 \leq x \leq 5$
    Thus, the solution set is {x: $-1 \leq x \leq 5$}

$-1 \leq x \leq 5$

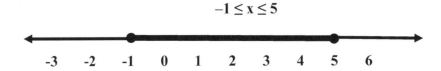

4) $|x - 4| \geq 1$
    $x - 4 \leq -1$ or $x - 4 \geq 1$
    $x - 4 + 4 \leq -1 + 4$ or $x - 4 + 4 \geq 1 + 4$…………(Add 4 to both sides)
    $x \leq 3$ or $x \geq 5$
    Thus, the solution set is {x: $x \leq 3$ or $x \geq 5$}

x ≤ 3 or x ≥ 5

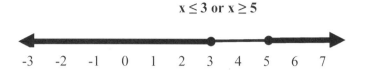

5) $-3|x - 1| + 6 \geq -12$

   $-3|x - 1| + 6 - 6 \geq -12 - 6$ ....(Subtract 6 from both sides)

   $-3|x - 1| \geq -18$

   $\dfrac{-3}{-3}|x - 1| \geq \dfrac{-18}{-3}$ ....(Divide both sides by –3)

   $|x - 1| \leq 6$..(The inequality sign is reversed)

   $-6 \leq x - 1 \leq 6$

   $-6 + 1 \leq x - 1 + 1 \leq 6 + 1$ ....(Add 1 to all sides of the inequalities)

   $-5 \leq x \leq 7$

   Thus, the solution set is {x: –5 ≤ x ≤ 7}

–5 ≤ x ≤ 7

6) $-2|x - 3| < -8$

   $\dfrac{-2}{-2}|x - 3| < \dfrac{-8}{-2}$ ....(Divide both sides by –2)

   $|x - 3| > 4$..(The inequality sign reversed)

   x – 3 < – -4 or x –3 > 4

   x – 3 + 3 <–4 + 3 or x – 3 + 3 > 4 + 3 ....(Add 3 on both sides)

   x <–1 or x > 7

   Thus, the solution set is {x: x <–1 or x > 7}

x <–1 or x > 7

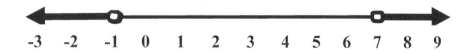

7) $|x - 1| \geq 0$

   x – 1 ≥ 0 or x – 1 ≤ 0

   x – 1 + 1 ≥ 0 + 1 or x – 1+1 ≤ 0 + 1 ....(Add 1 on both sides)

   x ≥ 1 or x ≤ 1

   Thus, the solution set is:................................. {x: x ≤ 1or x ≥ 1}

$(-\infty, 1] \cup [1, \infty)$

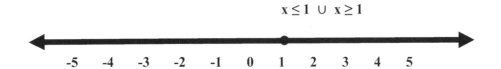

$$x \le 1 \ \cup \ x \ge 1$$

Note: - The solution set for $|x - 1| \ge 0$ is the set of real numbers.

i.e., $(-\infty, 1] \cup [1, \infty) = \mathbb{R}$

Exercise 4.6

Solve each of the following inequalities and indicate the solution set on the number line.

1) $|x - 3| < 2$
2) $|4 - 2x| > 3$
3) $|2x| \ge 8$
4) $|2x + 4| + 6 \ge 2$
5) $|x - 1| \le 4$
6) $|x + 4| < 3$
7) $|2x - 1| \ge 2$
8) $|x| \le 6$
9) $|2x| < -3$
10) $-|x| \le -2$

4.7. Graphs of Absolute Value Function on Coordinate Plane

Method of drawing the absolute value function $f(x) = |x|$ or $y = |x|$ on coordinate plane system is the same as linear function. To draw the graph of absolute value function, choose some values of x and find their corresponding values of y or their ordered pairs (x, y).

Examples
1) Graph $y = |x|$ on a coordinate plane system.

Solution: First make a table of values of x and y and use these values to draw the graph of the function $y = |x|$.

$y = |x|$	x	−3	−2	−1	0	1	2	3
	y	3	2	1	0	1	2	3

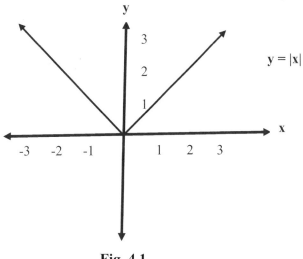

Fig. 4.1

2) Draw the graph of f(x) = |x + 1| on the rectangular coordinate plane system.
Solution: First make a table of values of x and y and use these values to draw the graph of f(x) = |x + 1|.

f(x) = |x + 1|	x	−4	−3	−2	−1	0	1	2	3	4
	f(x)	3	2	1	0	1	2	3	4	5

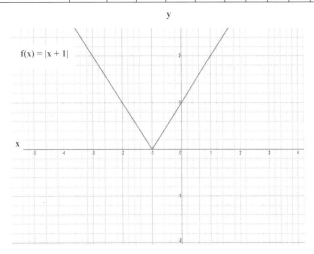

Fig. 4.2

3) Draw the graph of **f(x) = −|x|**
Solution: First make a table of values of x and y and use these values to draw the graph of the function: f(x) = |x|.

f(x)= −|x|	x	-3	-2	-1	0	1	2	3
	f(x)	-3	-2	-1	0	-1	-2	-3

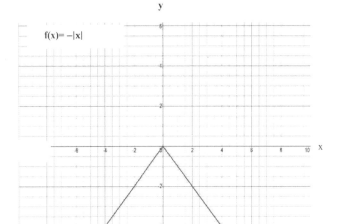

Fig. 4.3

4.8. Behaviors of Graphs of Parent Absolute Value Function: f(x) = |x|.

As we have seen above, graphs of absolute functions have different behaviors. Some of these are: -

1) The domain (value of x) of absolute value function is the set of all real numbers.
2) The range (value of y) of absolute value function is the set of all real numbers greater than or equal to zero.
3) The graphs of an absolute value functions generally have the shape of up right (up-side) V, or inverted (down-side) V.
4) An absolute value function graph has one axis of symmetry that passes through the vertex.

Note: -In mathematics, most of the time, the variable y is represented by f(x).
Example: $y = |x|$ can be rewritten as $f(x) = |x|$.

Exercise 4.7-4.8

Draw the graph of each of the following questions.

1) $f(x) = |x - 2|$
2) $f(x) = -|2x|$
3) $y = |x - 1| + 1$
4) $f(x) = |2x - 3|$

5) $y = -|x+2|$

4.9. The Graph of Linear Inequality in Two Variables

The graph of an equation in two variables is the set of all points whose coordinates, (x, y) makes the equation true. In the same way, the graph of inequality in two variables is the set of all points whose coordinates (x, y) makes the inequality true. The graph of a linear equality divides the coordinate plane in to two halves by a boundary line where one half represents the solutions of the inequality whereas the other not. This boundary line is dashed/ broken line for > and < signs or solid line for ≤ and ≥ signs.

Examples

1) Graph the inequality x + y > 3.
Solution; - Replace the inequality > by = and graph the linear equation.
When   x = 0
$$x + y = 3$$
$$0 + y = 3$$
$$y = 3$$
(0, 3)

When   y = 0
$$x + y = 3$$
$$x + 0 = 3$$
$$x = 3$$
(3, 0)

Here, (0, 3) is the y-intercept and (3, 0) is the x-intercept.

The x-intercept is 3 and it occurs at (3, 0) and the y-intercept is 3, and it occurs at (0, 3). Using the two intercepts, the boundary of its graph is shown as a dashed line in the figure below. This is because the inequality x + y > 3 contains a > sign/symbol. In other words, those points that make the equation true are not included in the solution set. This is because the true set of this equation excludes all points that satisfies x+y=3

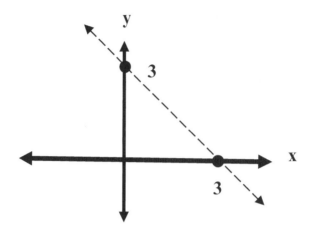

Fig. 4.4

As we have seen from the above graph, the plane is divided into two. Take any arbitrary test point either from the upper or lower part of a half-plane but not from the dashed line. A common test point is the origin, (0, 0). Then, substitute this chosen test point into the inequality $x + y > 3$. This test point belongs to the lower half plane of the graph. When we substitute $x = 0$ and $y = 0$ in $x + y > 3$, we get $0 + 0 > 3$ or $0 > 3$. This is false. Thus, shade the other half plane of the graph which does not include the point (0, 0). That is, shade the untested upper half plane. Indirectly, the solution set of the given inequality: $x + y > 3$ is the set of all points above the dashed line.

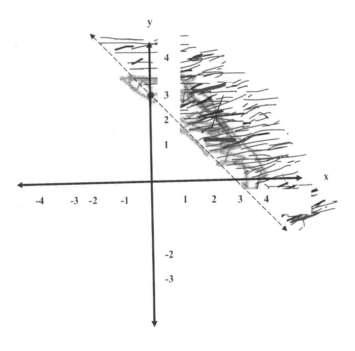

Fig. 4.5

2) Graph: $y \leq x$

    Solution: Replace the inequality sign/symbol $\leq$ by equality = sign/ symbol and graph the linear equation $y = x$. Since we are supposed to draw the graph of the inequality $y \leq x$, the

100

x and y intercepts for y = x are the same, the point (0, 0) or the origin. This shows that the intended graph passes through the point (0, 0).

Note: - Since the inequality symbol is less than or equal to, the boundary line of the graph is a solid line.

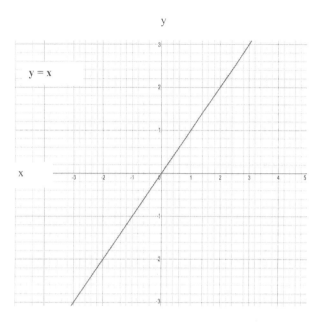

Fig. 4.6

Choose a test point; in this case we cannot take the test point (0, 0). This is because such point lies on the solution set of the graph. Let us take arbitrary test point (2, 3) which is on the upper half plane. This shows $y \leq x$, $3 \leq 2$. This is false, because it does not satisfy the inequality $y \leq x$.

This shows that all points above the boundary (the solid) line are not the solution set of the given inequality. Thus, shade the other half plane which does not include the point (2, 3). That is, shade the lower half plane. The solution set of the inequality: $y \leq x$ is the set of all points on the graph y=x and the rest of the regions that are found below the boundary solid line.

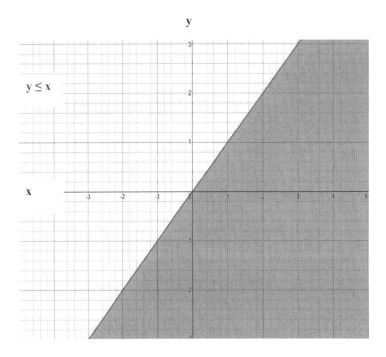

Fig. 4.7

3) Sketch the graph of the inequality $y \geq x + 2$.

Solution: Replace the inequality symbol $\geq$ by $=$ and graph the linear equation $y = x + 2$ to know the boundary reference. To draw the graph of the inequality $y \geq x + 2$, the x and y intercepts for $y = x + 2$ are $(-2, 0)$ and $(0, 2)$ respectively. Using these intercepts, we can draw the graph as shown below.

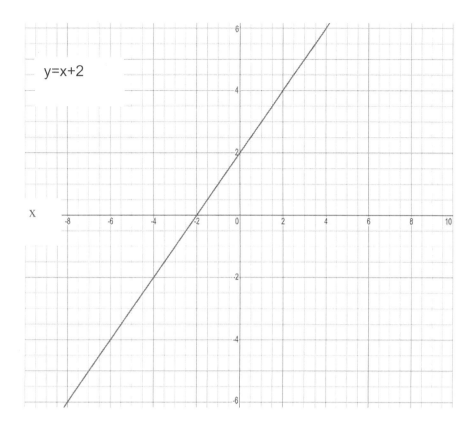

Fig. 4.8

Note: -Since the inequality symbol is greater than or equal to, the boundary reference line is solid line.

Choose any test point from upper or lower half plane but not from the boundary line. Let us take the common point (0, 0) and substituting in the inequality $y \geq x + 2$; $0 \geq 0 + 2$, $0 \geq 2$, which is false. Thus, shade the other half plane that does not include the point (0, 0). That is, shade the upper half plane. The solution set of the given inequality: $y \geq x + 2$ is the set of all points above and including the boundary solid line.

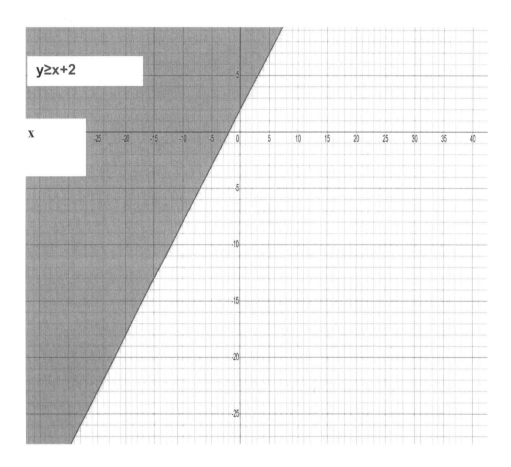

Fig. 4.9

4) Graph the inequality: $y \leq -x + 1$

Solution: -Replace the inequality symbol $\leq$ by = and graph the linear equation $y = -x + 1$, So as to know the boundary reference. Since we are going to draw the graph of the inequality $y \leq -x + 1$, the x and y intercepts for $y = -x + 1$ are (1, 0) and (0, 1) respectively. Using these intercepts, we can draw the graph as shown below

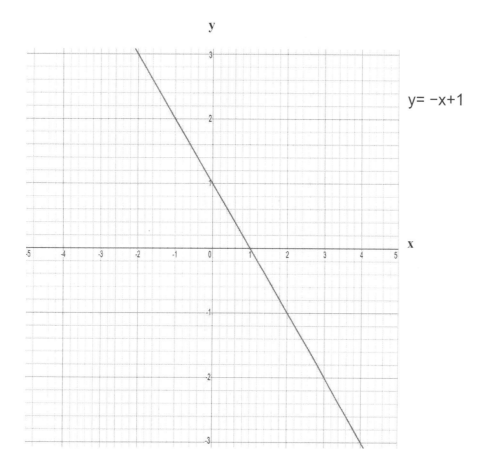

Fig. 4.10

Note: - Since the inequality symbol is less than or equal to, the boundary reference line is solid line.

Choose any test point from upper or lower half plane but not from the boundary line. Let us take the common point (0, 0) and substituting in the inequality $y \leq -x + 1$, $0 \leq 0 + 1$, $0 \leq 1$, which is absolutely true. Thus, shade the half plane which includes the point (0, 0). That is, shade the lower half plane. The solution set of the given inequality: $y \leq -x + 1$ is the set of all points below and including the boundary reference solid line.

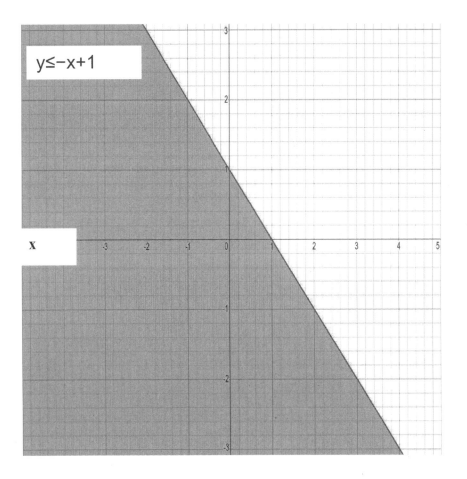

Fig. 4.11

5) Graph the inequality **y < x –1**

Solution: -Replace the inequality symbol < by = and graph the linear equation y = x – 1, for boundary reference. Since we are supposed to draw the graph of the inequality y < x –1, the x and y intercepts for y = x – 1 are (1, 0) and (0, –1) respectively. Using these intercepts, we can draw the graph of the reference boundary line as shown below.

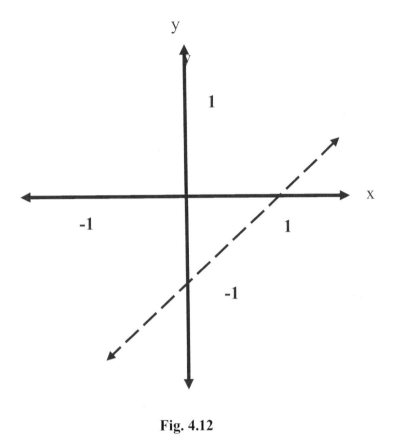

Fig. 4.12

Note: Since the inequality sign is less than (strict inequality), the boundary reference line is dashed line.

Remark: - A strict inequality equation is an inequality equation that has either > (greater than), or < (less than) sign in its term. In other words, a strict inequality is an inequality which does not have equal sign.

Choose any test point from above the half plane the point (0, 0) which is the common test point and substituting in the inequality $y < x - 1$, $0 < 0 - 1$, $0 < -1$, which is false. Thus, shade the half plane which does not include the test point (0, 0). That is, shade the lower half plane. The solution set of the given inequality: $y < x - 1$ is the set of all points below and not including the boundary reference dashed line.

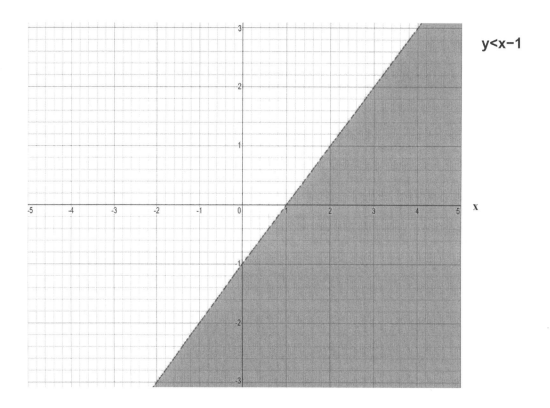

Fig. 4.13

Note: - We can graph linear inequalities (written in the form $y < mx + b$ or $y > mx + b$) without using test points. The inequality symbol $>$ or $<$ indicates which half-plane supposed to be shaded.

- If $y < mx + b$, we need to shade the lower half-plane of the line $y = mx + b$ (since the inequality symbol is strict ($<$), the boundary line must be dashed line.)
- For $y > mx + b$, we need to shade the upper half plane of the line $y = mx + b$ (since the inequality symbol is strict ($>$), the boundary line must be dashed line.)
- For $>$ or $<$ inequality signs, the dashed line is not in the solution set of the inequality.
- For the linear inequality $\leq$ or $\geq$, we need to draw solid line showing that the boundary line itself is in the solution set of the linear inequality.

i.e., - When we are graphing an inequality with two variables, for $<$ or $>$ sign, we use a dashed line and for $\leq$ or $\geq$ sign, we use solid line.

4.9.1. For Horizontal Line(s) $y \geq b$ or $y \leq b$

- If $y \geq b$, we need to shade the upper plane of the line $y = b$.
- If $y \leq b$, we need to shade the lower plane $y = b$.

Note: - Both graphs (the shaded parts) include those points that are found on the boundary line. No dashed line(s).

108

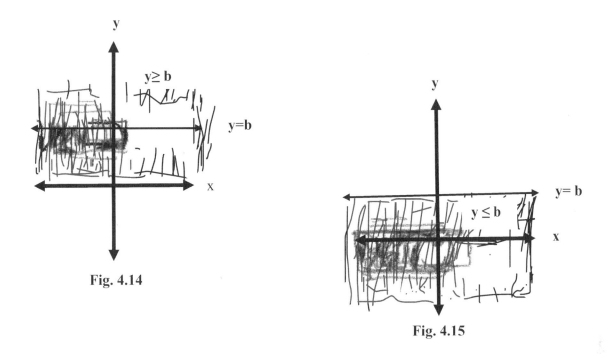

Fig. 4.14

Fig. 4.15

Note: - The rule holds true for the inequality y > b or y < b, but the boundary line must be dashed line and the dashed line is not in the solution set of the inequality line.

4.9.2. For Vertical Line(s) x ≥ a or x ≤ a:

- If x ≥ a, we need to shade the plane to the right side of the line x = a.
- If x ≤ a, we need to shade the plane to the left side of the line x = a.

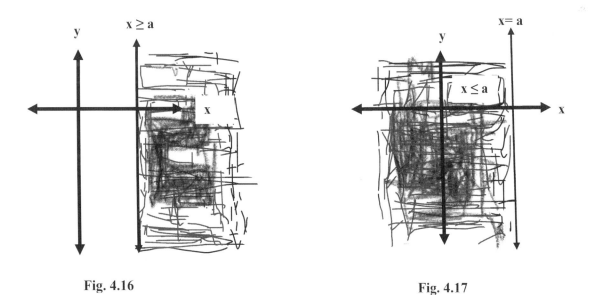

Fig. 4.16

Fig. 4.17

Note: - The rule holds true for the inequality x > a or x < a, but the boundary line must be dashed line and the dashed line is not in the solution set of the inequality line.

Examples: -

Graph each of the following inequalities in a rectangular coordinate plane without a test point.

a) x > 3
b) y < 4
c) y ≤ 1
d) x ≥ 0
e) y ≤ 0
f) y > –2
g) x < –3

Solution:

a) x > 3   b) y<4

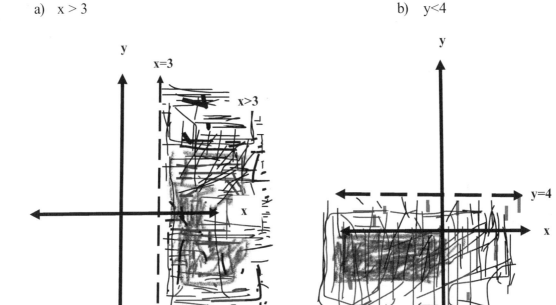

Fig. 4.18

Fig. 4.19

c) $y \leq 1$

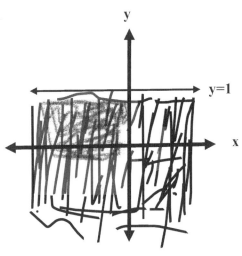

y=1

Fig. 4.20

Note: The graph of x = 0 is the y-axis.

d) $x \geq 0$

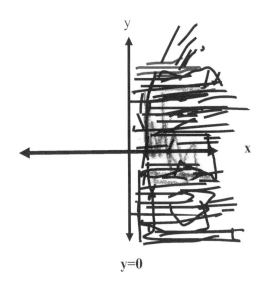

y=0

Fig. 4.21

e) $y \leq 0$

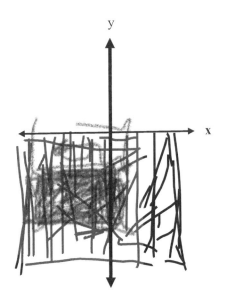

Fig. 4.22

Note: The graph of y = 0 is the x-axis.

f) $y > -2$

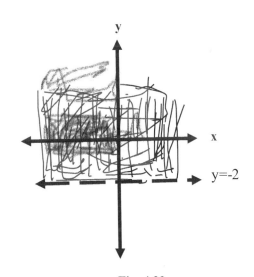

y=-2

Fig. 4.23

111

g) **x<-3**

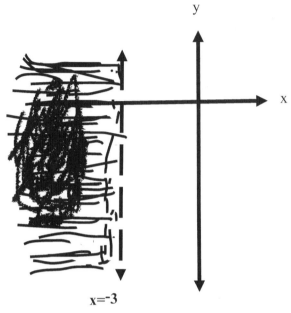

x=⁻3

Fig.4.24

Exercise 4.9 - 4.9.2

A) Draw the graph of each of the following inequality in a rectangular coordinate system.
1) $x + y < 2$
2) $x + y > -3$
3) $y \leq -x + 2$
4) $x - y \geq 4$
5) $y < 2x$
6) $y > -2x$
7) $x \leq 0$
8) $y \geq 0$

B) Draw the graph of each of the following inequality in a rectangular coordinate system.
1) $y \geq -2$
2) $x > 4$
3) $x \leq 5$
4) $x < -2$
5) $y < 6$
6) $y \geq 1$
7) $x > 2$
8) $y < -4$

CHAPTER 5

Simultaneous Equations

5.1. Definition of Simultaneous Equation

Simultaneous equations, also known as a system of equations, is a finite set of linear equations, which involve the same set of variables, for which common solutions that satisfy all the equations are searched.

Example

$$\begin{cases} x + y = 2 \\ x - y = 4 \end{cases}$$

The above example is a system of two variables x and y. A solution (truth set) to a linear system is an assignment of values to the variables x and y such that the two linear equations are simultaneously true. The solution set to the system of linear equation above is given by x = 3 and y = –1 since such values of x and y, when substituted in both linear equations, make both equations true. The phrase "systems of linear equation" indicates that the equations are to be considered collectively rather than one by one (individually).

5.2. How to solve systems of linear equations

There are different methods of solving simultaneous linear equations. In this lesson, we are going to see some of these methods like:

1) Elimination method.
2) Substitution method.
3) Graphical method

Examples

5.2.1. Solving Systems of Linear Equations by Elimination Method.

1) Solve: $\begin{cases} x + 2y = 5 \\ 2x - 2y = 4 \end{cases}$

From this equation, we can observe that both the first and the second linear equations have equal and opposite leading coefficient of y. To explain it more, the first linear equation has 2 while the second linear equation has –2 as the leading coefficient of y.

$$\begin{cases} x + 2y = 5 \\ 2x - 2y = 4 \end{cases}$$

Solution

In such kinds of equations, it is easy to solve using elimination method.

$$+ \begin{cases} x + 2y = 5 \\ 2x - 2y = 4 \end{cases}$$
$$\overline{3x + 0 = 9}$$
$$3x = 9$$
$$\frac{3x}{3} = \frac{9}{3}$$
$$x = 3$$

Add x with 2x, 2y with (–2y) and 5 with 4 so as to eliminate the variable y.

Substitute the value of x = 3, in either of the given linear equation so that you can get the value of y.

$$x + 2y = 5$$
$$3 + 2y = 5 \dots\dots\dots\dots\dots\dots\dots\dots\dots\text{(Substitute x = 3)}$$
$$3 - 3 + 2y = 5 - 3 \dots\dots\dots\dots\dots\text{(Subtract 3 from both sides)}$$
$$2y = 2$$
$$\frac{2y}{2} = \frac{2}{2} \dots\dots\dots\dots\dots\dots\dots\dots\dots\text{(Divide both sides by 2)}$$
$$y = 1$$

The solution set is {3, 1}

Check: $x + 2\ y = 5$
$$3 + 2(1) = 5$$
$$5 = 5 \checkmark$$

$$2x - 2y = 4$$
$$2(3) - 2(1) = 4$$
$$6 - 2 = 4$$
$$4 = 4 \checkmark$$

Note - 1) The solution set of the simultaneous equations is always written in bracket in order of the form {x, y}; that is, the value of 'x' is written first and the value of 'y' is written next.

2) There is no hard and stiff rule so as to eliminate first the variable x or y. So, one can eliminate either the variable x or y first depending up on his/her own choice. Usually the simple way to get (arrive at) the answer is preferred first. But whichever variable is eliminated first, the expected solution set is not changed. It is one and the same.

2) Solve: $\begin{cases} 3x - 2y = 4 \\ 4x - 3y = 1 \end{cases}$

In this equation, to eliminate one of the variables, we need to multiply the right and the left sides of both equations by suitable numbers to make sure that one of the two variables leading coefficient is equal and opposite. In this case, let us first eliminate the variable 'x.' In order to do this, multiply the right and left sides of the first and the second linear equations by –4 and 3, respectively. Then add like terms together. Thus,

$$\begin{cases} 3x - 2y = 4 \\ 4x - 3y = 1 \end{cases}$$

Solution

$$\begin{array}{l} (-4) \begin{cases} 3x - 2y = 4(-4) \\ (3) \end{cases} \quad 4x - 3y = 1(3) \qquad \text{(Add them)} \\ \underline{\quad -12x + 8y = -16} \\ + \quad 12x - 9y = 3 \\ \hline \quad\quad 0 - y = -13 \\ \quad\quad -y = -13 \\ \quad\quad \dfrac{-y}{(-1)} = \dfrac{-13}{(-1)} \quad \dots\dots\dots\dots\dots\dots\dots\dots\text{(Divide both sides by –1)} \\ \quad\quad y = 13 \end{array}$$

Substitute the values of y in either of the linear equation so that you can find the value of x.

$$3x - 2y = 4$$

Substitute y = 13, in the above equation.

$3x - 2(13) = 4$
$3x - 26 = 4$
$3x - 26 + 26 = 4 + 26$.............................(Add 26 on both sides)
$3x + 0 = 30$
$3x = 30$
$\dfrac{3x}{3} = \dfrac{30}{3}$ ....(Divided both sides by 3).
$x = 10$

Thus, the solution set of the system of equations is {10, 13}. You can check this answer by substituting the value of 10 for x and 13 for y in either equation.

3) Solve $\begin{cases} 3x - 2y = 3 \\ -2x + 4y = 6 \end{cases}$

Solution:

To solve this system of equations, multiply both sides of the first linear equation by 2 so that you can eliminate the variable 'y'.

$$(2)\begin{cases} 3x - 2y = 3(2) \\ -2x + 4y = 6 \end{cases}$$

$$(+)\begin{array}{l} 6x - 4y = 6 \\ -2x + 4y = 6 \end{array} \qquad \text{(Add them vertically)}$$

$$4x + 0 = 12$$

$$4x = 12$$

$$\frac{4x}{4} = \frac{12}{4} \qquad ....(\text{Divide both sides by 4})$$

$$x = 3$$

Substitute x = 3, in either of the equations, so that you can find the values of y.

$$3x - 2y = 3$$
$$3(3) - 2y = 3$$
$$9 - 2y = 3$$
$$9 - 9 - 2y = 3 - 9 ....(\text{Add 9 on both sides})$$
$$-2y = -6$$
$$\frac{-2y}{-2} = \frac{-6}{-2} \qquad ....(\text{Divide both sides by } -2)$$
$$y = 3$$

Thus, the solution for the systems of equations is {3, 3}

4) Solve $\begin{cases} 2x - 3y = 2 \\ \quad x = 7 \end{cases}$

Solution

$$\begin{cases} 2x - 3y = 2 \\ \quad x = 7 \end{cases}$$

Here, the value of x is already given as 7. To find the value of y, substitute this value of x in the first equation, Thus,

$$2x - 3y = 2$$
$$2(7) - 3y = 2$$
$$14 - 3y = 2$$
$$14 - 14 - 3y = 2 - 14 ....(\text{Subtract 14 from both sides})$$
$$0 - 3y = -12$$
$$-3y = -12$$
$$\frac{-3y}{-3} = \frac{-12}{-3} \qquad ....(\text{Divide both sides by } -3)$$
$$y = 4$$

Thus, the solution set for the systems of equations is {7, 4}.

5) Solve $\begin{cases} 2x - 5y = 1 \\ 4x + 3y = 4 \end{cases}$

Solution

$\begin{cases} 2x - 5y = 1 \\ 4x + 3y = 4 \end{cases}$

To solve this system of equations, multiply both sides of the first equation by –2, so that you can eliminate the variable 'x'.

$(-2)\begin{cases} 2x - 5y = 1(-2) \\ 4x + 3y = 4 \end{cases}$

$+\begin{cases} -4x + 10y = -2 \\ 4x + 3y = 4 \end{cases}$ (Add vertically)

$0 + 13y = 2$

$13y = 2$

$\dfrac{13y}{13} = \dfrac{2}{13}$ ....(Divide both sides by 13)

$y = \dfrac{2}{13}$

To find the value of x, substitute the value of $y = \dfrac{2}{13}$ in either of the original equations.

$2x - 5y = 1$

$2x - 5\left(\dfrac{2}{13}\right) = 1$

$2x - \dfrac{10}{13} = 1$

$2x - \dfrac{10}{13} + \dfrac{10}{13} = 1 + \dfrac{10}{13}$

$2x + 0 = \dfrac{23}{13}$

$2x = \dfrac{23}{13}$

$\dfrac{2x}{2} = \left(\dfrac{23}{13}\right)\left(\dfrac{1}{2}\right)$ ....(Divide both sides by 2)

$x = \dfrac{23}{26}$

Thus, the solution set of the system of equations is $\left\{\dfrac{23}{26}, \dfrac{2}{13}\right\}$.

Note: $\dfrac{a}{b} \div c = \left(\dfrac{a}{b}\right)\left(\dfrac{1}{c}\right)$, where b and c are not zero.

Exercise 5.1-5.2.1

Solve each of the following systems of linear equations using elimination method.

1) $x - 4y = 3$
   $2x - y = -1$

2) $3x + 2y = 1$
   $x - 4y = 3$

3) $8x - y = 12$
   $y = 4$

4) $\dfrac{1}{2}x - y = 7$
   $3x - 2y = 4$

5) $y = 2x - 3$
   $x - y = 5$

5.2.2. Solving Systems of Linear Equations Using Substitution Method

For solving systems of linear equations by substitution method, we have few steps, these are:

1) Step 1: Depending on the given variables in the equations, first, solve the value of one variable in terms of the other variable.
In this case, let the variables be x and y.
   • First, solve one of the systems of equations for y in terms of x.

Note: You can also solve for x in terms of y. There is no strict rule to solve either for x or for y first.

2) Step 2: Then substitute the newly found expression in the other systems of equations.

3) Step 3: Solve this equation and you have the value of x or y.

4) Then plug the newly found value of y to the other systems of equations to find the value of x or plug the newly found value of x to the other systems of equations to find the value of y.

Examples

Solve each of the following systems of equations using substitute method.

1) $x + y = 2$
   $2x + y = -1$

2) $3x - y = 3$
   $x - 2y = -4$

3) $-2x - y = 6$
   $2x + 4y = 3$

4) $x - 2y = 6$
   $y = 4$

5) $2a + b = 5$
   $4a - 2b = 6$

6) $w + 2z = 2$
   $2w - z = 4$

7) $2a - 3b = 8$
   $a = 3$

8) $\frac{1}{2}x - y = 3$
   $3x - \frac{3}{4}y = 1$

1) $x + y = 2$
   $\underline{2x + y = -1}$

Solution: -

Take the first systems of equations and solve for y in terms of x.

$x + y = 2$

$x - x + y = 2 - x$ ....(Subtract x from both sides)

$y = 2 - x$

Substitute $y = 2 - x$ in the second systems of equations and solve for x.

$2x + y = -1$

$2x + (2 - x) = -1$

$2x + 2 - x = -1$

$x + 2 = -1$

$x + 2 - 2 = -1 - 2$

$x = -3$

Then, substitute the value of x in $y = 2 - x$ so that you can find the value of y.

$y = 2 - (-3)$

$y = 2 + 3 = 5$

$y = 5$

Thus, the solution set of the systems of equations is $\{-3, 5\}$.

Remark: - To find the value of y, you can substitute the value of x in either of the original systems of equations.

2) $3x - y = 3$

    $x - 2y = -4$

Solution

Take the first systems of equations and solve for y in terms of x.

$3x - y = 3$

$3x - 3x - y = 3 - 3x$ ....(Subtract 3x from both sides)

$-y = 3 - 3x$

$(-1)(-y) = (-1)(3 - 3x)$

$y = -3 + 3x$

$\quad y = 3x - 3$

Note: $-a + b = b - a$

Substitute $y = 3x - 3$, in the second systems of linear equations and solve for x.

$x - 2y = -4$

$x - 2(3x - 3) = -4$

$x - 6x + 6 = -4$

$-5x + 6 = -4$

$-5x + 6 - 6 = -4 - 6$ ....(Subtract 6 from both sides)

$-5x = -10$

$\dfrac{-5x}{-5} = \dfrac{-10}{-5} = 2$ ....(Divide both sides by –5)

x=2

Substitute the value of x=2 in $y = 3x - 3$ and solve for y.

$y = 3x - 3$

$y = 3(2) - 3$

$y = 6 - 3$

$y = 3$

Thus, the solution set for the system of linear equations is $\{2, 3\}$.

3) $-2x - y = 6$

    $2x + 4y = 3$

Solution

Take the first systems of equations and solve for y in terms of x.

$-2x - y = 6$

$-2x + 2x - y = 6 + 2x$ ....(Add 2x on both sides)

$-y = 6 + 2x$

$(-1)(-y) = (-1)(6+2x)$ ....(Multiply both sides by -1)

$y = -6 - 2x$

Substitute $y = -6 - 2x$, in the second systems of equations and solve for x.

$2x + 4y = 3$

$2x + 4(-6 - 2x) = 3$

$2x - 24 - 8x = 3$

$2x - 8x - 24 = 3$

$-6x - 24 + 24 = 3 + 24$.....................................(Add 24 on both sides)

$-6x = 27$

$\dfrac{-6x}{-6} = \dfrac{27}{-6} =$ ....(Divide both sides by -6)

$x = \dfrac{-9}{2}$

Substitute the value of x in

$y = -6 - 2x$, and solve for y.

$y = -6 - 2\left(\dfrac{-9}{2}\right)$

$= -6 + 9$

$y = 3$

Thus, the solution set for the systems of linear equations is $\left\{\dfrac{-9}{2}, 3\right\}$.

4) $x - 2y = 6$

   $y = 4$

Solution

Observe that the value of y is already given that $y = 4$, substitute this value of y in the first systems of linear equations and solve for x.

$x - 2y = 6$

$x - 2(4) = 6$.....................................(Substituting value of y in the first equation)

$x - 8 = 6$

$x - 8 + 8 = 6 + 8$.................................(Add 8 on both sides)

$x = 14$

Thus, the solution set of the systems of linear equations is $\{14, 4\}$.

5) $2a + b = 5$

   $4a - 2b = 6$

Solution

Take the first systems of equations and solve for b in terms of a.

$2a + b = 5$

$2a - 2a + b = 5 - 2a$ ....(Subtract 2a from both sides)

$b = 5 - 2a$

Then, substitute $b = 5 - 2a$ in the second systems of equations and solve for a.

$4a - 2b = 6$

$4a - 2(5 - 2a) = 6$

$4a - 10 + 4a = 6$

$4a + 4a - 10 = 6$

$8a - 10 = 6$

$8a - 10 + 10 = 6 + 10$ ....(Add 10 on both sides)

$8a = 16$

$\dfrac{8a}{8} = \dfrac{16}{8}$ ....(Divide both sides by 8)

$a = 2$

Then, substitute the value of a in the expression $b = 5 - 2a$ and solve for b.

$b = 5 - 2(2)$......................................(Substituting value of a)

$b = 5 - 4$

$b = 1$

Thus, the solution set of the systems of equations is {2, 1}

6) $w + 2z = 2$

$2w - z = 4$

Solution

Take the first systems of equations and solve for z in terms of w.

$w + 2z = 2$

$w - w + 2z = 2 - w$ ....(Subtract w form both sides)

$2z = 2 - w$

$\dfrac{2z}{2} = \dfrac{2 - w}{2}$ ....(Divide both sides by 2)

$z = \dfrac{2 - w}{2}$

Substitute $z = \dfrac{2 - w}{2}$ in the second systems of linear equation and solve for w.

$2w - z = 4$

$2w - \left(\dfrac{2 - w}{2}\right) = 4$ ....(Substituting value of z)

$2w - \dfrac{2 + w}{2} = 4$ ....(LCM)

$\dfrac{4w - 2 + w}{2} = 4$ ....(LCM)

$\dfrac{5w - 2}{2} = 4$

$$2\left(\frac{5w-2}{2}\right) = 4(2) \quad \text{.........................(Multiplying both sides by reciprocal of } \frac{1}{2})$$

$$5w - 2 = 8$$

$$5w - 2 + 2 = 8 + 2 \quad \text{..........................(Add 2 on both sides)}$$

$$5w = 10$$

$$\frac{5w}{5} = \frac{10}{5} \quad \text{.................................(Divide both sides by 5)}$$

$$w = 2$$

Substitute the value of w in the expression $z = \dfrac{2-w}{2}$ and solve for z.

$$z = \frac{2-w}{2}$$

$$z = \frac{2-2}{2} = \frac{0}{2} \quad \text{.................................(Substituting the value of w)}$$

$$z = 0$$

Thus, the solution set for the systems of the equations is {2, 0}.

7) $2a - 3b = 8$
    $a = 3$

Solution

The value of 'a' is already given as a = 3; substitute a = 3, in the first systems of linear equations and solve for 'b'.

$$2a - 3b = 8$$

$$2(3) - 3b = 8 \quad \text{.................................(Substituting the value of a)}$$

$$6 - 3b = 8$$

$$6 - 6 - 3b = 8 - 6 \quad \text{.................................(Subtract 6 from both sides)}$$

$$0 - 3b = 2$$

$$-3b = 2$$

$$\frac{-3b}{-3} = \frac{2}{-3} \quad \text{.................................(Divide both sides by –3)}$$

$$b = \frac{-2}{3}$$

Thus, the solution set for the systems of the equations is $\left\{ 3, \dfrac{-2}{3} \right\}$.

8) $\dfrac{1}{2}x - y = 3$

$$3x - \frac{3}{4}y = 1$$

Solution

Take the first systems of linear equations and solve for y in terms of x.

$$\frac{1}{2}x - y = 3$$

$$\frac{1}{2}x - \frac{1}{2}x - y = 3 - \frac{1}{2}x \quad \text{.....................................(Subtracting } \frac{1}{2}x \text{ from both sides)}$$

$$0 - y = 3 - \frac{1}{2}x$$

$$-y = 3 - \frac{1}{2}x$$

$$-y = \frac{6 - x}{2}$$

$$(-1)(-y) = \left(\frac{6 - x}{2}\right)(-1) \quad \text{................................(Multiply both sides by } -1)$$

$$y = \frac{-6 + x}{2}$$

$$y = \frac{x - 6}{2}$$

Substitute the value of y in the second systems of linear equations.

$$3x - \frac{3}{4}y = 1$$

$$3x - \frac{3}{4}\left(\frac{x - 6}{2}\right) = 1$$

$$3x - \frac{3x + 18}{8} = 1$$

$$\frac{24x - 3x + 18}{8} = 1 \quad \text{..(LCM rule)}$$

$$\frac{21x + 18}{8} = 1$$

$$8\left(\frac{21x + 8}{8}\right) = 1 \times 8 \quad \text{..(Multiply both sides by reciprocal of } \frac{1}{8})$$

$$21x + 18 = 8$$

$$21x + 18 - 18 = 8 - 18$$

$$21x = -10$$

$$x = \frac{-10}{21}$$

Substitute this value of x in the expression of $y = \frac{x - 6}{2}$, and solve for y.

$$y = \frac{x-6}{2}$$

$$y = \frac{\frac{-10}{21}-6}{2}$$

$$y = \frac{\frac{-10-126}{21}}{2}$$

$$y = \frac{-136}{42}$$

$$y = \frac{-68}{21}$$

Thus, the solution set of the systems of linear equation is $\left\{\frac{-10}{21}, \frac{-68}{21}\right\}$

5.2.3. Solving Systems of Linear Equations by Graphing Method

To solve the systems of linear equations using graphing method, first we draw the graphs of the systems of linear equations on the same coordinate plane system. A solution to the systems of the equations is the point where the lines intersect. If the systems of linear equations have no point of intersection, then there is no solution set and if the two lines are exactly the same (coinciding), the equations have an infinite number of intersections; thus, they have infinite solutions.

Examples: -
Solve each of the following systems of liner equations using graphing method.

1) $2x - y = 3$
   $x - y = 1$

2) $2x + y = 4$
   $x + y = 3$

3) $x + y = 5$
   $x + y = 2$

4) $2x + 4y = 6$
   $x + 2y = 3$

Solutions
1) $2x - y = 3$
   $x - y = 1$

Draw the graphs of both of the given systems of linear equations on the same coordinate plane system.

First let us draw the graphs of the systems of linear equations on the same coordinate plane system. To draw the graphs, find x and y intercepts of each question.

For $2x - y = 3$,
x - intercept occurs, when $y = 0$.

$$2x - y = 3$$
$$2x - 0 = 3$$
$$2x = 3$$
$$2x/2 = 3/2$$
$$x = \frac{3}{2}$$
$$x = 1.5$$

Therefore, the x – intercept is (1.5, 0).
y - intercept occurs, when $x = 0$.

$$2x - y = 3$$
$$2(0) - y = 3$$
$$0 - y = 3$$
$$-y = 3$$
$$y = -3$$

Therefore, the y – intercept is (0, –3).
For $x - y = 1$,

x-intercept occurs, when $y = 0$.

$$x - 0 = 1$$
$$x = 1$$

Therfore, the x-intercept is (1, 0).

y-intercept occurs, when $x = 0$.

$$x - y = 1$$
$$0 - y = 1$$
$$-y = 1$$
$$-y/-1 = 1/-1$$
$$y = -1$$

Therefore, the y-intercept is (0, –1).

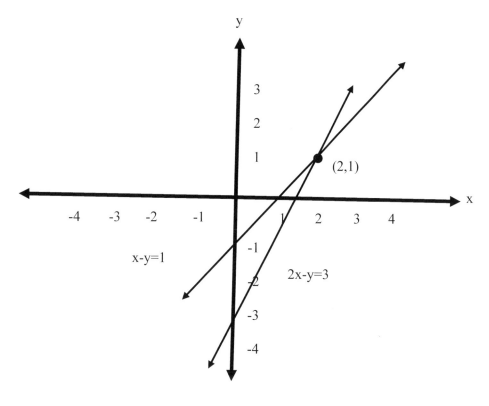

Fig. 5.1

In the figure above, the two lines intersect at a point (2, 1), which is the solution set of the systems of linear equations.

2) $2x + y = 4$
   $x + y = 3$

Draw the graphs of both of the systems of linear equations on the same coordinate plane system. To draw the graphs, find the x and y intercepts of the required systems of equations.

For $2x + y = 4$,
x-intercept occurs, when $y = 0$

$\quad 2x + 0 = 4$
$\quad 2x = 4$
$\quad \dfrac{2x}{2} = \dfrac{4}{2}$
$\quad x = 2$

Therfore, the x-intercept: (2, 0)
y-intercept occurs when $x = 0$.
$\quad 2x + y = 4$
$\quad 2(0) + y = 4$
$\quad y = 4$

Therfore,the y-intercept: (0, 4)

For x + y = 3,
x-intercept occurs when y = 0
    x + 0 = 3
    x = 3
Therefore, x-intercept: (3, 0)

y-intercept occurs when x = 0
    x + y = 3
    0 + y = 3
    y = 3
Therefore, the y-intercept: (0, 3).

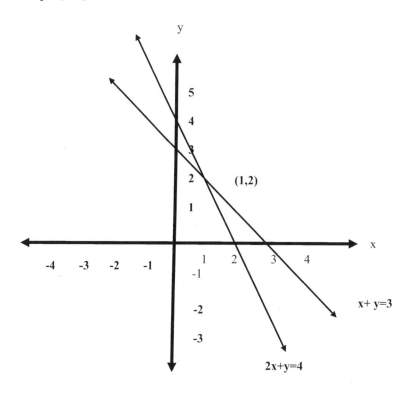

Fig. 5.2

In the figure above, the two lines intersect at a point (1, 2), which is the solution set of the system of linear equations.

3) x + y = 5
    x + y = 2

Draw the graphs of both of the systems of linear equations on the same coordinate plane system.

For x + y = 5,

x-intercept occurs when y = 0

$$x + 0 = 5$$
$$x = 5$$

Therefore, the x intercept: (5, 0)

y intercept occurs when x = 0

$$x + y = 5$$
$$0 + y = 5$$
$$y = 5$$

Therefore, the y-intercept: (0, 5)

For x + y = 2,

x-intercept occurs when y = 0

$$x + 0 = 2$$
$$x = 2$$

Therefore, the x-intercept: (2, 0)

x + y = 2

y-intercept occurs when x = 0

$$x + y = 2$$
$$0 + y = 2$$
$$y = 2$$

Therefore, the y-intercept: (0, 2)

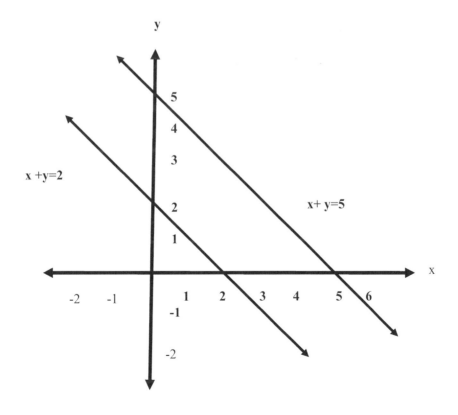

Fig. 5.3

In the figure above, the two lines are **parallel** and they do not intersect each other. Thus, there is **no** solution set. The solution set is **empty set.** S.S = { }

4) $2x + 4y = 6$
    $x + 2y = 3$

Draw the graphs of both of the systems of linear equations on the same coordinate plane system. To draw it, we have to find the x and y intercepts of both equations.
For $2x + 4y = 6$,
x-intercept occurs, when y = 0
    $2x + 4(0) = 6$
    $2x + 0 = 6$
    $2x/2 = 6/2$
    $x = 3$
Therefore, the x-intercept: (3, 0)

y-intercept occurs, when x = 0
    $2x + 4y = 6$
    $2(0) + 4y = 6$
    $4y = 6$
    $4y/4 = 6/4$
    $y = \dfrac{6}{4} = \dfrac{3}{2}$
    $y = 3/2$

Therefore, the y-intercept: $\left(0, \dfrac{3}{2}\right)$
For $x + 2y = 3$,
x-intercept occurs, when y = 0
    $x + 2(0) = 3$
    $x + 0 = 3$
    $x = 3$
Therefore, the x-intercept: (3, 0)

y-intercept occurs, when x = 0
    $x + 2y = 3$
    $0 + 2y = 3$
    $2y = 3$
    $y = \dfrac{3}{2}$

Therefore, the y-intercept: $\left(0, \dfrac{3}{2}\right)$

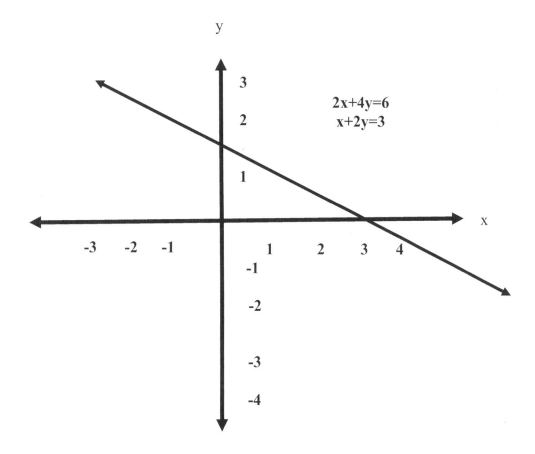

Fig. 5.4

In the figure above, the graphs of the two linear equations are **coinciding**. The lines are exactly the same, so that every coordinate pair on the line is a solution set to both equations. Thus, the lines have **infinitely many** solution sets.

5.3. Types of Systems of Linear Equations

There are three types of systems of linear equations in two variables having their own number of solutions. These are: -

- An independent system.
- An inconsistent system and
- A dependent system.

* If the graphs of the systems of equations intersect exactly at one point, then they have only one solution and such systems are called Independent Systems.

Example: $3x + 2y = 7$

$-x + y = 5$

* If the graphs of the systems of equations are parallel and never intersect, then the systems do not have a solution set. Such systems of equations are called Inconsistent Systems.

Example: $x + y = 3$
$\qquad\quad x + y = 5$

* If the graphs of the systems of equations are coincident, they have infinitely many solutions; such systems of equations are called Dependent Systems.

Example: $x - y = 4$
$\qquad\quad 2x - 2y = 8$

Moreover: * A consistent systems of linear equations is a system of equation with at least one set of values (at least one solution set) satisfying all questions. If a consistent system of equation has only one solution set, then it is considered to be an independent system.

* A consistent system of equations is considered to be dependent system if the equations have the same slopes and the same y-intercept. In other words, such lines are coincident lines.

Exercise 5.2.2-5.3

Find the solution set of each of the following systems of equations graphically.

a) $x - 2y = 4$
$\quad\; x + y = 2$

b) $x - y = 4$
$\quad\; 2x + y = 1$

c) $2x + y = 3$
$\quad\; x - y = 3$

d) $2x + y = 2$
$\quad\; x - y = 2$

e) $x - y = 5$
$\quad\; 2x - 2y = 10$

CHAPTER 6

Exponents and Exponential Functions

6.1. Definition of Exponent and Exponential Functions

Before we define the exponential function, let us start with simple examples. While operating mathematical problems, there are some occasions when we need to multiply a number by itself again and again. As an example, $2^4 = 2 \times 2 \times 2 \times 2 = 16$

In mathematics, we use the idea of exponents to represent a product involving the same factor. For example,

$2^4 = 2 \times 2 \times 2 \times 2$; Here, 2 occurs four times.
(We multiply 2 four times by itself).

Thus, 2 is called a base while 4 is called an exponent. We read as "Two raised to the power of four." Similarly, $5^3 = 5 \times 5 \times 5$, Here, 5 occurs three times, Thus, 5 is the base and 3 is the exponent. We read as "Five raised to the power of three."

In a real life, a simple example of exponential function is the growth of bacteria. Some bacteria doubled or tripled themselves every hour. To represent the number of bacteria that doubles itself every hour, we can show this fact exponentially as: $f(t) = 2^t$, where t is time in hour.

Let us see how a bacterium doubles itself every hour.

Before we start the experiment,
   At : $t = 0$, $f(0) = 2^0 = 1$

In 1 hour, $f(1) = 2^1 = 2$
In 2 hours, $f(2) = 2^2 = 4$
In 3 hours, $f(3) = 2^3 = 8$
In 4 hours, $f(4) = 2^4 = 16$
In 5 hours, $f(5) = 2^5 = 32$
In 6 hours, $f(6) = 2^6 = 64$

   .   .
   .   .
   .   .

In t hours $f(t) = 2^t$

In general, if 'a' is any number where a > 0 and a ≠ 1, then the exponential function is written in the form: **f(x) = aⁿ**

Where 'a' is called the base and 'n' is called the exponent of the equation.

('n' can be any real number)

Examples:

1) $6 \times 6 \times 6 \times 6 \times 6 = 6^5$

    Here, 6 is the base whereas 5 is the exponent of the equation.

2) $3 \times 3 \times 3 \times 3 \times 3 \times 3 = 3^6$

    Here, 3 is the base whereas 6 is the exponent of the equation.

3) $8 = 8^1$

    Here, 8 is the base whereas t is the exponent of the equation.

6.2. Properties of Exponents

6.2.1. Multiplication Properties of Exponents

Let a and b be non-zero real numbers and let the exponents m and n be integers, then

1) Product of power property

   $a^m \cdot a^n = a^{m+n}$

   To multiply powers of the same base. keep the base and add the exponents.

   Examples: - $4^3 \cdot 4^5 = 4^{3+5} = 4^8$

   $6^2 \cdot 6^9 = 6^{2+9} = 6^{11}$

2) Power of a power property.

   $(a^m)^n = a^{m \times n} = a^{mn}$

   To find a power, keep the base and multiply the exponents.

   Examples: $(3^4)^5 = 3^{4 \times 5} = 3^{20}$

   $(2^6)^4 = 2^{6 \times 4} = 2^{24}$

3) Power of a product property.

   $(a \times b)^n = a^n \times b^n$:

   Find the power of each factor and then multiply.

   Examples: $(6 \times 5)^4 = 6^4 \times 5^4$

   $(2 \times 3)^9 = 2^9 \times 3^9$

More examples

1) Product of power property
   a) $3^5 \cdot 3^6 = 3^{5+6} = 3^{11}$
   b) $4^8 \cdot 4^2 = 4^{8+2} = 4^{10}$

134

c) $x^6 \bullet x^3 = x^{6+3} = x^9$

d) $y \bullet y^5 = y^{1+5} = y^6$

e) $\left(\frac{1}{2}\right)^4 \cdot \left(\frac{1}{2}\right)^5 = \left(\frac{1}{2}\right)^{4+5} = \left(\frac{1}{2}\right)^9$

f) $(9)^6 \bullet (9)^4 = 9^{6+4} = 9^{10}$

g) $5^3 \bullet 5 = 5^3 \bullet 5^1 = 5^{3+1} = 5^4$

2) Examples of the power of power property

   a) $(4^3)^2 = 4^{3\times2} = 4^6$

   b) $(3^5)^4 = 3^{5\times4} = 3^{20}$

   c) $(x^2)^7 = x^{2\times7} = x^{14}$

   d) $(5^1)^2 = 5^{1\times2} = 5^2$

   e) $\left[\left(\frac{1}{3}\right)^4\right]^6 = \left(\frac{1}{3}\right)^{4\times6} = \left(\frac{1}{3}\right)^{24}$

   f) $\left[(x+2)^3\right]^5 = (x+2)^{3\times5} = (x+2)^{15}$

3) Examples of the power of a product property

   a) $(3 \times 4)^2 = 3^2 \times 4^2$

   b) $(5 \times 6)^4 = 5^4 \times 6^4$

   c) $\left(\frac{1}{2} \times \frac{1}{3}\right)^5 = \left(\frac{1}{2}\right)^5 \cdot \left(\frac{1}{3}\right)^5$

   d) $\left(\frac{1}{2} \cdot y\right)^9 = \left(\frac{1}{2}\right)^9 \cdot y^9$

   e) $(xy)^6 = x^6 \bullet y^6$

Exercise 6.2.1

A) Simplify each of the following expression using the product of power property.

   1) $m^2 \times m^5$

   2) $x^5 \bullet x^6$

   3) $6^4 \bullet 6^{11}$

   4) $\left(\frac{1}{2}\right)^9 \cdot \left(\frac{1}{2}\right)^4$

   5) $5^4 \bullet 5^{13}$

   6) $y^{16} \bullet y^{21}$

   7) $8^4 \bullet 8^{13}$

   8) $9^7 \bullet 9^6 \bullet 9^{40}$

   9) $m^6 \bullet m^4 \bullet m^8$

   10) $\left(\frac{1}{3}\right)^4 \cdot \left(\frac{1}{3}\right)^6 \cdot \left(\frac{1}{3}\right)^2$

   11) $n^{2014} \bullet n^{2001}$

B) Simplify each of the following expression using the power of a power property.

1) $(3^4)^2$
2) $[(8)^2]^3$
3) $(6^3)^4$
4) $(a^5)^4$
5) $\left[\left(\dfrac{1}{2}\right)^3\right]^6$
6) $(m^6)^9$
7) $(x^5)^4$
8) $\left[\left(\dfrac{1}{3}\right)^4\right]^{10}$
9) $(6^{1/2})^4$
10) $(y^{1/8})^8$
11) $\left(x^{3/2}\right)^6$

12) $(y^{3/5})^{25}$

C) Simplify each of the following expressions using power of a product property.

1) $(-3 \times y)^4$
2) $(2ab^2)^3$
3) $(5m^2n^3r^4)^2$
4) $(-4x^5y^3)^4$
5) $\left(\dfrac{1}{3}a^2b^6\right)^5$
6) $(2m^3n^4)^7$
7) $(6x^4y^4z^2)^{1/2}$
8) $\left(\dfrac{1}{4}x^2y\right)^3$
9) $(xyz)^4$
10) $(a^3b^2c^4)^5$
11) $(abc)^6$

6.2.2. Division Properties of Exponents

If the base 'a is a non-zero real number and the exponents m and n are integers, then

$$\frac{a^m}{a^n} = a^{m-n}$$

To divide powers that have the same base, keep the base and subtract the exponents.

Examples

1) $\dfrac{5^4}{5^3} = 5^{4-3} = 5$

2) $\dfrac{6^8}{6^2} = 6^{8-2} = 6^6$

3) $\dfrac{3^5}{3^5} = 3^{5-5} = 3^0$

4) $\dfrac{x^3 y^4 z^6}{xyz^5} = x^{3-1} y^{4-1} z^{6-5} = x^2 y^3 z$

5) $\dfrac{a^6 b^5 z^4}{a^2 bz^3} = a^{6-2} b^{5-1} z^{4-3} = a^4 b^4 z$

6) $\dfrac{m^4 n^6}{mn} = m^{4-1} n^{6-1} = m^3 n^5$

7) $\dfrac{x^6 y^6}{x^6 y^6} = x^{6-6} y^{6-6} = x^0 y^0 = 1$

Note: - Any non-zero number raised to zero is one.

That is, $a^0 = 1$, If $a \neq 0$.

Examples: -

a) **$6^0 = 1$**

b) **$(200)^0 = 1$**

c) $\left(\dfrac{1}{2}\right)^0 = 1$

For $a \neq 0$ and $x > 0$, any non-zero number to the power of a negative exponent is the reciprocal of the same power with positive exponent.

$a^{-x} = \dfrac{1}{a^x}$; For $a \neq 0$ and $x > 0$

Examples: -

1) $2^{-4} = \dfrac{1}{2^4} = \dfrac{1}{16}$

2) $3^{-2} = \dfrac{1}{3^2} = \dfrac{1}{9}$

3) $\left(\dfrac{4}{9}\right)^{-1} = \dfrac{9}{4}$

4) $\left(\dfrac{2}{3}\right)^{-2} = \left(\dfrac{3}{2}\right)^2 = \dfrac{9}{4}$

5) $\left(\dfrac{7}{10}\right)^{-3} = \left(\dfrac{10}{7}\right)^3 = \dfrac{1000}{343}$

6) $\dfrac{1}{6^4} = 6^{-4}$

Note: - For $a \neq 0$, $a^{-1} = \dfrac{1}{a}$

For a and b non-zero real numbers and n > 0,

$$\left(\frac{a}{b}\right)^{-n} = \left(\frac{b}{a}\right)^{n}$$

Examples: -

1) $\left(\dfrac{5}{4}\right)^{-2} = \left(\dfrac{4}{5}\right)^{2} = \dfrac{16}{25}$

2) $\left(\dfrac{3}{2}\right)^{-4} = \left(\dfrac{2}{3}\right)^{4} = \dfrac{16}{81}$

3) $\left(\dfrac{-5}{6}\right)^{-3} = \left(\dfrac{-6}{5}\right)^{3} = \dfrac{-216}{125}$

Examples

Solve each of the following equations.

a) $(x + 3)^{\frac{1}{3}} = 4$

b) $(27)^{x-2} = \dfrac{1}{27}$

c) $2^{5x-2} = 256$

d) $5^{x+7} = (125)^{2x-1}$

e) $(x^{2020}) \div (x^{2017}) = 27$

f) $3^{2x-1} = 3(3^{x-4})$

g) $4^{x^2-1} = 64$

h) $\begin{cases} 2^{x-y} = 4 \\ 3^{x+y} = 9 \end{cases}$

i) $5^{(x-2)} \cdot (25^{2x}) = \dfrac{1}{5}$

j) $(0.25)^{x-1} = \dfrac{1}{8}$

k) $\sqrt{\sqrt{x+2}} = 3$

l) $x - 5 = \left(\sqrt{11 - x}\right)$

m) $\sqrt{\sqrt{\sqrt{x}}} = 3$

n) $\sqrt{\sqrt{3^x}} = 27$

o) $2^x \cdot 16^x = 128$

Solutions: -

a) $(x+3)^{\frac{1}{3}} = 4$

$(x+3)^{\frac{1}{3}\times 3} = 4^{1\times 3}$ .... Multiply both sides of the exponents by reciprocal of the exponent of (x+3). i.e., Multiply by 3.

$(x+3) = 4^3$

$x+3 = 4\times 4\times 4$

$x+3 = 64$

$x+3-3 = 64-3$ .... Subtract 3 from both sides.

$x = 61$

Thus, the solution set is **{61}.**

b) $(27)^{x-2} = \dfrac{1}{27}$

$(27)^{x-2} = (27)^{-1}$

$3^{3(x-2)} = 3^{3(-1)}$ ....27 is written as in a power form with base 3.

$3(x-2) = 3(-1)$ ....The bases are cancelled out.

$3x-6 = -3$

$3x-6+6 = -3+6$..Add 6 on both sides of the equations.

$3x = 3$

$\dfrac{3x}{3} = \dfrac{3}{3}$ ....Divide both sides by 3.

$x = 1$

Thus, the solution set is: {1}.

c) $2^{5x-2} = 256$

$2^{5x-2} = 2^8$ ....$256 = 2^8$

$5x - 2 = 8$..(Recall: $a^x = a^y \Leftrightarrow x = y$)

$5x - 2+2 = 8 + 2$..Add 2 on both sides.

$5x = 10$

$\dfrac{5x}{5} = \dfrac{10}{5}$ ....Divide both sides by 5.

$x = 2$

Thus, the solution set is {2}

d) $5^{x+7} = (125)^{2x-1}$

$5^{x+7} = (5)^{3(2x-1)}$..$125 = 5^3$

$5^{x+7} = 5^{6x-3}$

$x + 7 = 6x - 3$..(Recall: $a^x = a^y \Leftrightarrow x = y$)

$x+7 - 7 = 6x-3 - 7$ ....(Subtract 7 from both sides)

$x = 6x - 10$

$x - 6x = 6x - 6x - 10$ .... Subtract 6x from both sides.

$-5x = -10$

$\dfrac{-5x}{-5} = \dfrac{-10}{-5}$ .... Divide both sides by –5.

$x = 2$

Thus, the solution set is: {2}

e) $(x^{2020}) \div (x^{2017}) = 27$

$\dfrac{x^{2020}}{x^{2017}} = 27$ .... Given

$x^{2020-2017} = 27$ .... Recall: $\dfrac{a^x}{a^y} = a^{x-y}$

$x^3 = 27$

$x^3 = 3^3$ .... $(27 = 3^3)$

$x^{3\left(\frac{1}{3}\right)} = 3^{3\left(\frac{1}{3}\right)}$ .... Multiply the exponent of both sides by reciprocal of 3.

$x = 3$

Thus, the solution set is: {3}.

f) $3^{2x-1} = 3(3^{x-4})$

$3^{2x-1} = 3^1(3^{x-4})$

$3^{2x-1} = 3^{1+x-4}$ .... $a^x \bullet a^y = a^{x+y}$

$3^{2x-1} = 3^{x-3}$

$2x - 1 = x - 3$ .... Recall: $a^x = a^y \Leftrightarrow x = y$

$2x - x - 1 = x - x - 3$ .... Subtract x from both sides.

$x - 1 = -3$

$x - 1 + 1 = -3 + 1$ .... Add 1 on both sides

$x = -2$

Thus, the solution set is {–2}.

g) $4^{(x^2-1)} = 64$

$2^{2\left(x^2-1\right)} = 2^6$ .... 4 and 64 written as exponent with base 2.

$2^{2x^2-2} = 2^6$ .... $2(x^2 - 1)$: Distributive property.

$2x^2 - 2 = 6$ .... Recall: $a^x = a^y \Leftrightarrow x = y$

$2x^2 - 2 + 2 = 6 + 2$ .... Add 2 on both sides.

$2x^2 = 8$

$\dfrac{2x^2}{2} = \dfrac{8}{2}$

$x^2 = 4$

$x = \pm\sqrt{4}$ .... Square root property.

$$x = \pm 2$$

Thus, the solution set is $\{-2, 2\}$.

The same answer can be found by expressing 64 as 4^3.

$x^2 - 1 = 3$

$x^2 - 1 + 1 = 3 + 1$ .... adding 1 to both sides

$x^2 = 4$

$x = \pm\sqrt{4}$ .... Square root property

Thus, the solution set is $\{-2, 2\}$.

h) $\begin{cases} 2^{x-y} = 4 \\ 3^{x+y} = 9 \end{cases}$

$\begin{cases} 2^{x-y} = 2^2 \\ 3^{x+y} = 3^2 \end{cases}$ .... The right side of the equations written as power form has the same base with the left side of the equations.

$\begin{cases} x - y = 2 \\ x + y = 2 \end{cases}$ .... Recall that $a^x = a^y \Leftrightarrow x = y$

Combining the two equations and solving for x and y,

$+\begin{cases} x - y = 2 \\ x + y = 2 \end{cases}$ Add downward

$\overline{\quad 2x + 0 = 4 \quad}$

$2x = 4$

$\dfrac{2x}{2} = \dfrac{4}{2}$ .... Divide both sides by 2.

$x = 2$

To find the value of y, substitute the value of x in either of the equations,

Thus,

$x - y = 2$

$2 - y = 2$ .... Substituting value of x = 2.

$2 - 2 - y = 2 - 2$ .... Adding 2 on both sides.

$0 - y = 0$

$-y = 0$

$(-1)(-y) = 0(-1)$ .... Multiplying both sides by –1.

$y = 0$

Thus, the solution set is: $\{2, 0\}$

Note: - When you list the solution set with two variables x and y, make sure to write in the form of $\{x, y\}$.

141

i) $5^{(x+4)} \cdot (25^{2x}) = \dfrac{1}{5}$

$5^{x+4} \cdot 5^{2(2x)} = 5^{-1}$ .... Writing in the same base.

$5^{x+4 + 2(2x)} = 5^{-1}$ .... Recall: $\mathbf{a^x \cdot a^y = a^{x+y}}$

$5^{x+4 + 4x} = 5^{-1}$

$x + 4 + 4x = -1$ .... Recall: $a^x = a^y \Leftrightarrow x = y$

$5x + 4 = -1$

$5x + 4 - 4 = -1 - 4$ .... Subtract 4 from both sides.

$5x = -5$

$\dfrac{5x}{5} = \dfrac{-5}{5}$

$x = -1$

Thus, the solution set is $\{-1\}$.

j) $(0 \cdot 25)^x = 64$

$\left(\dfrac{25}{100}\right)^x = 64$ .... $\left(0.25 = \dfrac{25}{100}\right)$

$\left(\dfrac{1}{4}\right)^x = 64$ .... $\left(\dfrac{25}{100} = \dfrac{1}{4}\right)$

$2^{-2x} = 2^6$, ... $(\dfrac{1}{4} = 2^{-2}$ and $64 = 2^6)$

$-2x = 6$ .... Recall that: $a^x = a^y \Leftrightarrow x = y$

$\dfrac{-2x}{-2} = \dfrac{6}{-2}$ .... Divide both sides by -2.

$x = -3$

Thus, the solution set is $\{-3\}$.

The same solution set can be found in another way.

$(0 \cdot 25)^x = 64$

$(25/100)^x = 4^3$

$1/4^x = 4^3$

$4^{-x} = 4^3$

$-x = 3$

$-x/-1 = 3/-1$

$X = -3$

Thus, the solution set is $\{-3\}$.

k) $\sqrt{\sqrt{x+2}} = 3$

$\left(\sqrt{\sqrt{x+2}}\right)^2 = (3)^2$ .... Squaring both sides

$\sqrt{x+2} = 9$

142

$$\left(\sqrt{x+2}\right)^2 = 9^2 \quad \text{......................................Again, squaring both sides.}$$

$x + 2 = 81$

$x + 2 - 2 = 81 - 2$

$x = 79$

Thus, S.S = {79}

l) $x - 5 = \left(\sqrt{11 - x}\right)$

    $(x - 5)^2 = \left(\sqrt{11 - x}\right)^2$ ....Squaring both sides

      $x - 5 = \left(\sqrt{11 - x}\right)$

      $(x - 5)^2 = \left(\sqrt{11 - x}\right)^2$ ......Squaring both sides.

      $x^2 - 10x + 25 = 11 - x$

      $x^2 - 10x + x + 25 - 11 = 11 - 11 - x + x$ ....(Add x on both sides and subtract 11 from both sides)

      $x^2 - 9x + 14 = 0$

      $x^2 - 7x - 2x + 14 = 0$ ....$[-7x + (-2x) = -9x$ and $(-7x)(-2x) = 14x^2]$

      $x(x - 7) - 2(x - 7) = 0$

      $(x - 2)(x - 7) = 0$ ......(Collecting outside terms and one of the other terms)

      $x - 2 = 0$ or $x - 7 = 0$

Check the answers by substituting in the original equation

$$x - 5 = \left(\sqrt{11 - x}\right)$$

Check when x=2

$$2 - 5 = \left(\sqrt{11 - 2}\right)$$

$$-3 = \sqrt{9}$$

$-3 \neq 3$ .... Invalid and thus, 2 cannot be the solution set of the equation.

Note that 2 is called **Extraneous solution**.

An **Extraneous solution** is a solution that come out from the process of solving the equation but not a valid solution to the equation.

Check when x = 7

$$7 - 5 = \left(\sqrt{11 - 7}\right)$$

$$2 = \sqrt{4}$$

$2 = 2$ .... Valid and thus,7 is the solution set of the equation.

Therefore, the solution set is {7}

m) $\sqrt{\sqrt{\sqrt{x}}} = 3$

Solution:

The square root of a number is the same as raising that number to the power of $\frac{1}{2}$, the expression $\sqrt{\sqrt{\sqrt{x}}} = 3$ has three radicals; thus, we can write as $(x)^{\frac{1}{2}}(x)^{\frac{1}{2}}(x)^{\frac{1}{2}} = 3$

Thus, $(x)^{\frac{1}{8}} = 3$

$x = 3^8$ ..... Recall that $(a)^{\frac{1}{b}} = c \Leftrightarrow a = c^b$

$x = 6,561$

n) $\sqrt{\sqrt{\sqrt{3^x}}} = 27$

Solution: $\sqrt{\sqrt{\sqrt{3^x}}} = 27$ can be written as $(3)^{\frac{x}{4}} = 3^3$

$(3)^{\frac{x}{4}} = 3^3$

$\frac{x}{4} = 3$ .... Recall that $a^x = a^y \Leftrightarrow x = y$

$x = 3 \times 4$

$x = 12$

Thus, the value of x is 12.

o) $2^x \cdot 16^x = 128$

Solution:

The expression $2^x \cdot 16^x = 128$ can be written in the form of $(2 \times 16)^x = 128$
(Recall that) $(a^x \times b^x) = (ab)^x$.... Power of a Product Rule.

Thus, $(2 \times 16)^x = 128$

$(32)^x = 128$

$(2)^{5x} = (2)^7$

$5x = 7$ .... (Recall that $(a)^x = (a)^y \Leftrightarrow x = y$)

$x = \frac{7}{5}$

Exercise 6.2.2

1) Solve each of the following equations.
   a) $x^{4/3} = 16$
   b) $y^{\frac{1}{3}} = 27$
   c) $2^x \cdot 3^x = 36$
   d) $3^{x-2} = 27$
   e) $\left(\frac{1}{4}\right)^{x-1} = 0.25$

f) $(27)^{x-2} = 3$

g) $8^{x-2} = 64$

h) $9^{x^2-2} = 81$

i) $5^{x-2} = (625)^{2x+3}$

j) $(0.25)^{x+1} = \dfrac{1}{8}$

k) $2^{x-1} = 2(4^{x+3})$

l) $\begin{cases} 2^{2x-y} = 8 \\ 3^{3x+y} = 27 \end{cases}$

m) $4^{(x-2)} \cdot 8^{(2x-1)} = \left(\dfrac{1}{32}\right)^{(x-1)}$

n) $25^{(x-3)} = \dfrac{1}{125}$

o) $2^{(0.05)x} \cdot 2^{(1.95)x} = 8$

p) $3^x \cdot 9^x = 81$

q) $\sqrt{2x+3} = \sqrt{x+5}$

r) $\sqrt{x-1} = x - 7$

s) $\sqrt{\sqrt{\sqrt{\sqrt{4^x}}}} = 8$

2) Give your answer in the simplest rational form.

a) $\left(\dfrac{3}{4}\right)^{-4}$

b) $\left(\dfrac{x^2 y^2}{xy}\right)^{-3}$

c) $\left(\dfrac{abc^2}{b^2 a}\right)^{-4}$

d) $\left(\dfrac{2}{9}\right)^{-3}$

e) $\left(\dfrac{mn^2 x^4}{xmn}\right)^{-5}$

145

f) $\left(\dfrac{9}{27}\right)^{-3}$

g) $\left(\dfrac{abc}{a^2b}\right)^{-5}$

h) $\left(\dfrac{27x^3y^3}{xy^2}\right)^{-3}$

i) $\left(\dfrac{x^3y^{-2}z^4}{xyz}\right)^{-3}$

j) $\left(\dfrac{32x^5y^5}{4x^{-1}y^{-1}}\right)^{\frac{1}{3}}$

k) $\left(\dfrac{144a^4b^4}{ab}\right)^{\frac{1}{2}}$

l) $\left(\dfrac{x^{\frac{1}{3}}y^{\frac{1}{3}}}{x^{\frac{2}{3}}y^{\frac{2}{3}}}\right)^3$

m) $\left(\dfrac{x^4y^6z^8}{x^2yz^2}\right)$

n) $\left(\dfrac{x^6y^4}{x^3yz^6}\right)^{\frac{-1}{3}}$

3) Write the simplified form of the following expression with positive exponent.

a) $\dfrac{x^5y^3z^{-1}}{xyz}$

b) $\dfrac{2ab}{8a^2b^2}$

c) $\left(\dfrac{m^5n^6}{mn}\right)\left(\dfrac{m^2n^3}{mn}\right)$

d) $(-1)^{-1}$

e) $\left(\dfrac{81x^2y^{-1}}{3xy}\right)^{-2}$

f) $4^{-1}+2^{-1}$

g) $(4^{-2})(2^{-3})$

h) $\left(\dfrac{1}{2}\right)^{-3}(8^{-2})$

146

i) $\left(\dfrac{2}{x^{-1}}\right)^3$

4) Simplify each of the following expressions

a) $\dfrac{6^2}{(12)^2}$

b) $\dfrac{r^2 s^2}{r^{-1} s}$

c) $\dfrac{25 x^4 y^3}{5^{-1} x^{-1} y}$

d) $\left(\dfrac{-3 x^3 y}{9 x y}\right)^{-3}$

e) $\left(\dfrac{5 x y^{-2} y}{3 x^{-1} y^5}\right) \cdot \left(\dfrac{4 x y}{3 x^{-1} y^{-1}}\right)$

5) Simplify each of the following expressions using one or more of the laws of exponents

a) $\dfrac{(6 x y)}{3 x y}$

b) $(3a^{-2} b^4)^{-3}$

c) $\left(\dfrac{x^2 y^2}{r s y}\right)\left(\dfrac{r s}{x^2 y}\right)$

d) $\left(\dfrac{m^3}{n^4}\right)^{-2}$

e) $\left(\dfrac{x^{-3} x^2}{y^{-2}}\right)^{-4}$

f) $\left(\dfrac{x^{-1} y^{-1}}{x^2 y^2}\right)^{-1}$

g) $\left(\dfrac{m^{-\frac{1}{2}}}{n^{-\frac{2}{3}}}\right)^{-4}$

h) $(3^{-2} + 4^{-3})^{-1}$

i) $[(-1)^{-1}]^{-1}$

j) $\dfrac{\left(3^{\frac{1}{2}}\right)^{2} \cdot \left(9^{-3}\right)}{27}$

6.3. Graphs of Exponential Functions

As explained before, depending on time x, a quantity of bacteria in laboratory dish begin to replicate itself by starting with small amount but increases rapidly to a very large number. We can observe that this pattern of growth of the bacteria is exponential. The variable x is the exponent of a function. Such function whose equations contain a variable in the exponent are called exponential functions. The exponential function is applicable in real life situations like population growth, growth of radioactive decay, growth of bacteria and others. The value of this function changes in an increasing or decreasing rate depending on the value of the given variable x. So, this phenomenon can be described using an exponential function.

6.3.1. Definition: If a > 0 and a ≠ 1, then the exponential function with the base 'a' and exponent 'x' is given by:

$$f(x) = a^x$$

To look into the properties and behavior of the exponential function, we sketch the graph of $f(x) = a^x$, when a > 1 and when 0 < a < 1

Examples:

1) Draw the graph of $f(x) = 2^x$.

Solution: We begin by setting up a table of coordinate by choosing a few positive and negative values of x and finding their corresponding values of f(x).

Examples: $f(x) = a^x$
$f(0) = 2^0 = 1$
$f(1) = 2^1 = 2$
$f(2) = 2^2 = 4$
$f(x) = 2^3 = 8$
$f(-1) = 2^{-1} = \frac{1}{2}$
$f(-2) = 2^{-2} = \frac{1}{4}$
$f(-3) = 2^{-3} = \frac{1}{8}$

	x	-3	-2	-1	0	1	2	3
y = 2ˣ	f(x)	$\frac{1}{8}$	$\frac{1}{4}$	$\frac{1}{2}$	1	2	4	8

Let us plot these points on the co-ordinate system and connect them by a smooth curve to obtain the graph of $f(x) = 2^x$

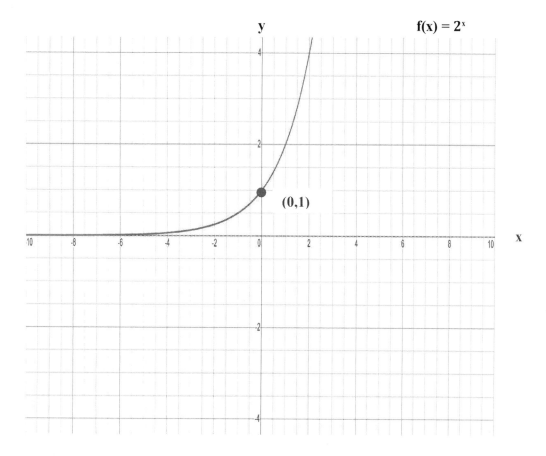

Fig. 6.1

We plot these points, connecting them with a continuous curve and we get the graph as shown in the figure above. Observe that the graph of **f(x) = 2^x** approaches but never touches the negative portion of the x-axis. Thus, the x-axis or y=0 is a horizontal asymptote. The range or value of f(x), (i.e., value of 'y') is the set of all positive real numbers.

Note:- An **asymptote** is a line that continually approaches a given curve but does **not** touches it at any **finite** distance.

1) Draw the graph of $g(x) = \left(\dfrac{1}{2}\right)^x$

$g(x) = \left(\dfrac{1}{2}\right)^x$	x	-3	-2	-1	0	1	2	3
	g(x)	8	4	2	1	$\dfrac{1}{2}$	$\dfrac{1}{4}$	$\dfrac{1}{8}$

Note: - The graph of f(x) = a^x, for any 0<a<1 has similar shape to the graph of $\mathbf{y} = \left(\dfrac{1}{2}\right)^x$.

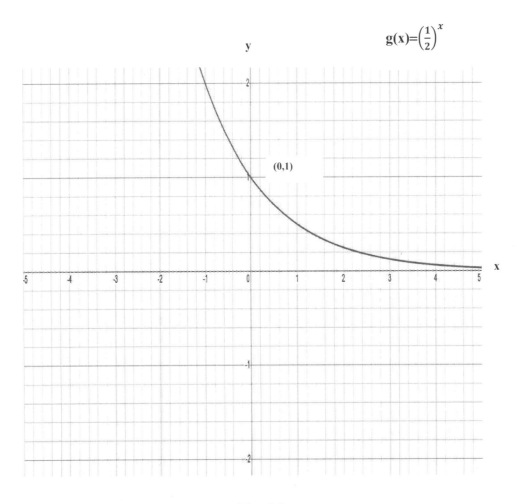

$$g(x)=\left(\frac{1}{2}\right)^x$$

(0,1)

Fig. 6.2

6.4. Basic Properties of Graphs of Exponential Functions

- The graph of f(x) = a^x, a > 1 has the following basic properties.

1) The domain (value of x) is the set of all real numbers.
2) The range (value of y) is the set of all positive real numbers.
3) The function is increasing.
4) The y-intercept of the graph is 1.
5) The function is one-to-one and has an inverse function.
6) The values of the function are greater than 1 for x > 0 and between 0 and 1 for x < 0.
7) The graph approaches the x-axis as an asymptote on the left and increases without limit on the right.

- The graph of f(x) = a^x, 0 < a < 1 has the following basic properties.

1) The domain (value of x) is the set of all real numbers.
2) The range is the set of all positive real numbers.

150

3) The y-intercept of the graph is 1.
4) The function is decreasing.
5) The function is one-to-one and has an inverse function.
6) The values of the function are greater than 1 for x < 0 and between 0 and 1 for x > 0.
7) The graph approaches the x-axis as an asymptote on the right and increases without limit on the left.

Exercise 6.3-6.4

Answer each of the following questions.

1) Construct tables of values and draw the graphs of:
   a) $f(x) = 3^x$

   b) $g(x) = \left(\dfrac{1}{3}\right)^x$

   c) $f(x) = 3^x$ and $g(x) = \left(\dfrac{1}{3}\right)^x$ using the same coordinate system.

2) Answer each of the following questions by referring the functions: $f(x) = 3^x$ and $g(x) = \left(\dfrac{1}{3}\right)^x$.
   a) Find the domain of each function.
   b) Find the range of each function.
   c) What is the y-intercepts of each function?
   d) Determine which function is increasing and which function is decreasing.

   e) Determine which function is one -to- one function and which is not one -to-one.
   f) Find the asymptote for each graph.

6.5. Scientific Notation

Scientific notation is a way of writing very large or very small numbers that uses exponents. It is used by mathematician and scientists when they are working with very large or very small numbers. Using exponential notation, large or small numbers can be rewritten in a way that is easy to read. The scientific notation is also called Standard Form. In scientific notations, non-zero numbers are written in the form of: -

$M \times 10^n$, where 'n' is an integer, and the coefficient 'M' is a non-zero real number (usually between 1 and 10).

The integer 'n' is called the exponent and the real number 'M' is called the significand.

Below are some examples of the way how to convert decimal notation to scientific notation.

Decimal notation	Scientific notation
3	3×10^0
2000	2×10^3
400000	4×10^5
2343578	2.343578×10^6
−48000	-4.8×10^4
68300000	6.83×10^7
0.5	5×10^{-1}
0.05	5×10^{-2}
0.645	6.45×10^{-1}
0.0986	9.86×10^{-2}
0.0000435	4.35×10^{-5}

More Examples

1) Write the decimal notation 0.000004 in scientific notation.

Answer: Move the decimal point to the right of the first non-zero digit so that you have a number between 1 and 10. Here, the first non-zero digit is 4. So, you should move the decimal to the right of 4 to get 4 which is a number between 1 and 10.

Multiply the number 4 by 10 raised to the power of a negative number by counting the number of places the decimal moved. In this case we had to move the decimal 6 places to the right so that the answer will be 4×10^{-6}
Thus, $0.000004 = 4 \times 10^{-6}$.

Conversely, to write any positive number whose value is greater than 1 in scientific notation, multiply this number by 10 raised to the power of a positive number found by counting the number of places the decimal moved to the left.

2) Write 4000000 in scientific notation.

Here, the number of places the decimal moved to the left is six. Therefore,
   $4000000 = 4 \times 10^6$
3) Write 0.0000046 in scientific notation.
   $0.0000046 = 4.6 \times 10^{-6}$
4) Write 0.0000000084 in scientific notation.
   $0.0000000084 = 8.4 \times 10^{-9}$
5) Write 378000 in scientific notation.
   $378000 = 3.78 \times 10^5$
6) Write 62486 in to scientific notation.
   $62486 = 6.2486 \times 10^4$

6.6. Converting Scientific Notation to Decimal Notation

To convert a number from scientific notation to decimal notation, first remove the multiplication sign and the number 10^n. Then, shift the decimal separator 'n' digits to the right (for positive n) or to the left (for negative n).

Examples:
1) Convert the scientific notation 1.4305×10^6 to decimal notation.
Answer: The number 1.4305×10^6 has its decimal separator shifted six digits to the right and become 1430500.

2) Convert the scientific notation -5.0431×10^{-3} to decimal notation.
Answer: The number -5.0431×10^{-3} has its decimal separator moved three digits to the left and become: -0.0050431

3) Convert the scientific notation 3.045×10^7 to decimal notation.
Answer: 30,450,000

4) Convert the scientific notation 8.63×10^{-8} to decimal notation.
Answer: 0.0000000863

5) Convert the scientific notation -4.3×10^{-6} to decimal notation.
Answer: -0.0000043

Exercise 6.5-6.6

1) Convert the following decimal notation to scientific notation.
   a) 0.0000048
   b) -0.00003054
   c) 0.865403
   d) 20.5436
   e) -0.563408
   f) 0.0432
   g) -5.86000
   h) 467.83020
   i) -40.6321

2) Convert the following numbers to scientific notation.
   a) 6042000
   b) 54.321
   c) 400006
   d) 1000000
   e) 47000000

3) Write the following numbers in decimal form.
   a) 4.32×10^6
   b) -4.3×10^{-4}
   c) -5×10^{-6}
   d) 6.0083×10^{-3}
   e) 8.24×10^7
   f) 7×10^{-5}
   g) 5.65×10^{-4}
   h) 6.02×10^{-6}
   i) -3×10^{-8}
   j) -56×10^{-4}
   k) 6.0321×10^5
   l) 943.214×10^4

CHAPTER 7

SETS

7.1. Introduction to Sets

In mathematics, a set is a well-defined collection of objects. The phrase "well defined" means that there is a rule which can be used to identify the members of the set. The member that makes up a set can be anything: people, mathematical objects, such as numbers, lines or other geometrical objects. There are, however; some collections whose members may be difficult to identify. For example, the collection of all students in a class who have eight eyes. This collection cannot be an example of a set because it is not well-defined (in reality, there is no student who has eight eyes). Once in conclusion, to be a set, a collection must be a well-defined.

It should have a meaningful collection of members or elements in it.

Examples of Sets

* The set of all even numbers.
* The set of all prime numbers.
* The set of all English alphabets.

Moreover,

* A set is represented by brace, {   } sometimes called ' curly bracket'.
* A set is usually denoted by a capital letter such as A, B, R, S, D, …
* The generic elements of a set are denoted by lowercase letters a, x, c,.m.d.y.h, …

Examples: A = {1, 2, 3, ...}
We read this as "Set A", **not** as "Set a".

This form of representation is called Roster Method or Listing Method. We use commas to separate the elements of a set. The three dots (dot-dot-dot) or **(…)** after an element in a set is called an ellipsis. It indicates that there is no final element and thus the listing is an endless.

* R is used to represent the set of all real numbers.
* N is used to represent the set of all natural numbers.
* A distinct object that belongs to that set is called an element or a member.

Examples: A = {1, 2, 3}
    1, 2, and 3 are members or elements of set A.

Note: - If 'x' is an element of set A, we usually express it as 'x∈A', where ∈ stands for 'belongs to' or 'is an element of' whereas ∉ stands for 'doesn't belong to' or 'is **not** an element of'.

Example: M = {0, 1, 2, 6}

0 ∈ M, 1 ∈ M, 2 ∈ M, 6 ∈ M, 8 ∉ M, 5 ∉ M, {2, 6}∉M, {1, 2, 3}∉M

7.2. Description of Sets

In mathematics, a set can be described in three ways:
1. Verbal Method.
2. Listing Method. (The Roster Method).
3. Set Builder Method.

7.2.1. (1) Verbal Method

This method is used when there is large number of elements and such elements cannot be named or listed.

Examples:
- The set of even numbers.
- The set of odd numbers.
- The set of integers.

7.2.2. (2) Listing Method (The Roster Method)

This method is used to describe a set by the listing method. We list all elements of the set and enclose them in a pair of braces.

Examples: -
A. {1, 2, 3, 4, 5}
B. {1, 3, 5, 7, 11, . . . 43}
C. {0, 2, 4, 6, 8, . . .}

7.2.3. (3) Set Building Method

In set builder method, the elements of the set are described, but not listed.

Examples

A = {x|x is an integer between 2 and 16}

M = {x|x is an odd number}

S = {x|x is a counting number less than 13}

Note: - When we list elements of a well-defined set, it is not necessary to repeat the element(s).

- When we list the elements of a set, the order does not matter.
- A set which contains no element in it is called an empty set or null set. Sometimes, it is also called a void set. It is denoted by {   } or ∅.

Examples:

A = {x|x ∈ N and 2 < x < 3}

M = {x|x ∈ Z and 5 < x < 6}

7.3. Finite and Infinite Set

7.3.1. Finite Set

A finite set is a set having a countable number of elements.

Examples:

{1, 5, 7, 8} is a finite set with four elements.

{2,6,7,10,14, 9,4} is a finite set with seven elements.

The set of even numbers less than 55. This is a finite set.

7.3.2. Infinite Set

An infinite set is a set whose elements cannot be counted.

Examples:

1) {1, 2, 3, ...} is an infinite set whose elements cannot be counted. Listing all members of this set has no end. In other words, it has no countable numbers of elements.

2) {2, 4, 6, 8,10,...} is an infinite set whose elements cannot be counted. Listing all members of this set has no end.

3) The set of odd numbers. This is an infinite set.

4) The set of real numbers. This is an infinite set.

Note: An empty set is a finite set with cardinal number 0.

7.4. Operation on Sets

In mathematics, there are several fundamental operations for constructing new sets from a given two or more sets. Thus, if two or more sets combine together to form a single set under the given conditions, the operations on sets are accomplished. On this lesson, we define the following four rules of operations on sets.

1) Union of sets.

2) Intersection of sets.

3) Difference of two sets.

4) Cartesian product of sets.

7.4.1. Union of Sets.

If A and B are two sets, then their union, denoted by A ∪ B, is the set of all elements that are found either in A or in B or in both.

The union of a set is denoted by the symbol 'U'.

In a set builder notation,

A $\cup$ B = {x|x $\in$ A or x $\in$ B}

Examples:

   A = {1, 2, 4, 6}

B = {a, e, i, o, u, 1, 2}

Find A $\cup$ B

Answer:

A $\cup$ B = {1, 2, 4, 6, a, e, i, o, u}

b) If M = {0, 2, 4, 6, 8}

N = {0, 1, 3, 4, 6}

Find M $\cup$ N

Answer

M $\cup$ N = {0, 1, 2, 3, 4, 6, 8}

Note: List duplicate elements only once while you are listing union of set of elements.

Example:

R= {1,3,5,5,7,8,8}

S= {2,8, a, b, b,7,6,9}

Find R $\cup$ S

R $\cup$ S= {1,3,5,7,8,2, a, b,6,9}

= {1,2,3,5,6,7,8,9,a,b}

7.4.2. Intersection of Sets

If A and B are two sets, then their intersection, denoted by A $\cap$ B, is the set of all elements which are common to both set A and B.

The intersection of a set is denoted by the symbol $\cap$.

In a set builder notation,

A $\cap$ B = {x| x$\in$ A and x $\in$ B}

Examples:

a)   If   A = {0, 2, 4, 6, 8}

         B = {2, 4, 7, 9, 10}

   Find A $\cap$ B

   Answer

   A $\cap$ B = {2, 4}

b) If R= {a, e, i. o, u,3,4}

   S= {a, u,3,8,9}

   Find R $\cap$ S

   Answer

   R $\cap$ S = {a, u,3}

7.4.3. Difference of two sets

Let A and B be sets. The difference of A and B denoted by A – B, is the set containing those elements that are in A but **not** in B.

In set builder notation

A – B {x: x∈ A and x∉ B}

Note: A – B can be written as A/B and red as "A complement B". or "A less B".

Examples

1) If A= {2,3,4, a, b}
   B= {1,2,5, a, c}

   Find
   a) A – B   b) B – A
       A/B read as set A less set B. It is the same as set A – B
   Answer
   a) A – B = A/B = {3, 4, b}
   b) B – A = B/A = {1, 5, c}

2) If R = {a, b, c, 3, 6, 8}
   S = {4, 3, 9}
   Find
   a) R – S
   b) S – R

   Answer
   a) R – S = {a, b, c, 6, 8}
   b) S – R = {4, 9}

7.4.4. Cartesian Product (Cross Product) of a Set.

The Cartesian product (or cross product) of A and B, denoted by A × B, is the set of all possible pair of elements of set A × B = {(a, b)/a ∈ A and b ∈ B}. Thus, the elements (a, b) are ordered pairs.

Examples:

1) If   A = {1, 2, 3} and
        B = {c, d} then, the Cartesian product of set A and B is given by:
        A × B = {(1, c), (1, d), (2, c), (2, d), (3, c), (3, d)}
        B × A= {(c,1), (c, 2), (c, 3), (d, 1), (d, 2), (d, 3)}

2) If   A = {1, 2, 5} and
        B = {x, y, 3} then, the Cartesian product of B × A is
        B × A = {(x, 1), (x, 2), (x, 5), (y, 1), (y, 2), (y, 5), (3, 1), (3, 2), (3, 5)}

3) If   A = {Alex, Maria} and

B = {Teresa, Robert} then the Cartesian product of A and B is
A × B = {(Alex, Teresa), (Alex, Robert), (Maria, Teresa), (Maria, Robert)}

7.5. Properties of Cartesian Product

- The Cartesian product is non- commutative, A × B ≠ B × A
- A × B = B × A, if only A = B
- A × B = Ø, if either A = Ø or B = Ø
- The Cartesian product is non-associative. (A × B) × C ≠ A × (B × C).

7.5.1. Distributive Property over Set of Intersection.

A × (B ∩ C) = (A × B) ∩ (A × C)

7.5.2. Distributive Property over Set of Union.

A × (B ∪ C) = (A × B) ∪ (A × C)

7.6. Cartesian Product of three Sets

For three sets A, B and C, the Cartesian product of A, B and C is denoted by A × B × C and defined by:

A × B × C = {(x, y, z)/x ∈ A and y ∈ B and z ∈ C}
Example
If A = {1, 3}, B = {4, 5} and C = {6}, then find A × B × C
Answer
    A = {1, 3}
    B = {4, 5}
    C = {6}
Thus,
A × B × C = {(1, 4, 6), (1, 5, 6), (3, 4, 6), (3, 5, 6)}

7.7. Symmetric Difference between Two Sets

Let A and B be any sets. The symmetric difference between A and B denoted by A Δ B is the set of all elements which belongs either in set A or set B but not to both.
In set builder notation,
A Δ B = {x| x ∈ A ∪ B and x∉ A ∩ B}
Note: A Δ B can also be expressed by (A ∪ B) – (A ∩ B). In other words, A Δ B = A/B ∪ B/A.

Examples:

1) If A = {1, 2, 3, a, b, c} and
        B = {1, 5, b, c}, then

A Δ B = {2, 3, a, 5}
Or A/B = {2, 3, a}
B/A = {5}
Therefore, A/B ∪ B/A = A Δ B = {2, 3, a, 5}.

2) If R = {x, y, 2, 5, 9} and
S = {x, 2, 5, 8, 10}, then
R Δ S = {y, 9, 8, 10}
R/S = {y, 9}
S/R = {8,10}
Therefore, R/S ∪ S/R = R Δ S = {y, 9, 8, 10}.

Note: A Δ B = A ∪ B, if A ∩ B = ∅
A Δ B = B Δ A

7.8. Venn Diagram

A Venn diagram consists of a closed rectangular region that may consists of another circular sub region(s) in it. Such regions represent sets.

Examples

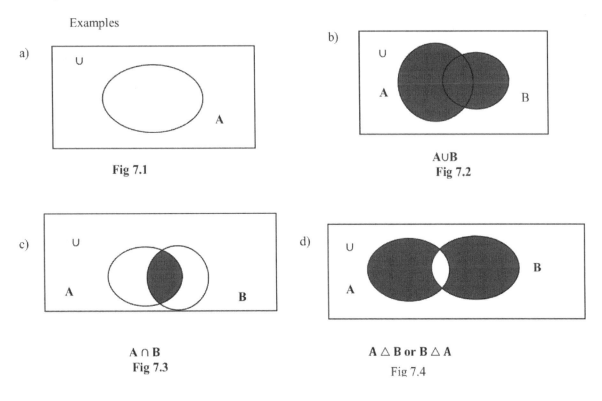

Examples

a)

Fig 7.1

b)

AUB
Fig 7.2

c)

A ∩ B
Fig 7.3

d)

A △ B or B △ A
Fig 7.4

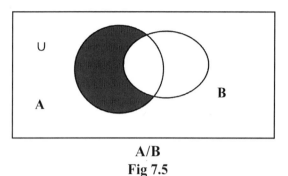

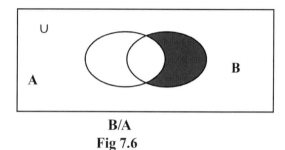

A/B	B/A
Fig 7.5	**Fig 7.6**

Note: The procedure for creating a Venn-diagram is that:
1) Draw a rectangle and label it U to represent the universal set.
2) Draw circles with in the rectangle to represent the subsets of the universe.
3) Label the circles and write the applicable elements in each circle.
4) Enter the differences: - Inside each of the given circle, place the elements that are unique to that specific item and that do not belong to any of the other topic.
5) Enter the similarities: - If two or more sets have common elements, place that element in the section in which all such circular shapes overlap.

7.9. Universal Set

In mathematics, a universal set denoted by U is the set containing all objects or elements including itself and of which all other sets are its subsets or proper subsets.

Examples:

1) Let U = {1, 2, 3, 4, 5, 6, 7, 8}
        A = {1, 3, 5, 6}
        B = {1, 3, 4, 5}
   Universal set Venn-diagram representation

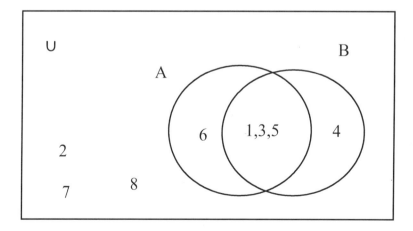

Fig. 7.7

In Example 1) above, set A and B are overlapping sets. The intersections of set A and B consists of elements 1, 3 and 5. Thus,

A ∩ B = {1, 3, 5}.

2) Let U= {1, 2, 3, 4, 5, a, b, c}
   A = {1, 3, 5, a}
Universal set Venn-diagram representation

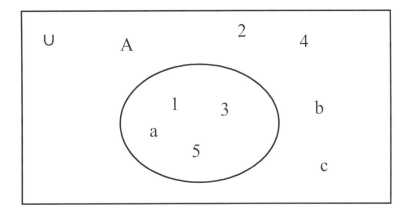

Fig. 7.8

3) Let U = {1, 3, 5, 6, 8, 9, a, e, f, n}
   A = {1, 5, 6}
   B = {1, 8, 9, a, e}
   C = {1, 9, f}
Universal set Venn-diagram representation

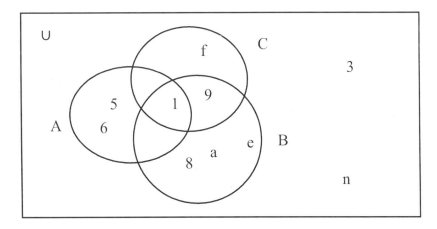

Fig. 7.9

7.10. Complement of a Set

The complement of a set, denoted by A is the set of all elements in the given universal set U that are not in set A.

In set builder notation

A = {x/x∈U and x∉ A}

Thus, A = U – A

Below is the Venn-diagram for the complement of A.

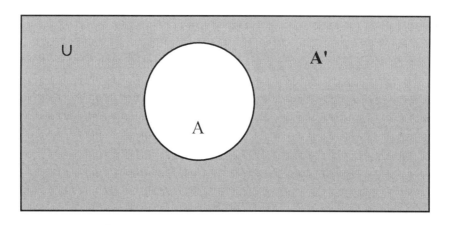

Fig. 7.10

Examples

1) If A = {1, 2, a, b} and U= {1, 2, a, b, 4, 5, 6, 7} then find A complement (A')
   Answer: Complement of set A is the set containing all elements that belong in universal set but not in set A.
   Here, the elements 4, 5, 6, and 7 are members of U but not belongs to set A.
   Thus, A' = {4, 5, 6, 7}

2) Let U = {4, 5, a, b, c, 8, 9} and A = {4, b, 8}.
   Then, find A' and show your answer using Venn-diagram.
   Answer:
   As it is mentioned before, complement of set A (i.e., A') is the set containing all elements present in the universal set but not in set A. For the given example, elements 5, a, c and 9 are found in universal set U but not in set A. Thus,
   A' = {5, a, c, 9}
   Using Venn-diagram

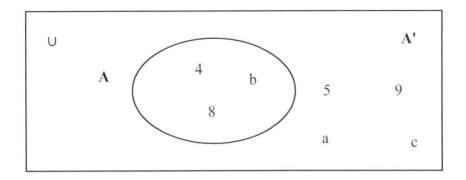

Fig. 7.11

7.10.1. Some Properties of Complement Sets

    a) For every set A, there exists a unique set A with respect to U.

    b) A∪A=U

    c) A∩A=∅

    d) (A∪B)= A∩ B ……………….DeMorgan's First Law

    e) (A∩B) = A∪B ……………….DeMorgan's Second Law

    f) ∅= ∪

    g) ∪=∅

    h) (A) =A

7.11. SUBSET

A set 'A' is said to be a subset of set 'B' if every element of set A is also an element of set B. Thus, x ∈ A implies x∈ B.

Notation: If A is a subset of set B, then we write A ⊆ B and read as "A is a subset of B". If some elements x in A are not in set B, then set A is not a subset of set B and it is denoted by

A ⊄ B. (Read as "A is not a subset of B").

That is, ⊄ is a symbol for "not subset"

Examples

1) If A = {1, 2, a, b}

      B = {1, 2, a, b, 5, 6}

    Then, A ⊆ B, because every element in set A is also an element of set B.

    But recall that set B is not a subset of set A. This is because some of the elements in set B are not belong in set A.

    Mathematically, B ⊄ A.

2) If A = {a, b, c, 1, 2}

      B = {a, b, c, 5, 6}

Since some of the elements in set A are not found in set B, that is; $1 \notin B$ and $2 \notin B$, we say $A \nsubseteq B$.

3) If $A = \{1, 3, 5, 7\}$ and
$B = \{1, 3, 5, 7\}$, then
$A \subseteq B$ or $B \subseteq A$, because every element in set A is also in set B and vice versa.

4) The set of natural numbers (N) is a subset of the set of integers (I). Thus, $N \subseteq I$
$N = \{1, 2, 3, 4, ...\}$
$I = \{..., -3, -2, -1, 0, 1, 2, 3, ...\}$
i.e., $N \subseteq I$

Note: Every set is a subset of itself because $x \in A$ means $x \in A$ and this is always true.
An empty set is a subset of every set.
Thus, $\varnothing \subseteq A$, for every set A.

7.12. Proper Subset

Set A is a proper subset of a set B, if and only if every element of A also in B, and there exists at least one element in B that is not in A.

- Proper subset is represented by the symbol "$\subset$".
  $A \subset B$: read as "Set A is a proper subset of set B"

Example: -

If $A = \{1,2,3\}$ and
$B = \{1,2,3,4,5\}$, then
$A \subset B$, (Set A is a proper subset of set B).
Note: Set B is not a proper subset of set A. i.e., ($B \not\subset A$).
- Any set is **not** a proper subset of itself.

Example: If $A = \{1, 2, 3\}$
$B = \{1, 2, 3\}$
Thus, Set A is **not** a proper subset of set B.
- An empty set is a proper subset of every set except for the empty set, itself.

7.13. Equal Sets and Equivalent Sets.

7.13.1. Equal sets are two or more sets having the same types (identical) and equal number of elements. Two sets A and B can be equal only if each element of set A is the element of set B or else each element of set B is also the element of set A.

Note: The arrangement or the order of the elements in the given set does not matter so as to determine whether they are equal or equivalent sets.
*   Equal sets can be represented by a sign "=".

Examples: -

1. If   A = {1, 2, 3, a, b} and
         B = {1, 2, 3, a, b},
    Since all elements in set A are also in set B and their number of elements are equal to each other, we say set A and set B are equal sets. Thus, A = B

2. If   A = {1, 2, 3, a, b, c} and
         B = {1, 2, a, c, 3, b}
    Since every element in set A is also in set B, and its converse is also true, we say set A and set B are equal sets. Thus, A = B
    (The order does not matter)
    Note: If A ⊆ B and B ⊆ A, then A = B.

7.13.2. Equivalent Sets: Two sets are said to be equivalent sets if they have the same number of elements in each set which are not necessarily identical.

*   Equivalent sets are represented symbolically as "~".

Examples: -

1) If   A = {1, 2, 3, a, b, c} and
         B = {5, 6, 7, x, y, z}
    Since set A has six elements and also set B has six elements, we say set A and set B are equivalent sets. Thus, A ~ B.
2) If   R = {1, 2, 3, 4} and
         S = {2, 1, 4, 3}
    Both set R and set S have 4 elements. As the number of elements in set R is the same as that of set S, set R and set S are equivalent sets. Hence, R ~ S
    In addition to this, set R and set S have the same types of elements. They are identical. In short, they are equal sets. Thus, R = S.
    Note: - Set R and set S are also equal sets.
    *   Equal sets are always equivalent sets; but its converse is not necessarily true.
        i.e., Equivalent sets are not always equal sets.

7.14. Disjoint Sets

Two sets A and set B are said to be disjoint sets if they have no element in common. Disjoint sets can be shown in a Venn-diagram as follows:

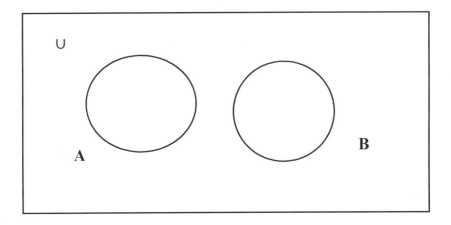

Fig. 7.12
Disjoint sets

Examples:

1) If   A = {1, 2, 3, a, b} and
        B = {5, 8, 4, 7}
   Since set A and set B do not have any common elements, we say set A and set B are disjoint sets.

2) If E is the set of even numbers and O is the set of odd numbers, then set E and set O are disjoint sets.

7.15. Cardinality of a Set

The cardinality of a set is the number of elements in that set.
The cardinality of the set A is denoted by |A| or n(A).

Examples: -

1) If   A = {1, 2, 3, a, b} and
        B = {a, 2, 4}
   The cardinality of set A is 5 and the cardinality of set B is 3. Thus, n(A) = 5 and n(B) = 3
   • Cardinality of A∩B = 2, that is n (A∩B) = 2.

7.16. Ordered Pairs

An ordered pair is a composition of the x coordinate and the y coordinate, having two values written in a fixed order within the parentheses. Thus, a pair of elements a, b (in which the first number is 'a' and the second number is 'b') is called an ordered pair denoted by (a, b).

Examples:

1) (4, 3) is an ordered pair, for which the first entry 4, belongs to the x-coordinate in the Cartesian plane and the second entry 3, belongs to the y-coordinate in the same plane. Thus,

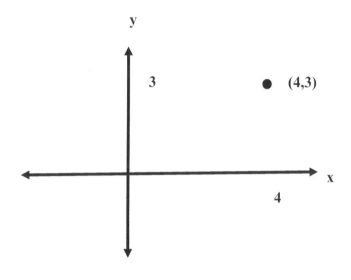

Fig. 7.13

2) (–5, 6),

Here, the first entry **–5** is the x-coordinate whereas the second entry 6 is the y-coordinate.

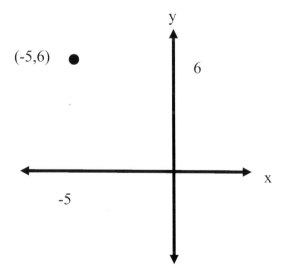

Fig. 7.14

7.17. Power Set

All possible subsets of set A are known as the power set of set A. It includes an empty set and the given set itself. It is denoted by P(A).

Example

Let A = {2, 4, 6}, then
The possible subsets of set A are: Ø, {2}, {4}, {6}, {2,4}, {2, 6}, {4, 6}, {2, 4, 6}. Thus, the power set of set A, P(A) = {Ø, {2}, {4}, {6}, {2, 4}, {2, 6}, {4, 6}, {2, 4, 6}}.
- The cardinality of the set A is 8.
- The power set of set A is given by P(A) = 2^n, where n is the number of elements in the given set A.

In the above example, set A has 3 elements; i.e., n = 3, the power set of set A is
P(A) = 2^n, P(A) = 2^3, P(A) = $2 \times 2 \times 2 = 8$, which means set A has 8 number of possible subsets.

Note: If set A has 'n' number of elements, then set A has 2^n-1 proper subset.

Example: - Let A = {1, 3}, then set A has: $2^n-1 = 2^2 - 1 = 2 \times 2 - 1 = 4 - 1 = 3$ proper subsets.

These proper subsets are: Ø, {1}, and {3}
Thus, the Possible proper subset of set A = Ø, {1}, {3}.

More examples related to the topic

1) If the ordered pairs: (5x-2, 4) = (3, 2y-6), then find the value of x,y and the original ordered pairs.
   Solution:
   Equate the first entry of the first ordered pairs with the first entry of the second ordered pairs and solve for x.
   (5x-2) is the first entry of the first ordered pairs and 3 is the first entry of the second ordered pairs. Thus,

   5x-2=3
   5x-2+2=3+2 ....Add 2 on both sides.
   5x=5
   $\frac{5x}{5} = \frac{5}{5}$ .... Divide both sides by 5.

   $x=1$

   Equate the second entry of the first ordered pairs with the second entry of the second ordered pairs and solve for y.
   4 is the second entry of the first ordered pairs and 2y-6 is the second entry of the second ordered pairs. Thus,

$4=2y-6$

$4+6=2y-6+6$ ....Add 6 on both sides.

$10 = 2y$

$\dfrac{10}{2} = \dfrac{2 \cdot y}{2}$ .... Divide both sides by 2.

$5=y$

$y=5$

Thus, x= 1 and y = 5.

To find the original ordered pairs, substitute the value of x and y in

$(5x - 2, 4) = (3, 2y - 6)$

$(5(1) - 2, 4) = (3, 2(5) - 6)$

$(3, 4) = (3, 4)$

Thus, the original ordered pair is (3, 4).

2) In the medical examination of 120 persons, it was known that 25 persons are anemic and 16 are diabetic, and 6 are both anemic and diabetic of the 120 persons examined. How many persons are neither anemic nor diabetics?

Solution

Let U be the universal set or the total number of persons. i.e U = 120

Let A be the number of anemic persons = 25

Let D be the number of diabetic persons = 16

From your previous knowledge: $n(A) + n(B) = n(A \cup B) + n(A \cap B)$

In this case,

$n(A) + n(D) = n(A \cup D) + n(A \cap D)$

$25 + 16 = n(A \cup D) + 6$

$41 = n(A \cup D) + 6$

$n(A \cup D) = 41 - 6$

$n(A \cup D) = 35$

Thus, out of 120 examined persons, 35 persons are either diabetic or anemic patients.

Hence: $n(U|A \cup D) = n(U) - n(A \cup D)$

$= 120 - 35$

$= 85$

Therefore, the number of persons who are neither anemic nor diabetics is 85.

This can also be easily solved using universal set Venn-diagram representation.

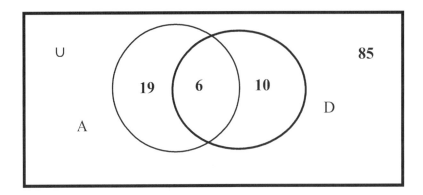

Fig. 7.15

In generl,19 persons anemic only,10 persons diabetic only,6 persons are both anemic and diabetic, whereas 85 of them are neither anemic nor diabetics.

3) In a class of 80 students, 37 like physics and 54 likes Biology and each student likes at least one of the two subjects. How many students like both physics and Biology?

Solution

Let 'P' be the number of students who like physics.

'B' be the number of students who like biology.

'U' be the total number of students in a class.

Given

$U = 80$, $\cap(P) = 37$, $\cap(B) = 54$, then $\cap(P\cap B) = \cap(P) + \cap(B) - \cap(P\cup B)$

$\cap(P\cap B) = 37 + 54 - 80$

$\cap(P\cap B) = 91 - 80$

$\cap(P\cap B) = 11$

Thus, 11 students like both physics and biology.

Universal Venn-diagram representation

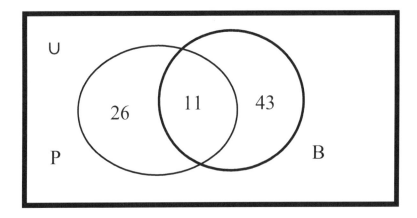

Fig. 7.16

26 students like only Physics.

43 students like only Biology.

11 students like both Physics and Biology.

4) In a group of 110 persons, 76 people can speak English and 51 can speak Spanish.
   a) How many people can speak English only?
   b) How many people can speak Spanish only?
   c) How many people can speak both English and Spanish?

   Solution

   Let E be the set of people who speak English

   S be the set of people who speak Spanish

   E − S be the set of people who speak English and not Spanish.

   S − E be the set of people who speak Spanish and not English.

   E∩S be the set of people who speak both English and Spanish

   Given: $n(E) = 76$, $n(S) = 51$, and $n(E∪S) = 110$

   Thus, $n(E∩S) = n(E) + n(S) − n(E∪S)$

   $$= 76 + 51 − 110$$

   $$= 127 − 110$$

   $$n(E∩S) = 17$$

   Therefore, the numbers of people who speaks both English and Spanish is 17.

   $n(E) = n(E − S) + n(E∩S)$

   $n(E − S) = n(E) − n(E∩S)$

   $$= 76 − 17$$

   $n(E − S) = 59$

   and $n(S − E) = n(S) − n(E∩S)$

   $n(S − E) = 51 − 17$

   $$= 34$$

   Hence,

   a) The number of people who speak only English = 59
   b) The number of people who speak only Spanish = 34
   c) The number of people who speak both English and Spanish = 17

   Universal Venn-diagram representation.

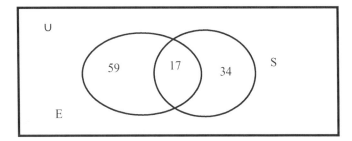

Fig. 7.17

173

5) Let A and B are two sets such that n(A) = 56, n(A∩B) = 12, n(A∪B) = 73, then find n(B)

Solution

Given: n(A) = 56, n(A∩B) = 12,

n(A∪B) = 73, n(B) =?

Using the general formula n(A) + n(B) = n (A∩B) + n(A∪B)

56 + n(B) = 12 + 73

56 + n(B) = 85

56 – 56 + n(B) = 85 – 56

$\qquad$ n(B) = 29

Thus, n(B) = 29

6) Let A = {1, 2}, B = {3, 4} and c = {a, b} then find
   a) A × B
   b) B × A
   c) B × C
   d) A × C
   e) A × (B∩C)
   f) (A × B)∩(A × C)
   g) A × (B ∪ C)
   h) (A × B) ∪ (A × C)

Solutions
   a) A × B = {(1, 3), (1, 4), (2, 3), (2, 4)}
   b) B × A = {(3, 1), (3, 2), (4, 1), (4, 2)}

Remark: A × B ≠ B × A
   c) B×C = {(3, a), (3, b), (4, a), (4, b)}
   d) A × C = {(1, a), (1, b), (2, a), (2, b)}
   e) A × (B∩C) = (A × B) ∩ (A × C)
       {(1, 3), (1, 4), (2, 3), (2, 4)} ∩ {(1, a), (1, b), (2, a), (2, b)}
   Thus, A × (B∩C) = Ø
   f) (A × B) ∩ (A ×C) =
       {(1, 3), (1, 4), (2, 3), (2, 4)} ∩ {(1, a), (1, b), (2, a), (2, b)}
   Thus, (A × B) ∩ (A × C) = Ø
   g) A × (B∪C) = (A × B) ∪ (A × C)
       {(1, 3), (1, 4), (2, 3), (2, 4)} ∪ {(1, a), (1, b), (2, a), (2, b)}
       = {(1, 3), (1, 4), (2, 3), (2, 4)}, (1, a), (1, b), (2, a), (2, b)}
   h) (A × B) ∪ (A × C) =
       {(1, 3), (1, 4), (2, 3), (2, 4), (1, a), (1, b), (2, a), (2, b)}

7) If A = {1, 2, 3, a, b} and B = {4, 5, a, b} then

Determine each of the following sets.
   a) A∩B
   b) A∪B

c) A|B
d) B/A
e) A Δ B
f) B Δ A
g) n(A)
h) n(B)

Solutions
a) A∩B = {a, b}
b) A∪B = {1, 2, 3, 4, 5, a, b}
c) A|B = {1, 2, 3}
d) B|A = {4, 5}
e) A Δ B = A/B∪B/A = {1, 2, 3, 4, 5}
f) B Δ A = B/A∪A/B = {1, 2, 3, 4, 5}
g) n(A) = 5
h) n(B) = 4

8) If   U = {1, 2, 3, x, y, z, a, b, c, d}
        A = {1, 3, x, y}
        B = {1, 3, z, a, b} and
        C = {1, b, x, y}, then
(i) Draw and label a Venn diagram to represent the set U, A, B and C. Indicate all the elements of each set in the Venn-diagram.
(ii) Determine each of the following.
    a) A'
    b) B'
    c) C'
    d) A∩(B∪C)
    e) A Δ B
    f) B Δ C
    g) U'
    h) A|B
    i) B|C
    j) Cardinal number of A, B and C.
    k) Number of proper subsets of set A, B and C.
    l) Write all possible subsets of set A.
    m) Write all possible proper subsets of set A.
    n) Mention the power set of set A.

175

Solutions

(i)

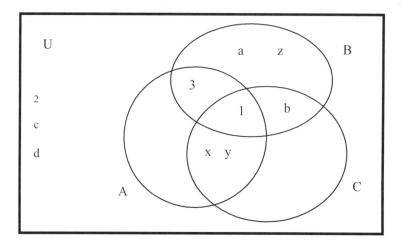

(ii) a) A' = {2, c, d, a, z, b}

b) B' = {2, c, d, x, y}

c) C' = {2, c, d, 3, a, z}

d) A∩ (B∪C) = {1, 3, x, y}

e) A Δ B = {x, y, z, a, b}

f) B Δ C = {3, z, a, x, y}

g) U' = ∅

h) A|B = {x, y}

i) B|C = {3, z, a}

j) Cardinal number of A = 4, B = 5 and C = 4.

k) Number of proper subsets of set A = $2^n- 1 = 2^4 - 1 = 2$ x2 x 2 x 2 – 1= 16 – 1= 15.
   Number of proper subsets of set B= $2^n - 1 = 2^5 - 1 = 2$ x2 x 2 x 2x 2 – 1= 32 – 1=31,
   and Number of proper subsets of set C = $2^n - 1 = 2^4 - 1 = 2$ x2 x 2 x 2 – 1= 16 – 1= 15.

l) Set A = {1, 3, x, y}
   Possible subsets of set A = $2^n = 2^4$ =2x2x2x2= 16

   These are {∅, {1}, {3}, {x}, {y}, {1, 3}, {1, x}, {1, y}, {3, x}, {3, y}, {x, y}, {1, 3, x},
   {1, 3, y}, {3, x, y}, {1, x, y}, {1, 3, x, y}}.

m) Set A = {1, 3, x, y}
   Possible proper subsets of set A = $2^n - 1 = 2^4 -1$=2x2x2x2 - 1= 16 – 1 = 15
   These are {∅, {1}, {3}, {x}, {y}, {1, 3}, {1, x}, {1, y}, {3, x}, { 3 ,y}, {x, y}, {1, 3, x},
   {1, 3, y}, {3, x, y}, {1, x, y}}.

n) The power set of set A = {1, 3, x, y}
   P(A) = {, {1}, {3}, {x}, {y}, {1, 3}, {1, x}, {1, y}, {3, x}, {3, y}, {x, y}, {1, 3, x},
   {1, 3, y}, {3, x, y}, {1, x, y}, {1, 3, x, y}}.

9) If $n(A-B) = 22$, $n(A\cup B) = 74$ and $\cap(A\cap B) = 29$, then find $\cap(B)$

Solution

Using the formula: -

$$\cap(A\cup B) = \cap(A - B) + \cap(A\cap B) + \cap(B - A)$$
$$74 = 22 + 29 + \cap(B - A)$$
$$\cap(B - A) = 74 - (22 + 29)$$
$$= 74 - (51)$$
$$\cap(B - A) = 23$$

Thus, $\cap(B) = \cap(A \cap B) + \cap(B - A)$
$$= 29 + 23$$
$$= 52$$

Universal Venn-diagram representation

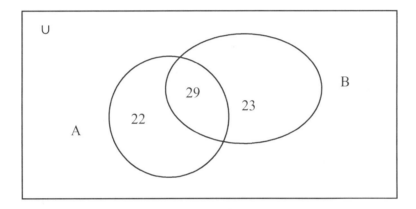

10) In a survey of Collage students, 68 had taken Biology course, 98 had taken Mathematics course, 62 had taken Chemistry course, 32 had taken Biology and chemistry, 30 students had taken Biology and Mathematics courses, 26 had taken mathematics and chemistry courses, and 18 had taken all the three courses. How many students had taken only one course?

Solution: -

Let B, M, and C represent sets of students who had taken Biology, Mathematics and Chemistry course respectively. Then, from the given information, we have:

$n(B) = 68$, $n(M) = 98$, $n(C) = 62$
$n(B \cap C) = 32$, $n(B\cap M) = 30$,
$n(M \cap C) = 26$, and $n(B \cap M\cap C) = 18$

Thus,

Number of students who had taken only Biology
$$= n(B) - [n(B \cap C) + n(B \cap M) - n(B \cap M \cap C)]$$
$$= 68 - [32 + 30 - 18]$$
$$= 68 - [62 - 18]$$
$$= 68 - 44$$
$$= 24$$

Number of students who had taken only mathematics:

= n(M) – [n (B ∩ M) + n (M ∩ C) – n (B ∩ M ∩ C)]

= 98 – [30 + 26 – 18]

= 98 – [56 – 18]

= 98 – 38

= 60

Number of students who had taken only chemistry

= n(C) – [n (B ∩ C) + n (M ∩ C) – n(B∩M∩C)]

= 62 – [32 + 26 – 18]

= 62 – [58 – 18]

= 62 – [40]

= 22

The total number of students who had taken only one course:

n(B) + n(M) + n(C)

= 24 + 60 +22

= 106

Therefore, the total number of students who had taken only one course is 106.

Remark: -The above question can also be solved alternatively using Venn-diagram method.

11) Below is the representation of set U, set A and set B. With Universal Venn-diagram.

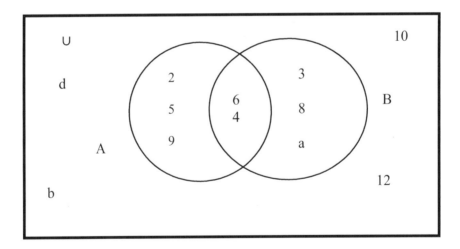

Then, from the above given Venn-diagram information, Find: -

a) n(U)

b) n(A)

c) n(B)

d) n(A ∩ B)

e) $n(A \cup B)$
f) $A \cap B$
g) $A \cup B$
h) A|B
i) B|A
j) U'
k) $A \Delta B$
l) $B \Delta A$
m) A'
n) B'

Solution
a) $n(U) = 12$
b) $n(A) = 5$
c) $\cap(B) = 5$
d) $n(A \cap B) = 2$
e) $n(A \cup B) = 8$
f) $A \cap B = \{4,6\}$
g) $A \cup B = \{2,5,9,4,6,3, a,8\}$
h) $A|B = \{2,5,9\}$
i) $B|A = \{3, a,8\}$
j) $U' = \emptyset$
k) $A \Delta B = \{2,5,9,3, a,8\}$
l) $B \Delta A = \{2,5,9,3, a,8\}$
m) $A' = \{3, a,8, d, b,10,12\}$
n) $B' = \{2,5,9, d, b,10,12\}$

Exercise 7.1-7.17

1) Write each of the following sets using listing method.|
   a) {x: x is a vowel in English alphabet}
   b) {x: x is an even number between 1 and 20}
   c) {x: x is a latter in the word "World"
   d) {x: x a natural number and multiple of 5}
   e) {y: y is non-negative integer}
   f) {y: y is an integer between 0 and 10}
   g) {x: x is a natural number between 3 and 14 and divisible by 2}
   h) {x: x is a natural number and divisible by 4}
   i) {x|x is a non-positive integer}
   j) {x|x is a letter in the word 'march'}

2) Write each of the following sets using set-builder method.
   a) {2, 4, 6, 8, 10, 12}
   b) {7, 14, 21, 28, ...}
   c) {Monday, Tuesday, Wednesday, Thursday}
   d) {..., –2, –1,0, 1, 2, 3, ...}
   e) {0, 1, 2, 3, ...}

3) If A = {–1, –2, 1, 2, 3, 5, a, b} and
   B = {–1, 1, x, y, 5, b}, then
   Compute: -
   a) A ∪ B
   b) A ∩ B
   c) A|B
   d) B|A

4) For each of the following sets (i-iv) compute:
   a) A∩B
   b) A ∪ B
   c) A|B
   d) B|A
       (i)   A = {x: x is an even number}
             B = {x: x is an odd number}
       (ii) A = {x: x is an integer}
             B = {0, 1, 2, 3, ...}
       (iii) A = {..., –3, –2, –1, 0, 1, 2, 3, ...}
             B = {0, 1, 2, 3, ...}
       (iv) A = {a, x, y, 1,2}
             B = {a, 1, 2, 4}

5) Let U = {–5, –2, –1, 0, 3, 6, 2, a, b, x}
   A = {–2, –1, 0, 3, x}
   B = {0, –5, 6, 2, a, b}
   C = {x, –5, 6}
   (i) Draw the Universal Venn diagram representation for the above sets U, A, B and C, and
        label all elements in the appropriate set.
   (ii) Compute
        a) A∩B
        b) A ∪ B
        c) A ∩ C
        d) B ∩ C
        e) A ∪ C
        f) B ∪ C
        g) A∩B∩C
        h) A∩(B∪C)

i) U'
j) A Δ B
k) A Δ C
l) A|B
m) A|C
n) B|C
o) A × B
p) A × (B∩C)
q) A × (B∪C)
r) A'
s) B'
t) C'

(iii) Find the power set of the set
a) A
b) B
c) C
d) U
e) A∩C
f) A∩C
g) A ∪ B
h) B∩C
i) A∪(B∩C)
j) A∩(B∪C)
k) A∩(B∩C)

(iv) Find the number of proper subsets of set
a) A
b) B
c) C
d) A∩B
e) A∩C
f) B∩C
g) A∪(B∩C)
h) A∩(B∪C)

(v) Find the number of subsets of set
a) A
b) B
c) C

6) In a class of 120 students, 36 like chemistry and 42 like math, 16 like both subjects. How many students like either of them and how many like neither?

7) Use the letters in the Venn-diagram below to describe the region for each of the sets given below.

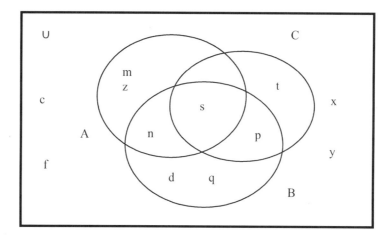

a) U
b) A
c) B
d) C
e) A∩B
f) A∩C
g) B∩C
h) A∪(B∪C)
i) A∩(B∩C)
j) A'
k) B'
l) C'
m) U'
n) A|B
o) B|C
p) A|C
q) A Δ B
r) B Δ C

8) In a university, 50 students enrolled in Biology, 60 students in Physics, 35 in Mathematics, 20 in Biology and Physics, 16 in Mathematics and Physics, 5 in Mathematics and Biology. No one enrolled in all the three subjects, find how many are enrolled at least one of the subjects.

9) Which of the following pairs of sets are disjoint sets?
   a) A = {2, 4, 6, a, b}, B = {5, 6, x, y}
   b) A = {x, y, z}, B = {0, 2, y}
   c) A = {x, y, z}, B = {z, y, m}

d)   A = {xy, x, z}, B = {xzy, y}

10) Which of the following set is an example of finite set?
   a)   A = {..., –2, –1, 0, 1, 2, ...}
   b)   A = {x|x is a positive number}
   c)   A = {x|x is a vowel letters in English alphabets}
   d)   A = {0, 1, 2, 3, ...}

11) Let A = {1, 3, 5}, B = {a, b, c}, C = {x, y, z}
   Compute
   (i)   A × B
   (ii) A × C
   (iii) A × (B∪C)

12) Using Venn diagrams, verify the following identities
   a)   A Δ B
   b)   A|B
   c)   A∩B
   d)   A∪B
   e)   A'
   f)   B'

13) Let A and B be two finite sets such that n(A) = 30, n(B) = 38 and n(A∩B) = 13, then find n(A∪B).

14) If   U = {a, b, c, 4, 5, 6, 9}
   A = {a, b, 5}
   B = {b, c, 4, 5}, then
   Compute:
   a)   A∩B
   b)   A'
   c)   B'
   d)   (A∩B)'
   e)   (A∪B)'
   f)   A∪B

CHAPTER 8

Polynomial Functions

8.1. Definition of a Polynomial Function

Let **n** be a non-negative integer and let $a_n, a_{n-1}, ..., a_2, a_1, a_0$, be real numbers, with $a_n \neq 0$. The function defined by:

$f(x) = a_n x^n + a_{n-1} x^{n-1} + ... + a_2 x^2 + a_1 x + a_0$ is called a polynomial function of degree **n.**

Note that in the definition of a polynomial function $p(x) = a_n x^n + a_{n-1} x^{n-1} + ... + a_1 x + a_0$

(i) $a_n, a_{n-1}, a_{n-2}, ..., a_0$ are called the coefficients of the polynomial function.
(ii) The number a_n is called the leading coefficient of the polynomial function and $a_n x^n$ is the leading term.
(iii) The number a_0 is called the constant term of the polynomial.
(iv) The number **n** (the exponents of the highest power of x) is the degree of the polynomial.

Note: In a polynomial, the leading term is the term with the highest power of x. For example, in the expressions $f(x) = 3x^4 + 2x^3 + 2x^2 + 3x + 9$, the leading term is $3x^4$ because it has the highest power.

- In $f(x) = 6x^3 + 7x^2 + 5x^4 + 2x - 8x^6 - 4$, the leading term is $-8x^6$, because it has the highest power.

Note: The power expressions $a_n x^n, a_{n-1} x^{n-1}, ..., a_1 x$, and a_0 are called terms of the polynomial $a_n x^n + a_{n-1} x^{n-1} + ... + a_1 x + a_0$.

- A polynomial with only one term is called a Monomial.

Examples: $2x^2, 3x^3, 12x$

A polynomial with only two non-zero terms is called a Binomial.
Examples: $2x-1, 3y+4, 2x^4 - 3x, -x^2 + 9$
- A polynomial with only three non-zero terms are called Trinomial.
  Examples: $2x^2 + 5x-1, x^3 - 3x^2 + 6, -x^4 + 8x^2 - 9$

Examples:

1) $p(x) = 6x^4 + 3x^3 - 2x^2 + 8x - 4$, this is a polynomial.

(i)   whose leading term is 6x^4.
(ii) whose coefficients are 6, 3, –2, 8 and –4.
(iii) whose leading coefficient is 6.
(iv) whose constant term is –4.
(v)   whose degree is 4.

2) $p(x) = -3x^5 - 6x^4 + 8x^2 - 9$ is a polynomial of:
(i)   leading term: $-3x^5$.
(ii) coefficients: –3, –6, 0,8, and -9.
(iii) leading coefficient –3.
(iv) constant term –9.
(v)   Degree 5.

3) $p(x) = 2x^4 - 3x + 8$ is a polynomial of
(i)   leading term: $2x^4$
(ii) coefficients: 2, 0, 0, –3, and 8.
(iii) leading coefficient 2.
(iv) constant term 8.
(v)   Degree 4.

4) $p(x) = -6x^2 + 9$ is a polynomial of
(i)   leading term: $-6x^2$
(ii) coefficients: –6, 0 and 9.
(iii) leading coefficient –6.
(iv) constant term 9.
(v)   Degree 2

5) $p(x) = 2x^3$ is a polynomial of
(i)   leading term: $2x^3$
(ii) coefficient 2
(iii) leading coefficient 2.
(iv) constant term 0.
(v)   Degree 3.
      Note: $2x^3 = 2x^3 + 0x^2 + 0x + 0$

6) f(x)=1
(i)   Leading coefficient: 1
(ii) Constant term: 1
(iii) Degree:0 as $1x^0 = 1$
      Recall that $x^0 = 1$, where $x \neq 0$

7) $K(x) = \dfrac{x^4+2}{x^4+2} = 1$

(i)   Leading coefficient: 1

(ii) Constant term :1

(iii) Degree: 0

8) f(x)=0

    (i)   It does not have degree (Not defined)

    (ii) Zero function

    (iii) Leading coefficient: 0

Note: A polynomial whose coefficients are all equal to zero is called Zero polynomial.

- The domain of a polynomial function is the set of all real numbers.
- The range of a polynomial function is the set of all real numbers.

Remark: In some cases, we may have to set a restriction on the domain because of situations. Examples: - In population function, the domain is always the set of positive numbers.

- In geometrical application, if the radius of a circle is x feet long, and A(x) is the area of the circle, the domain of the function A is the set of positive real numbers.

- Sometimes a polynomial may have more than one variable.

Examples:

1) $p(x) = 6x^2y^3 + 2xy^2 - xy + y + 3$, in these expressions the degree of each term is the sum of its exponents. In this case: In the polynomial: $6x^2y^3 + 2xy^2 - xy + y + 3$:

Degree of: $6x^2y^3$ is $2 + 3 = 5$

Degree of: $2xy^2$ is $1 + 2 = 3$

Degree of: $-xy$ is $1 + 1 = 2$

Degree of: y is 1

Degree of 3 is 0.

Thus, in the example above the degree of $6x^2y^3 + 2xy^2 - xy + y + 3$, is the **highest exponent** of the various terms that constitute the polynomial. $6x^2y^3$ has the highest exponents and thus, the degree of this polynomial is the sum of the degree of x and y, which is, $6x^2y^3 = 2+3=5$

2) $p(x) = 3x^2y^3z^5 + 5xy^2 - 6xy + 2$ is a polynomial of:

    (i)   Coefficients: 3, 5, –6, 2

    (ii) Leading coefficient: 3

    (iii) Constant term: 2

    (iv) Degree: $2 + 3 + 5 = 10$

        Note: - The degree of a polynomial in more than one variable is the sum of the exponents in each term and the degree of the polynomial is the largest sum in that polynomial.

3) $p(x) = 3x^5y^4z^2 + 8xy^2z - 4xy + 9$

    (i)   Coefficients: 3, 8, –4, 9.

    (ii) Leading coefficient: 3

    (iii) Constant term: 9.

    (iv) Degree: $5 + 4 + 2 = 11$

4) $p(x) = 5x^7y^4 + 6xy^2z^3 - 8xy + 8$

    (i)   Coefficients: 5, 6, –8, 8.

    (ii) Leading coefficient: 5.

    (iii) Constant term: 8

    (iv) Degree: $7 + 4 = 11$

5) What is the degree of $8x^2y^7z^{10} + 2x^5y^{13}z^2 - x^3y + y + 3$

    Since the term $2x^5y^{13}z^2$ has the highest exponents, the degree is the sum of the degree of x, y, and z which is $2x^5y^{13}z^2 = 5 + 13 + 2 = 20$, the degree is 20.

8.2. Operation on Polynomial Functions

In mathematics, the basic algebraic operations are addition, subtraction, multiplication and division. When we perform operation on polynomial functions, most of the time, we use the commutative, associative and distributive laws in order to combine like terms together. This helps someone to make the intended operation easy.

To revise algebraic operations once more, let a, b and c be real numbers. Thus, the following are some rules of algebra which are applicable to operate polynomials.

8.2.1. Commutative Law for Addition

$$a + b = b + a$$

8.2.2. Commutative Law for Multiplication

$$(a \cdot b) = (b \cdot a)$$

8.2.3. Associative Property of Addition.

$$a + (b + c) = (a + b) + c$$

8.2.4. Associative Property of Multiplication.

$$a \cdot (b \cdot c) = (a \cdot b) \cdot c$$

8.2.5. Distributive Property

$$a \cdot (b + c) = (a \cdot b) + (a \cdot c)$$

8.3. Addition of Polynomial Functions

The method of adding polynomial functions is the same as the method of adding real numbers. Adding polynomials involve combining like terms and adding their coefficients. Note that like terms are terms having the same variables with the same powers perhaps different coefficients.

Examples:

1) If $f(x) = 2x^2$ and $g(x) = 3x^2 + 3$, then find $f(x) + g(x)$
   Solution: $f(x) + g(x) = (2x^2) + (3x^2 + 3)$
   $$= 5x^2 + 3$$

2) If $g(x) = 3x^3 + 2x^2 - 5$ and $h(x) = 4x^3 - 4x^2 + 7$, then find $g(x) + h(x)$
   Solution: $g(x) + h(x) = (3x^2 + 2x^2 - 5) + (4x^3 - 4x^2 + 7)$
   $$= (3x^3 + 4x^3) + [2x^2 + (-4x^2)] + (-5 + 7)$$
   $$g(x) + h(x) = 7x^3 - 2x^2 + 2$$

3) Find the sum of each of the following polynomial functions.
   a) $f(x) = x^4 - 2x^3 + \dfrac{1}{2}x^2 - 4$ and $g(x) = x^4 + 2x^3 + 6$
   b) $f(x) = x^5 + 3x^2 - 2x - 1$ and $g(x) = 3x^5 + 6x^2 + 4x + 5$
   c) $g(x) = -3x^6 + 4x^2 - 3x + 6$ and $h(x) = 3x^6 - 4x^2 + 3x + 9$
   d) $g(x) = x^4 + x^3 - 2x^2 + x + 1$ and $h(x) = x^3 - 3x^2 + x + 6$
   Solution:

   a) $f(x) = x^4 - 2x^3 + \dfrac{1}{2}x^2 - 4$ and $g(x) = x^4 + 2x^3 + 6$
   $$f(x) + g(x) = (x^4 - 2x^3 + \dfrac{1}{2}x^2 - 4) + (x^4 + 2x^3 + 6)$$
   $$= (x^4 + x^4) + (-2x^3 + 2x^3) + \dfrac{1}{2}x^2 + (-4 + 6)$$
   $$= 2x^4 + 0 + \dfrac{1}{2}x^2 + 2$$
   $$\therefore f(x) + g(x) = 2x^4 + \dfrac{1}{2}x^2 + 2$$

   b) $f(x) = x^5 + 3x^2 - 2x - 1$ and $g(x) = 3x^5 + 6x^2 + 4x + 5$
   $$f(x) + g(x) = (x^5 + 3x^2 - 2x - 1) + (3x^5 + 6x^2 + 4x + 5)$$
   $$= (x^5 + 3x^5) + (3x^2 + 6x^2) + (-2x + 4x) + (-1 + 5)$$
   $$= 4x^5 + 9x^2 + 2x + 4$$
   $$\therefore f(x) + g(x) = 4x^5 + 9x^2 + 2x + 4$$

   c) $g(x) = -3x^6 + 4x^2 - 3x + 6$ and $h(x) = 3x^6 - 4x^2 + 3x + 9$
   $$g(x) + h(x) = (-3x^6 + 4x^2 - 3x + 6) + (3x^6 - 4x^2 + 3x + 9)$$
   $$= (-3x^6 + 3x^6) + [4x^2 + (-4x^2)] + (-3x + 3x) + (6 + 9)$$

$$= 0 + 0 + 0 + 15$$
$$= 15$$
$$\therefore g(x) + h(x) = 15$$

d) $g(x) = x^4 + x^3 - 2x^2 + x + 1$ and $h(x) = x^3 - 3x^2 + x + 6$
$$g(x) + h(x) = (x^4 + x^3 - 2x^2 + x + 1) + (x^3 - 3x^2 + x + 6)$$
$$= x^4 + (x^3 + x^3) + [-2x^2 + (-3x^2)] + (x + x) + (1 + 6)$$
$$= x^4 + 2x^3 - 5x^2 + 2x + 7$$
$$\therefore g(x) + h(x) = x^4 + 2x^3 - 5x^2 + 2x + 7$$

In general, the sum of two polynomial functions f and g is written as f + g, and it is defined by: f + g: (f + g) (x) = f(x) + g(x), for all x ∈ |R

Note: The sum of any two or more polynomial function is always a polynomial function.

•   The domain of the sum of any two or more polynomial function is the set of real numbers.

8.4. Subtraction of Polynomial Functions

To subtract a polynomial from a polynomial, convert signs of terms of a polynomial that is subtracted and then perform addition instead of subtraction.

Examples

1) If $f(x) = 3x^4 + 6x^3 - x^2 + 4x + 2$ and
$g(x) = x^4 + 2x^3 + x^2 - 2x + 1$, then
find $f(x) - g(x)$
Solution:
$$f(x) - g(x) = (3x^4 + 6x^3 - x^2 + 4x + 2) - (x^4 + 2x^3 + x^2 - 2x + 1)$$
$$= 3x^4 + 6x^3 - x^2 + 4x + 2 - x^4 - 2x^3 - x^2 + 2x - 1$$
$$= 3x^4 - x^4 + 6x^3 - 2x^3 - x^2 - x^2 + 4x + 2x + 2 - 1$$
$$f(x) - g(x) = 2x^4 + 4x^3 - 2x^2 + 6x + 1$$

2) If $f(x) = 2x^5 + 4x^3 + 2x^2 + x + 1$ and
$g(x) = 2x^5 + 3x^3 + 2x + 4$, then
find $f(x) - g(x)$
Solution:
$$f(x) - g(x) = (2x^5 + 4x^3 2 2x^2 + x + 1) - (2x^5 + 3x^3 + 2x + 4)$$
$$= 2x^5 + 4x^3 + 2x^2 + x + 1 - 2x^5 - 3x^3 - 2x - 4$$
$$= 2x^5 - 2x^5 + 4x^3 - 3x^3 + 2x^2 + x - 2x + 1 - 4$$
$$f(x) - g(x) = x^3 + 2x^2 - x - 3$$

3)   If $f(x) = 10x^4 + x^2 - 4x + 8$ and $g(x) = 4x^3 - 2x + 1$, then find $f(x) - g(x)$

Solution:

$$f(x) - g(x) = (10x^4 + x^2 - 4x + 8) - (4x^3 - 2x + 1)$$

$$= 10x^4 - 4x^3 + x^2 - 4x - (-2x) + 8 - 1$$

$$= 10x^4 - 4x^3 + x^2 - 4x + 2x + 8 - 1$$

$$f(x) - g(x) = 10x^4 - 4x^3 + x^2 - 2x + 7$$

4)   If $g(x) = 5x^4 + 4x^3 - 2x^2 + 7x - 13$ and $h(x) = 6x^4 - 7x^3 + 3x^2 + 9x + 16$, then find
$2h(x) - g(x)$

Solution: $2h(x) = 2(6x^4 - 7x^3 + 3x^2 + 9x + 16)$

$$2h(x) = 12x^4 - 14x^3 + 6x^2 + 18x + 32$$

$$2h(x) - g(x) = (12x^4 - 14x^3 + 6x^2 + 18x + 32) - (5x^4 + 4x^3 - 2x^2 + 7x - 13)$$

$$= 12x^4 - 5x^4 - 14x^3 - 4x^3 + 6x^2 + 2x^2 + 18x - 7x + 32 - 13$$

$$2h(x) - g(x) = 7x^4 - 18x^3 + 8x^2 + 11x + 19$$

5)   If $f(x) = 7x^4 + 18x^3 - 7x + 4$ and $g(x) = 7x^4 + 18x^3 + 7x + 15$, then find,

$f(x) - g(x)$

Solution:

$$f(x) - g(x) = (7x^4 + 18x^3 - 7x + 4) - (7x^4 + 18x^3 + 7x + 15)$$

$$7x^4 - 7x^4 + 18x^3 - 18x^3 - 7x - 7x + 4 - 15$$

$$= 0 + 0 - 14x - 11$$

$$= -14x - 11$$

Note that the degree of f(x) – g(x) is 1.

Thus, the difference of two polynomial function f and g is written as f – g, and it is defined by:
$(f - g) = (f - g)(x) = f(x) - g(x)$, for all $x \in |R$

- The domain of the difference of any two polynomial functions is the set of real numbers.
- When we operate the difference of two polynomials, the degree of the result depends on the degree of the polynomial functions f and g. If the degree of f is not equal to the degree of g, then the degree of (f – g) (x) is the degree of f(x) or the degree of g(x), which ever has the highest degree. If they have the same degree; however, the degrees of (f – g) (x) might be lower than this common degree when they have the same leading coefficients. (See examples 1 - 5 above).

8.5. Multiplication of Polynomial Functions

To multiply two polynomial functions, multiply each term in one polynomial by each term in the other polynomial turn by turn, collect like terms and then add or subtract, if necessary.
Examples

1) If $f(x) = 2x^3 + 3x - 2$ and $g(x) = 2x - 4$, then find the product of $f(x)$ and $g(x)$.

   Solution

   $f(x) \bullet g(x) = (2x^3 + 3x - 2)(2x - 4)$
   $= 2x^3(2x - 4) + 3x(2x - 4) + (-2)(2x - 4)$
   $= 4x^4 - 8x^3 + 6x^2 - 12x - 4x + 8$

   $f(x) \bullet g(x) = 4x^4 - 8x^3 + 6x^2 - 16x + 8$

   Thus, the product of $f(x)$ and $g(x)$ is: $4x^4 - 8x^3 + 6x^2 - 16x + 8$.

2) Multiply: $f(x) = (3x^3 - 2x^2 + 4)$ by
   $g(x) = (x^4 - 2x^3 - 3x^2 + 8)$

Solution: $f(x) \bullet g(x) = 3x^3(x^4 - 2x^3 - 3x^2 + 8) + (-2x^2)(x^4 - 2x^3 - 3x^2 + 8) + 4(x^4 - 2x^3 - 3x^2 + 8)$
   $= 3x^7 - 6x^6 - 9x^5 + 24x^3 - 2x^6 + 4x^5 + 6x^4 - 16x^2 + 4x^4 - 8x^3 - 12x^2 + 32$
   $= 3x^7 - 6x^6 - 2x^6 - 9x^5 + 4x^5 + 6x^4 + 4x^4 + 24x^3 - 8x^3 - 16x^2 - 12x^2 + 32$
   $= 3x^7 - 8x^6 - 5x^5 + 10x^4 + 16x^3 - 28x^2 + 32$

   Thus; $f(x) \bullet g(x) = 3x^7 - 8x^6 - 5x^5 + 10x^4 + 16x^3 - 28x^2 + 32$

   Its degree is the highest degree in the given term, which is 7.

   Note: - The product of two polynomial functions f and g is written as:

$f \bullet g = (f \bullet g)(x) = f(x) \bullet g(x)$, for all $x \in$ |R.

- The degree of the product of two non-zero polynomials are the sum of the highest degrees of the two polynomials in the product.

  Thus, for $a > 0$, and m and n are non-negative numbers.

  If $f(x) = a^m$ and $g(x) = a^n$, then $f(x) \bullet g(x) = a^m \bullet a^n = a^{m+n}$, for $a > 0$, and m and n are non-negative numbers.

  Examples: -

  1) If $f(x) = x^6$, for $x > 0$ and $g(x) = x^8$, for $x > 0$, then $f(x) \bullet g(x)$ is given by: -
     $f(x) \bullet g(x) = x^6 \bullet x^8 = x^{6+8} = x^{14}$

  2) For $x > 0$, if $f(x) = x^5$ and $g(x) = x^8$ then, $f(x) \bullet g(x)$ is given by:
     $f(x) \bullet g(x) = x^5 \bullet x^8 = x^{5+8} = x^{13}$

8.6. Division of Polynomial Functions

Using a long division process, similar to that used in arithmetic, it is possible to divide a polynomial by another polynomial of the same or lower degree. A generalized type of this common arithmetic technique is called Long Division.

8.6.1. The Division Algorithm

If $P(x)$ and $D(x)$ are polynomials, such that $D(x) \neq 0$, and the degree of $D(x)$ is less than or equal to the degree of $P(x)$, then there exist unique polynomials $Q(x)$ and $R(x)$ such that:

   $P(x) = D(x) \bullet Q(x) + R(x)$

   Dividend = Divisor $\bullet$ Quotient + Remainder

The remainder, $R(x)$ equals zero or it is of degree less than the degree of $D(x)$. If $R(x) = 0$, we say that $D(x)$ divides exactly into $P(x)$ and that $D(x)$ and $Q(x)$ are factor of $P(x)$.

Thus, $\dfrac{P(x)}{D(x)} = Q(x) + \dfrac{R(x)}{D(x)}$

Look at the elementary division below, where 643 is being divided by 11.

$$\begin{array}{r} 58 \\ 11\overline{)643} \\ -55 \\ \hline 93 \\ -88 \\ \hline 5 \end{array}$$

- Divide 64 by 11.
- Multiply the quotient 5 by 11.
- Subtract 55 from 64.
- Bring down 3.
- Again divide 93 by 11.
- Multiply the quotient 8 by 11.
- Subtract 88 from 93.
- Now 5 is a remainder.

Thus, $643 = (58 \times 11) + 5$

Here, 643 is the dividend, 11 is a divisor, 58 is the quotient and 5 is the remainder of the division.

Examples:

1) Divide: $2x^2 + 3x + 2$ by $x - 1$

Solution:

$$\begin{array}{r} 2x + 5 \\ x-1\overline{)2x^2 + 3x + 2} \\ -(2x^2 - 2x) \\ \hline 5x + 2 \\ -(5x - 5) \\ \hline 7 \end{array}$$

$\leftarrow \dfrac{2x^2}{x} = 2x$, multiply $(2x)$ by $(x-1)$

$\leftarrow$ subtract

$\leftarrow \dfrac{5x}{x} = 5$, multiply 5 by $(x-1)$

$\leftarrow$ subtract

$\leftarrow$ Remainder

7 is no longer divided by $(x - 1)$ and it is a remainder.

Thus,

Quotient: $q(x) = 2x + 5$ and remainder: $R(x) = 7$

Since the remainder is not zero, $(x - 1)$ is not a factor of $2x^2 + 3x + 2$.

More over; $2x^2 + 3x + 2 = (x - 1)(2x + 5) + 7$

That is, $\dfrac{2x^2 + 3x + 2}{(x-1)} = 2x + 5 + \dfrac{7}{(x-1)}$

2) Divide: $2x^3 - 5x^2 + x + 2$ by $2x + 1$

Solution:

$$
\begin{array}{r}
x^2 - 3x + 2 \\
(2x+1)\overline{)\, 2x^3 - 5x^2 + x + 2} \\
-(2x^3 + x^2) \\
\hline
-6x^2 + x + 2 \\
-(-6x^2 - 3x) \\
\hline
4x + 2 \\
-(4x + 2) \\
\hline
0
\end{array}
$$

$\leftarrow \dfrac{2x^3}{2x} = x^2$

$\leftarrow$ subtract

$\leftarrow -\dfrac{6x^2}{2x} = -3x$

$\leftarrow$ subtract

$\leftarrow \dfrac{4x}{2x} = 2$

$\leftarrow$ subtract

$\leftarrow$ Remainder

Thus, Quotient: $Q(x) = x^2 - 3x + 2$ and remainder: $R(x) = 0$
Since the remainder is zero, $(2x + 1)$ is a factor of $2x^3 - 5x^2 + x + 2$.
Moreover, $(2x^3 - 5x^2 + x + 2) = (2x + 1)(x^2 - 3x + 2)$

3) Divide: $x^2 + 10x + 24$ by $x + 4$

Solution:

$$
\begin{array}{r}
x + 6 \\
(x+4)\overline{)\, x^2 + 10x + 24} \\
-(x^2 + 4x) \\
\hline
6x + 24 \\
-\ (6x + 24) \\
\hline
0
\end{array}
$$

$\leftarrow \dfrac{x^2}{x} = x$, multiply it by $(x + 4)$

$\leftarrow$ subtract

$\leftarrow \dfrac{6x}{x} = 6$, multiply it by $(x + 4)$

$\leftarrow$ subtract

$\leftarrow$ Remainder

Since the remainder is 0, $x + 4$ is a factor of $x^2 + 10x + 24$.
Thus, $x^2 + 10x + 24 = (x + 4)(x + 6)$

4) Divide: $4x^4 + 2x^3 + 6x^2 + 13x + 11$ by $2x + 1$

Solution:

$$
\begin{array}{r}
2x^3 + 3x + 5 \\
(2x+1)\overline{)\, 4x^4 + 2x^3 + 6x^2 + 13x + 11} \\
-\ 4x^4 + 2x^3 \\
\hline
6x^2 + 13x + 11 \\
-\ 6x^2 + 3x \\
\hline
10x + 11 \\
-\ 10x + 5 \\
\hline
6
\end{array}
$$

...... $\dfrac{4x^4}{2x} = 2x^3$, multiply by $(2x+1)$

$\longleftarrow$ Subtract

$\longleftarrow$ Bring down and $\dfrac{6x^2}{2x} = 3x$

$\longleftarrow$ Subtract

$\longleftarrow$ Bring down 11 and $\dfrac{10x}{2x}$

$\longleftarrow$ Subtract

$\longleftarrow$ Remainder

6 is no longer divide by 2x and thus, it is a remainder.

Thus, Quotient: Q(x) is $2x^3 + 3x + 5$ and the remainder is 6.

Since the remainder is different from zero, (2x+1) is not a factor of $4x^4 + 2x^3 + 6x^2 + 13x + 11$

8.7. The Remainder Theorem

If f(x) is any polynomial and f(x) is divided by (x-r), then the remainder is f(r).

The validity of this theorem can be proved in example number 1 above.

1) In this example, when $2x^2 + 3x + 2$ was divided by (x − 1) the remainder was 7.

   Based on the Remainder Theorem, the remainder can be found by substituting x=1 in f(x).

   i.e., f(1)

   Thus,

   $f(x) = 2x^2 + 3x + 2$
   $f(1) = 2(1)^2 + 3(1) + 2$
   $f(1) = 2 + 3 + 2$
   $f(1) = 7$

2) Find the remainder when $x^2 + 5x + 6$ is divided by x+4.

   Solution: Based on the Remainder Theorem, the remainder can be found by substituting x= -4 in f(x). i.e., f(-4)

   Thus,

   $f(x) = x^2 + 5x + 6$
   $f(-4) = (-4)^2 + 5(-4) + 6$
   $f(-4) = 16 + (-20) + 6$
   $f(-4) = 16 + 6 - 20$
   $f(-4) = 22 - 20$
   $f(-4) = 2$

   Thus, the remainder is 2

3) Find the remainder when $2x^4 + 3x^3 - x^2 - 1$ is divided by (x-2)

   Solution: Based on the Remainder Theorem, the remainder is found by substituting x=2, in f(x).

   i.e., f(2)

   Thus,

   $f(x) = 2x^4 + 3x^3 - x^2 - 1$
   $f(2) = 2(2)^4 + 3(2)^3 - (2)^2 - 1$
   $\quad = 2(16) + 3(8) - 4 - 1$
   $\quad = 32 + 24 - 4 - 1$
   $\quad = 56 - 5$

$f(x) = 51$

Thus, the remainder is 51

4) Find the remainder when $x^2 - 8x + 6$ is divided by $(x+3)$

Solution: Based on The Remainder Theorem, the remainder is found by substituting x=-3 in f(x), i.e., f (-3)

Thus,

$f(-3) = x^2 - 8x + 6$

$f(-3) = (-3)^2 - 8(-3) + 6$

$f(-3) = 9 + 24 + 6$

$f(-3) = 39$

Thus, the remainder is 39

8.8. The Factor Theorem

If f(x) is any polynomial function and f(x) is divided by (x-r) and the remainder f(r)=0, then (x-r) is a factor of f(x).

Examples:

1) Show that (x-1) is a factor of $f(x) = 3x^3 + 2x^2 - 2x-3$

Solution: By the Factor Theorem, if (x-1) is a factor of f(x), when substituting x=1, i.e., f (1) must be zero.

$f(x) = 3x^3 + 2x^2 - 2x-3$

$f(1) = 3(1)^3 + 2(1)^2 - 2(1)-3$

$f(1) = 3+2-2-3$

$f(1) = 5-5$

$f(1) = 0$

Substituting x=1 in f(x) gives 0, that means the remainder is zero. Thus (x-1) is a factor of $f(x) = 3x^3 + 2x^2 - 2x-3$

2) Show that (y+4) is a factor of $f(y) = y^2 + 10y + 24$

Solution: By the Factor Theorem, if (y+4) is a factor of f(y), when substituting y=-4, i.e., f(-4) must be zero.

$f(y) = y^2 + 10y + 24$

$f(y) = (-4)^2 + 10(-4) + 24$

$f(-4) = 16 + (-40) + 24$

$= 16+24-40$

$= 40-40$

$f(y) = 0$

Substituting $y = -4$ in f(y) yields 0. That means the remainder is zero. Thus, (y+4) is a factor of $f(y) = y^2 + 10y+24$

8.8.1. The Remainder and The factor Theorem for the divisor (ax+b)

If the divisor is not in the form of (x-r) and given in the form of linear equation (ax+b),
We cannot use (x-r) to find the remainder because the coefficient of x, i.e., "a" is different form one. Thus, to find the remainder when the divisor is of the form (ax+b). we use the following method.

Assume: ax+b=0, then
     ax+b-b=0-b ....(Subtracting 'b' from both sides)
     ax=-b

$$\frac{ax}{a} = \frac{-b}{a}$$ ....(Dividing both sides by 'a')

$$x = \frac{-b}{a}$$

Thus, when a polynomial is divided by the divisor (ax+b) the remainder is $f\left(\frac{-b}{a}\right)$.

From the above assumption, we can restate The Remainder Theorem when the divisor is of the form (ax+b).

If f(x) is any polynomial and f(x) is divided by (ax+b), then the remainder is $f\left(\frac{-b}{a}\right)$. Moreover, if

$f\left(\frac{-b}{a}\right) = 0$, then the divisor (ax+b) is a factor of f(x).

Examples:

1) Is (3x+2) factor of a polynomial function $f(x) = 3x^3 - 13x^2 + 8x + 12$?

Solution: To show whether (3x+2) is a factor of f(x) or not, we apply the above rule.

Consider: 3x+2=0
     3x+2-2=0-2
     3x=-2
     $x = \frac{-2}{3}$

Thus, substitute, $x = \frac{-2}{3}$ in $f(x) = 3x^3 - 13x^2 + 8x + 12$ and find the reamainder.

$$f(x) = 3x^3 - 13x^2 + 8x + 12$$

$$f\left(\frac{-2}{3}\right) = 3\left(\frac{-2}{3}\right)^3 - 13\left(\frac{-2}{3}\right)^2 + 8\left(\frac{-2}{3}\right) + 12$$

$$f\left(\frac{-2}{3}\right) = 3\left(\frac{-8}{27}\right) - 13\left(\frac{4}{9}\right) - \frac{16}{3} + 12$$

$$= \frac{-24}{27} - \frac{52}{9} - \frac{16}{3} + 12$$

$$= \frac{-8}{9} - \frac{52}{9} - \frac{16}{3} + 12 \quad \ldots\ldots\ldots\ldots\ldots\left(\frac{-24}{27} = \frac{-8}{9}, \text{ simplified by } 3\right)$$

$$= \frac{-8 - 52 - 48 + 108}{9}$$

$$f\left(\frac{-2}{3}\right) = \frac{-108 + 108}{9}$$

$$f\left(\frac{-2}{3}\right) = 0$$

Since $f\left(\frac{-2}{3}\right) = 0$, the divisor $(3x+2)$ is a factor of $f(x) = 3x^3 - 13x^2 + 8x + 12$
i.e., When we substitute, $x = \frac{-2}{3}$ in $f(x)$, the result is zero, showing that the remainder is zero.
Thus, the divisor $(3x+2)$ is a factor of $f(x)$.

2) Is $(2x-1)$ a facto of a polynomial function $g(x) = 2x^4 + 3x^3 - x^2 - 2x + 8$?
   Solution: Consider: $2x-1=0$

$$2x-1+1=0+1 \quad \ldots\ldots \text{ Add 1 on both sides}$$
$$2x=1$$
$$x = \frac{1}{2}$$

Now, substitute $x = \frac{1}{2}$ in $g(x)$ and find the remainder.

$$g(x) = 2x^4 + 3x^3 - x^2 - 2x + 8$$

$$g\left(\frac{1}{2}\right) = 2\left(\frac{1}{2}\right)^4 + 3\left(\frac{1}{2}\right)^3 - \left(\frac{1}{2}\right)^2 - 2\left(\frac{1}{2}\right) + 8$$

$$= \frac{1}{8} + \frac{3}{8} - \frac{1}{4} - \frac{2}{2} + 8$$

$$= \frac{4}{8} - \frac{5}{4} + 8$$

$$= \frac{4-10+64}{8} \quad \dots\dots\dots\dots\dots(\text{ Taking LCM of the denominators})$$

$$g\left(\frac{1}{2}\right) = \frac{58}{8}$$

$$g\left(\frac{1}{2}\right) = \frac{58}{8}$$

$$g\left(\frac{1}{2}\right) = \frac{29}{4}$$

Since, $g\left(\frac{1}{2}\right) = \frac{29}{4}$, which is different from zero, the divisor (2x-1) is **not** a factor of g(x). meaning the remainder is $\frac{29}{4}$, which is not zero, therefore, the divisor (2x-1) is **not** a factor of g(x).

Supplementary Examples: -

1) Find the remainder using The Remainder Theorem when $x^3 - 2x^2 + 3x - 2$ is divided by
   a) x − 2
   b) x + 2
   c) x − 1
   d) x + 1
   e) x − 3
   f) x + 3

Solutions: -
$x^3 - 2x^2 + 3x - 2$
a) x − 2
   x − 2 = 0
   x = 2
Substitute x = 2 in $x^3 - 2x^2 + 3x - 2$
   $2^3 - 2(2)^2 + 3(2) - 2$
   $8 - 8 + 6 - 2$
   $0 + 4 = 4$
Thus, when $x^3 - 2x^2 + 3x - 2$ is divided by x − 2, the remainder is 4.
Therefore, x − 2 is not a factor of $x^3 - 2x^2 + 3x - 2$.

b) x + 2
   x + 2 = 0
   x = −2
Substitute −2 in $x^3 - 2x^2 + 3x - 2$
   $(-2)^3 - 2(-2)^2 + 3(-2) - 2$
   $-8 - 8 - 6 - 2$

–24

When $x^3 - 2x^2 + 3x - 2$ is divided by $x + 2$, the remainder is –24. Therefore, x-2 is not a factor of $x^3 - 2x^2 + 3x - 2$.

c) x – 1

    x – 1 = 0

    x = 1

Substitute x = 1 in $x^3 - 2x^2 + 3x - 2$

    $1^3 - 2(1)^3 + 3(1) - 2$

    1 – 2 + 3 – 2

    4 – 4

    0

Thus, the remainder is zero and x – 1 is a factor of $x^3 - 2x^2 + 3x - 2$.

d) x + 1

    x + 1 = 0

    x = –1

Substitute x = –1 in $x^3 - 2x^2 + 3x - 2$.

    $(-1)^3 - 2(-1)^2 + 3(-1) - 2$

    –1 – 2 – 3 – 2

    –8

Thus, the remainder is –8, and x + 1 is not a factor of $x^3 - 2x^2 + 3x - 2$.

e) x – 3

    x – 3 = 0

    x = 3

Substitute x = 3, in $x^3 - 2x^2 + 3x - 2$.

    $3^3 - 2(3)^2 + 3(3) - 2$

    27 – 18 + 9 – 2

    36 – 20

    16

Thus, the remainder is 16, and x – 3 is not a factor of $x^3 - 2x^2 + 3x - 2$.

f) x + 3

    x + 3 = 0

    x = –3

Substitute x = –3 in $x^3 - 2x^2 + 3x - 2$.

    $(-3)^3 - 2(-3)^2 + 3(-3) - 2$

    –27 – 18 – 9 – 2

    –56

Thus, when $x^3 - 2x^2 + 3x - 2$ is divided by x + 3, the remainder is –56. Therefore; x + 3 is not a factor of $x^3 - 2x^2 + 3x - 2$.

2) Find the value of 'm' such that when $x^3 - 4x^2 + 2mx + 6$ is divided by $x - 2$, where the remainder is 26.

Solution

Let $f(x) = x^3 - 4x^2 + 2mx + 6$

$\quad x - 2 = 0$

$\quad x = 2$

$\quad f(2) = 26$ (By remainder theorem)

Thus, $2^3 - 4(2)^2 + 2m(2) + 6 = 26$

$\quad 8 - 16 + 4m + 6 = 26$

$\quad -8 + 4m + 6 = 26$

$\quad -2 + 4m = 26$

$\quad 4m = 26 + 2$

$\quad 4m = 28$ …………..(Divide both sides by 4)

$\quad m = \dfrac{28}{4}$

$\quad m = 7$

The value of m is 7.

3) When the polynomial $f(x) = ax^4 + bx^3 - 3x - 4$ is divided by $x + 1$ and $x - 1$, the remainders are both 0. Find the values of a and b.

Solution

Given: $f(x) = ax^4 + bx^3 - 3x - 4$

Both $f(1)$ and $f(-1) = 0$

Thus, $f(-1) = a(-1)^4 + b(-1)^3 - 3(-1) - 4$

$\quad f(-1) = a - b + 3 - 4$

$\quad f(-1) = a - b - 1$

$\quad a - b - 1 = 0 - - - (*)$ Since $f(-1) = 0$

$\quad f(1) = a(1)^4 + b(1)^3 - 3(1) - 4$

$\quad = a + b - 3 - 4$

$\quad a + b - 7 = 0 - - - (**)$ Since $f(1) = 0$

Combine (*) and (**)

$$\begin{cases} a - b - 1 = 0 \\ a + b - 7 = 0 \end{cases}$$

$\quad \underline{\begin{aligned} a - b &= 1 \\ a + b &= 7 \end{aligned}}$ ……….. Adding vertically

$\quad 2a + 0 = 8$

$\quad 2a = 8$

$\quad a = \dfrac{8}{2}$

$\quad a = 4$

$\quad a - b = 1$

$\quad 4 - b = 1$

$-b = 1-4$

$-b = -3$

$-1(-b) = -3(-1)$

$b = 3$

Thus, the value of a is 4 and that of b is 3.

4) Find the remainder when $2x^{31} - 4x^2 + 4$ is divided by $(x - 1)$

Solution:

   $x - 1 = 0$

   $x = 1$

To find the remainder, substitute $x = 1$ in the equation $2x^{31} - 4x^2 + 4$

   $2(1)^{31} - 4(1)^2 + 4$

   $2 - 4 + 4 = 2$

$\therefore$ The remainder is 2

8.9. Special Formulas

   1. $(a - b)(a + b) = a^2 - b^2$
   2. $(a + b)^2 = (a + b)(a + b) = a^2 + 2ab + b^2$
   3. $(a - b)^2 = (a - b)(a - b) = a^2 - 2ab + b^2$
   4. $(a + b)^3 = a^3 + 3a^2b + 3ab^2 + b^3$
   5. $(a - b)^3 = a^3 - 3a^2b + 3ab^2 - b^3$

Examples:

Multiply each of the following with appropriate formula

   a) $(2x + 3y)(2x - 3y)$
   b) $(6x^2 - 2y)^2$
   c) $(4x + 3)^2$
   d) $(3x + 5)(3x + 5)$
   e) $(x - 3)^3$
   f) $(y + 2)^3$
   g) $(2x - 4)^2$

Solutions:

   a) $(2x + 3y)(2x - 3y)$

      $(a + b)(a - b) = a^2 - b^2$

      $a = 2x$ and $b = 3y$

      Thus, $(2x + 3y)(2x - 3y) = (2x)^2 - (3y)^2$

      $\qquad\qquad\qquad\qquad\qquad = 4x^2 - 9y^2$

b) $(6x^2 - 2y)^2$

$(a - b)^2 = a^2 - 2ab + b^2$

$a = 6x^2$ and $b = 2y$

$= a^2 - 2ab + b^2$

$= (6x^2)^2 - 2(6x^2)(2y) + (2y)^2$

$= 36x^4 - 24x^2y + 4y^2$

c) $(4x + 3)^2$

$(a + b)^2 = a^2 + 2ab + b^2$

$a = 4x$ and $b = 3$

$(4x + 3)^2 = (4x)^2 + 2(4x)(3) + 3^2$

$= 16x^2 + 24x + 9$

d) $(3x + 5)(3x + 5)$

$(a + b)^2 = a^2 + 2ab + b^2$

$a = 3x$ and $b = 5$

$(3x + 5)(3x + 5) = (3x)^2 + 2(3x)(5) + 5^2$

$9x^2 + 30x + 25$

e) $(x - 3)^3$

$(a - b)^3 = a^3 - 3a^2b + 3ab^2 - b^3$

$a = x$ and $b = 3$

$(x - 3)^3 = x^3 - 3x^2(3) + 3x(3)^2 - 3^3$

$= x^3 - 9x^2 + 27x - 27$

f) $(y + 2)^3$

$(a + b)^3 = a^3 + 3a^2b + 3ab^2 + b^3$

$(y + 2)^3 = y^3 + 3y^2(2) + 3(y)(2^2) + 2^3$

$= y^3 + 6y^2 + 12y + 8$

g) $(2x - 4)^2$

$(a - b)^2 = (a - b)(a - b) = a^2 - 2ab + b^2$

$(2x - 4)^2 = (2x - 4)(2x - 4) = (2x)^2 - 2(2x)(4) + 4^2$

$= 4x^2 - 16x + 16$

Thus, $(2x - 4)^2 = 4x^2 - 16x + 16$

Additional note for The Factor Theorem

Let f(x) be a polynomial of degree greater than or equal to one, and let 'r' be any real number, then

(i) $(x - k)$ is a factor of f(x), if f(k) = 0, and

(ii) f(k) = 0, if x − k is a factor of f(x).

Examples:

1) Let $f(x) = 2x^2 - x - 1$
   a) Evaluate $f(1)$ and show that $(x - 1)$ is a factor of $f(x)$.
   b) Show that $2x + 1$ is a factor of $f(x)$.

   Solutions
   a) $f(x) = 2x^2 - x - 1$
      $f(1) = 2(1)^2 - 1 - 1$
      $\quad\ = 2 - 1 - 1$
      $\quad\ = 2 - 2$
      $f(1) = 0$
      Since $f(1) = 0$, $(x - 1)$ is a factor of $f(x)$.

   b) $2x + 1$
      $2x + 1 = 0$
      $2x = -1$
      $x = \dfrac{-1}{2}$

      To show that $2x + 1$ is a factor of $f(x)$, substitute $x = \dfrac{-1}{2}$, in $f(x) = 2x^2 - x - 1$.

      $$f\left(\tfrac{-1}{2}\right) = 2\left(-\tfrac{1}{2}\right)^2 - \left(\tfrac{-1}{2}\right) - 1$$

      $$= 2\left(\tfrac{1}{4}\right) + \tfrac{1}{2} - 1$$
      $$= \tfrac{1}{2} + \tfrac{1}{2} - 1$$
      $$= 1 - 1$$
      $$= 0$$
      $$f\left(\tfrac{-1}{2}\right) = 0$$

      Thus, $f\left(\dfrac{-1}{2}\right) = 0$, hence $(2x + 1)$ is a factor of $f(x)$.

8.10. Zero of Polynomial Function

For a polynomial function f and a real number k, If
$f(k) = 0$, then k is a zero of 'f'.
Note that if $x - k = 0$ is a factor of $f(x)$, then k is a zero of $f(x)$.
Examples: -
Find the zeros of each of the following:
a) $f(x) = 2x + 3$
b) $f(x) = x^2 - 1$

c) $f(x) = x^2 - x - 2$

Solutions:

a) $f(x) = 2x + 3$

To find the zero of f(x), make f(x) = 0.

$2x + 3 = 0$

$2x + 3 - 3 = 0 - 3$

$2x = -3$

$\dfrac{2x}{2} = \dfrac{-3}{2}$

$x = \dfrac{-3}{2}$

Thus, $\dfrac{-3}{2}$ is the zero of f (x).

b) $f(x) = x^2 - 1$

To find the zero of f (x), make f(x) = 0.

$x^2 - 1 = 0$

$(x + 1)(x - 1) = 0$ .... factorizing

$x + 1 = 0$ or $x - 1 = 0$

$x = -1$ or $x = 1$

Thus, the zeros of f (x) are –1 and 1.

c) $f(x) = x^2 - x - 2$

To find the zeros of f (x), make f(x) = 0.

$x^2 - x - 2 = 0$

$x^2 - 2x + x - 2 = 0$

$x(x - 2) + 1(x - 2) = 0$

$(x + 1)(x - 2) = 0$ .... factorizing

$x + 1 = 0$ or $x - 2 = 0$

$x = -1$ or $x = 2$

Thus, the zeros of f (x) are –1 and 2.

More examples

Solve each of the following equations:

a) $x^2 - 25 = 0$

b) $x^3 - 49x = 0$

c) $x^2 - 14x + 49 = 0$

d) $x^2 - 2 = 0$

e) $x^3 - 25x = 0$

Solutions:

a) $x^2 - 25 = 0$

$a^2 - b^2 = (a + b)(a - b)$

$x^2 - 25 = 0$

$x^2 - 5^2 = (x + 5)(x - 5) = 0$

$= x + 5 = 0$ or $x - 5 = 0$

$x + 5 = 0$ or $x - 5 = 0$

$x = -5$ or $x = 5$

Thus, S.S. = $\{-5, 5\}$

b) $x^3 - 49x = 0$

$x^3 - 49x = 0$

$x(x^2 - 49) = 0$

$x(x + 7)(x - 7) = 0$

$x = 0$ or $x + 7 = 0$ or $x - 7 = 0$

$x = 0$ or $x = -7$ or $x = 7$

Thus, S.S = $\{-7, 0, 7\}$

c) $x^2 - 14x + 49 = 0$

$x^2 - 7x - 7x + 49 = 0$

$x(x - 7) - 7(x - 7) = 0$

$(x - 7)(x - 7) = 0$

$x - 7 = 0$ or $x - 7 = 0$

$x = 7$

Thus, S.S = $\{7\}$

d) $x^2 - 2 = 0$

$x^2 - \left(\sqrt{2}\right)^2 = 0$, since $\left(\sqrt{2}\right)^2 = 2$

$\left(x + \sqrt{2}\right)\left(x - \sqrt{2}\right) = 0$ ....Factorizing

$x + \sqrt{2} = 0$ or $x - \sqrt{2} = 0$

$x = -\sqrt{2} = 0$ or $x = \sqrt{2}$

S.S $= \left\{-\sqrt{2}, \sqrt{2}\right\} = \left\{\pm \sqrt{2}\right\}$

e) $x^3 - 25x = 0$

$x(x^2 - 25) = 0$ ....Taking common factor.

$x(x + 5)(x - 5) = 0$ ....Factorizing

$x = 0$ or $x + 5 = 0$ or $x - 5 = 0$

$x = 0$ or $x = -5$ or $x = 5$

S.S = $\{-5, 0, 5\}$

Exercise 8.1 - 8.10

1) Determine the leading term, leading coefficient, the degree and constant term of each of the following polynomial functions.
   a) $p(x) = -2x^5 + 4x - 8$
   b) $p(x) = 6x^2 - 4x^6 + 3x^3 + 1$
   c) $p(x) = 3x^4 - 2x^3 - 6x + 5$
   d) $p(x) = 4x - 1$
   e) $p(x) = 0$
   f) $p(x) = \sqrt{3}\, x^5 - 4x^2 + 6x + \sqrt{8}$
   g) $p(x) = \sqrt{5}\, x^2 - 2x + 1$
   h) $p(x) = 4x^4 - 8x^{14} + 7x + 6$
   i) $p(x) = 6x^4 - 2x^3 + 3x^2 - 5x + 9$

2) If $f(x) = x^3 - 2x^2 + 3x - 1$ and
   $g(x) = 4x^3 + 6x^3 + 4x + 5$, then find: -
   a) $f(x) + g(x)$
   b) $f(x) - g(x)$
   c) $g(x) - f(x)$

3) If $f(x) = x^2 - 3$ and $g(x) = x - 4$, then find: -
   a) $f(x) + g(x)$
   b) $f(x) - g(x)$
   c) $g(x) - f(x)$
   d) $g(x) + f(x)$
   e) $f(x) \cdot g(x)$

4) Determine each of the following whether they are polynomial or not
   a) $p(x) = 2$
   b) $p(x) = x^{-3} + 2x^2 + 3x + 1$
   c) $p(x) = \sqrt{3}\, x^5 + 2x^4 + 3x + 4$
   d) $p(x) = 3x - 1$
   e) $p(x) = \pi x^4 + 3x^2 - 1$
   f) $p(x) = \dfrac{x+1}{x-1}$
   g) $p(x) = 0$
   h) $p(x) = \sqrt{2x - 3}$
   i) $p(x) = \dfrac{x^4 + 1}{x^4 + 1}$
   j) $p(x) = \dfrac{x^3 + 1}{x + 1}$

5) Divide
   a) $x^2 - 2x + 1$ by $x - 1$
   b) $x^3 - 2x^2 + 1$ by $x + 2$
   c) $x^4 + 2x^2 + 3x$ by $x^2 + 1$
   d) $3x^3 + 2x^2 + 3x + 1$ by $3x + 2$
   e) $x^3 + 3x + 1$ by x^2

6) If $f(x) = x^2 + 1$ and $g(x) = 2x^2 - 2x + 3$, then find: -
   a) $f(-2)$
   b) $g(0)$
   c) $f(1) + g(2)$
   d) $2f(3)$
   e) $g(-3)$
   f) $3g(1)$
   g) $f(3) - g(1)$
   h) $g(4) - f(3)$
   i) $f(1) \cdot g(3)$

7) Find the remainder when $p(x) = x^3 - 3x^2 + 5$ is divided by $x - 2$.

8) If $p(x) = 3x^3 - 2x^2 - 4k + 3$ is divided by $x - 1$ the remainder is 14, then find the value of k.

9) Find the value of m such that $p(x) = -2x^3 + 3x^2 + 2mx + 3$ divided by $x + 1$, the remainder is 0.

10) When $f(x) = ax^3 + bx^2 + 3x + 1$ is divided by $(x + 2)$ and $(x - 2)$, the remainders are both zero. Find the values of a and b.

11) If $p(x) = x^{101} + 1$ is divided by $x + 1$, then the remainder is _____

12) Multiply each of the following using appropriate formula.
   a) $(3x - 2y)(3x + 2y)$
   b) $(5x + 3y)(5x - 3y)$
   c) $(3a - 4b)(3a - 4b)$
   d) $(2y - 1)(2y + 1)$
   e) $(x + 2)^3$
   f) $(x - 2)^3$
   g) $(3x^2 - 2)^2$
   h) $(4x + 3)^3$

13) Find the zeros of each of the following equations.
   a) $3x^2 - 9 = 0$
   b) $2x^2 - 8 = 0$
   c) $y^2 - 5 = 0$
   d) $y^2 - 2 = 0$
   e) $x^2 = 27$
   f) $2x^2 - 49 = 0$
   g) $3x^2 - 27 = 0$

CHAPTER 9

Review of Algorithm for Simplifying and Solving Equations

9.1. Rational Expressions

In mathematics, an algebraic expression $\dfrac{p(x)}{q(x)}$, where p(x) and q(x) are polynomial expressions and q(x) ≠ 0 is called a rational expression. The domain of a rational function is the set of all real numbers except the values of x- that make the denominator zero.

$$f(x) = \frac{p(x)}{q(x)}, \; q(x) \neq 0$$

p(x) is a numerator whereas q(x) is a denominator of a function.

Examples:

a) $f(x) = \dfrac{x^2 - 1}{x - 2}, \; x \neq 2$

Here, p(x) = x² – 1 and q(x) = x – 2 are both polynomials. So that, $f(x) = \dfrac{x^2 - 1}{x - 2}, \; x \neq 2$, is a rational expression.

Note: - Remember to be a rational expression, f(x) should have the restriction x ≠ 2; otherwise, it is **not** a rational expression.

b) $f(x) = \dfrac{2x}{x^2 + 4}$ is a rational expression since the domain of f(x) is the set of all real numbers and for all x∈ ℝ, the denominator x² + 4 never be equal to 0.

c) $f(x) = \dfrac{\sqrt{2x - 1}}{3x + 2}, \; x \neq \dfrac{-2}{3}$ is **not** an example of rational expressions. (Why?)
   - Because the numerator $\sqrt{2x - 1}$ is **not** a polynomial, its domain (values of x) is **not** the set of all real numbers.

Note: - The domain of $\sqrt{2x - 1}$ is $\left\{ x : x \geq \dfrac{1}{2} \right\}$.

d) $f(x) = \dfrac{x^2 - 9}{x + 3}, \; x \neq -3$

f(x) is a rational expression since the domain (values of x) for the denominator is already restricted, which is x ≠ –3.

e) $f(x) = \dfrac{x}{\sqrt{x-5}}$

   $f(x)$ is not a rational expression. (Why?) because the denominator $\sqrt{x-5}$ is not a polynomial since its domain is not the set of all real numbers. Thus, $\sqrt{x-5}$ is a radical expression.

f) $f(x) = 2^x$, $x \in R$. This is not an example of rational expression, because it is not written in the form of $f(x) = \dfrac{p(x)}{q(x)}$, $q(x) \neq 0$

9.2. Reduction of a Rational Expression

In mathematics, a rational expression can be reduced to its lowest form. A rational expression is reduced to its lowest term if all common factors from the numerator and denominator are cancelled. To simplify (reduce) a rational expression to lowest term, first factorize the numerator and denominator as much as possible and then cancel their common factors if there exist.

Examples
Simplify each of the following expressions:

1) $\dfrac{x^2 - 1}{x+1}$, $x \neq -1$

2) $\dfrac{a^2 - 4}{a - 2}$, $a \neq 2$

3) $\dfrac{x^4 + 1}{x^4 + 1}$

4) $\dfrac{x^2 - 5x + 6}{x - 3}$, $x \neq 3$

5) $\dfrac{x^2 - 5x}{x}$, $x \neq 0$

6) $\dfrac{x^2 - 2x + 1}{\dfrac{(x-1)^2}{x^2 - 1}}$, $x \neq -1, 1$

7) $\dfrac{x^2 - 8x + 15}{x^3 + 5x^2 + 6x}$, $x \neq 0, -2, -3$

8) $\dfrac{x^2 - 4}{x^2 + 2x}$, $x \neq 0, -2$

9) $\dfrac{y + 3}{y^3 - 9y}$, $y \neq 0, -3, 3$

10) $\dfrac{x^5 + 6x}{x^4 + 6}$

Solutions

1) $\dfrac{x^2 - 1}{x + 1}$, $x \neq -1$

$\dfrac{(x+1)(x-1)}{(x+1)}$ ....Factorizing

$= x - 1$

$\dfrac{(x+1)}{(x+1)}$ is cancelled, thus $\dfrac{(x+1)}{(x+1)} = 1$

The Simplified form is $(x - 1)$.

2) $\dfrac{a^2 - 4}{a - 2}$, $a \neq 2$

$\dfrac{(a+2)\,(a-2)}{a-2}$ ....$[a^2 - 4 = (a+2)(a-2)]$

$a + 2$..$(\dfrac{a-2}{a-2}$ is cancelled, thus $\dfrac{a-2}{a-2} = 1$ $)$

The simplified form is $a + 2$.

3) $\dfrac{x^4 + 1}{x^4 + 1}$

Since the domain of the numerator and denominator is the set of all real numbers, $\dfrac{x^4 + 1}{x^4 + 1}$ cancelled each other, thus

$\dfrac{x^4 + 1}{x^4 + 1} = 1$

The simplified form is 1.

4) $\dfrac{x^2 - 5x + 6}{x - 3}$, $x \neq 3$

$\dfrac{x^2 - 3x - 2x + 6}{x - 3}$ ....factorizing $x^2 - 5x + 6$

$\dfrac{x(x-3) - 2(x-3)}{x - 3}$ ....Taking the outside terms.

$\dfrac{(x-2)(x-3)}{(x-3)}$

$$\frac{(x-2)(x-3)}{(x-3)} \qquad \left(\frac{x-3}{x-3} \text{ cancelled each other}\right)$$

$$(x-2)$$

The simplified form is $x - 2$.

5) $\dfrac{x^2 - 5x}{x}, x \neq 0$

$$\frac{x(x-5)}{x} \qquad \dots\dots\dots\dots\dots\dots\dots\dots\dots\dots\dots\dots\dots \left(\text{Take the common factor x and cancel } \frac{x}{x}\right)$$

$x - 5$

The simplified form is $(x - 5)$

6) $\dfrac{\dfrac{x^2 - 2x + 1}{(x-1)^2}}{x^2 - 1}, x \neq -1, 1$

$$\frac{\frac{x^2-2x+1}{(x-1)^2}}{\frac{x^2-1}{1}}, x \neq 1, -1 \dots\dots\dots\dots\dots \frac{\frac{a}{b}}{c} = \frac{ac}{b} \text{ where } b, c \neq 0$$

$$\frac{(x^2-2x+1)}{(x-1)^2}(x^2 - 1)$$

$$\frac{(x-1)(x-1)}{(x-1)(x-1)} \cdot (x+1)(x-1) \dots\dots\dots \frac{(x-1)(x-1)}{(x-1)(x-1)} = 1, \text{ cancelled each other}$$

$(1) (x + 1) (x - 1)$

$(x^2 - 1)$

Thus, the simplified form is $(x^2 - 1)$

7) $\dfrac{x^2 - 8x + 15}{x^3 - 5x^2 + 6x}, x \neq 0, -2, -3$

212

$$\frac{x^2 - 3x - 5x + 15}{x(x^2 - 5x + 6)}$$ ....Factorizing
................Taking common factor

$$\frac{x(x-3) - 5(x-3)}{x(x^2 - 3x - 2x + 6)}$$

$$\frac{(x-5)(x-3)}{x[x(x-3) - 2(x-3)]}$$

$$\frac{(x-5)(x-3)}{x(x-2)(x-3)}$$ ....($\frac{x-3}{x-3}$ is cancelled each other)

$$\frac{(x-5)}{x(x-2)}$$

The simplified form is $\dfrac{(x-5)}{x(x-2)}$

8) $\dfrac{x^2 - 4}{x^2 + 2x}$, $x \neq 0, -2$

$$\frac{(x+2)(x-2)}{x(x+2)}$$ ....Factorizing
................Taking common Factor

$$\frac{x-2}{x}$$ ($\frac{x+2}{x+2}$ is cancelled each other)

The simplified form is $\dfrac{x-2}{x}$.

9) $\dfrac{y+3}{y^3 - 9y}$

$$\frac{y+3}{y(y^2 - 9)}$$ ....(Taking common factor).

$$\frac{(y+3)}{y(y+3)(y-3)}$$

$$\frac{1}{y(y-3)}$$ ....($\frac{y+3}{y+3}$ is cancelled each other)

The simplified form is $\dfrac{1}{y(y-3)}$

10) $\dfrac{x^5 + 6x}{x^4 + 6}$

$$\frac{x(x^4 + 6)}{x^4 + 6}$$ ....Taking common factor x.

213

$$\frac{x(\cancel{x^4 + 6})}{\cancel{x^4 + 6}} \quad \dots\dots\dots\dots\dots\dots\dots\dots\dots\dots\dots\dots\dots(\frac{x^4 + 6}{x^4 + 6} \text{ is cancelled})$$
$$x$$

The simplified form is x

Note: In question number 10 above, $\frac{x^4 + 6}{x^4 + 6}$ is cancelled each other. Because the domain of $x^4 + 6$ is the set of all real numbers.

9.3. Complex Rational Expressions

Complex rational expressions are expressions which have numerators or denominators containing one or more rational expressions. Complex rational expressions are also called complex fractions.

Examples

a) $\dfrac{2 + \dfrac{2}{x}}{2 - \dfrac{2}{x}}$

b) $\dfrac{\dfrac{1}{x + y} - \dfrac{1}{x}}{y}$

A simple method for simplifying a complex rational expression is to combine its numerator into a single expression and combine its denominator into one expression. Then operate the terms by inverting the denominator.

Examples:

1) Simplify: $\dfrac{2 + \dfrac{2}{x}}{2 - \dfrac{2}{x}}$

Solution

First, add the rational expression in the numerator to get a single rational expression.
$$2 + \frac{2}{x} = \frac{2}{1} + \frac{2}{x} = \frac{2x + 2}{x}$$

The Least Common Multiple (LCM) for 1 and x is x.

Second, subtract the rational expression in the denominator to get a single rational expression.
$$2 - \frac{2}{x} = \frac{2}{1} - \frac{2}{x} = \frac{2x - 2}{x}$$

The LCM for 1 and x is x.

The next step is to perform the division shown by the main fraction bar: to do this, invert the denominator and multiply it with the numerator. If they have common factor, simplify further.

Thus, $\dfrac{2+\dfrac{2}{x}}{2-\dfrac{2}{x}} = \dfrac{\dfrac{2x+2}{x}}{\dfrac{2x-2}{x}} = \left(\dfrac{2x+2}{\cancel{x}}\right)\left(\dfrac{\cancel{x}}{2x-2}\right) = \dfrac{2x+2}{2x-2}$

$= \dfrac{\cancel{2}(x+1)}{\cancel{2}(x-1)},$ ....2 is a common factor.

$= \dfrac{x+1}{x-1}$

2) Simplify: $\dfrac{\dfrac{1}{x}+\dfrac{3}{4}}{\dfrac{1}{x}-\dfrac{5}{2}}$

Solution

First add the numerator to get a single rational expression.

$\dfrac{1}{x}+\dfrac{3}{4} = \dfrac{4+3x}{4x}$

The LCM of x and 4 is 4x

Second step: subtract the denominator to get a single rational expression.

$\dfrac{1}{x}-\dfrac{5}{2} = \dfrac{2-5x}{2x}$

The LCM of x and 2 is 2x

The third step is to operate the division shown by the main fraction bar.

$\dfrac{\dfrac{1}{x}+\dfrac{3}{4}}{\dfrac{1}{x}-\dfrac{5}{4}} = \dfrac{\dfrac{4+3x}{4x}}{\dfrac{2-5x}{2x}} = \left(\dfrac{4+3x}{4x}\right)\left(\dfrac{2x}{2-5x}\right)\cdots\cdots\left(\dfrac{2x}{4x}\right) = \dfrac{1}{2}$

$= \dfrac{4+3x}{2(2-5x)}$

$= \dfrac{4+3x}{4-10x}$

3) Simplify: $\dfrac{\dfrac{1}{x+y} - \dfrac{1}{x}}{y}$

Solution

First perform subtraction in the numerator to get a single expression

$$\frac{1}{x+y} - \frac{1}{x} = \frac{x-(x+y)}{(x+y)\cdot x} = \frac{x-x-y}{(x+y)\cdot x} = \frac{-y}{(x+y)\cdot x} \quad \ldots\ldots\text{The LCM of } (x+y) \text{ and } x \text{ is}$$

$(x + y).x$, Now combines the result $\dfrac{-y}{(x+y)x}$ with the original denominator and write in the form shown below.

$$\frac{\dfrac{-y}{(x+y)\cdot x}}{y}$$

Finally, invert the original denominator y and multiply it with the new simplified terms.

$$\frac{-y}{(x+y)\cdot x} \cdot \left(\frac{1}{y}\right) \quad \ldots\ldots \text{ where } x \neq 0,\ y \neq 0,\ x \neq -y \text{ and } y \text{ is cancelled out by -y and the final result is: -}$$

$$\frac{-1}{(x+y)\cdot x} \quad \ldots\ldots\ldots\ldots \text{ where } x \neq 0,\ y \neq 0,\ x \neq -y$$

4) Simplify: $\dfrac{\dfrac{1}{x+6} - \dfrac{1}{x}}{6}$

Solution

Before we simplify the given expression, we have to restrict the domain, the domain of the given expression is {x: x $\neq$ 0 and x $\neq$ −6}

First, subtract to get a single rational expression in the numerator.

$$\frac{1}{x+6} - \frac{1}{x} = \frac{x-(x+6)}{x(x+6)} = \frac{x-x-6}{x(x+6)} = \frac{-6}{x(x+6)} \quad \ldots\ldots\text{The LCM of } x \text{ and } x+6 \text{ is } x(x+6)$$

Second, perform the division indicated by the main fraction bar. To do this, invert, and multiply and then simplify further, if possible.

$$\frac{\dfrac{1}{x+6} - \dfrac{1}{x}}{6} = \frac{\dfrac{-6}{x(x+6)}}{6} = \frac{-6}{x(x+6)} \cdot \frac{1}{6} \quad \ldots\ldots\ldots\text{6 by -6 is cancelled out and the final result is:}$$

$$= \frac{-1}{x(x+6)}$$

9.4. Evaluating Rational Expression

To evaluate a rational expression $\frac{p(x)}{q(x)}$, for x = k, we substitute k for x in the expression and reduced to its lowest term.

Examples:

1) Evaluate: $\frac{x^2-6x+10}{x-3}$ for

   a) $x = 2$
   b) $x = 4$
   c) $x = 3$
   d) $x = -1$

   Solutions
   a) $x = 2$

   Substitute x = 2 in $\frac{x^2-6x+10}{x-3}$

   $$\frac{2^2-6(2)+10}{2-3}$$

   $$\frac{4-12+10}{-1}$$

   $$\frac{-8+10}{-1}$$

   $$\frac{2}{-1}$$

   $$-2$$

   b) $x = 4$

   Substitute x = 4 in $\frac{x^2-6x+10}{x-3}$

   $$\frac{4^2-6(4)+10}{4-3}$$

   $$\frac{16-24+10}{1}$$

   $$\frac{26-24}{1}$$

   $$= 2$$

c) x = 3

   Substitute x = 3 in $\dfrac{x^2 - 6x + 10}{x - 3}$

   $$\dfrac{3^2 - 6(3) + 10}{3 - 3}$$

   $$\dfrac{9 - 18 + 10}{0}$$

   $$\dfrac{-18 + 19}{0}$$

   $$\dfrac{1}{0}$$

   Here, $\dfrac{1}{0}$ is undefined.

d) x = −1

   Substitute x = −1 in $\dfrac{x^2 - 6x + 10}{x - 3}$

   $$\dfrac{(-1)^2 - 6(-1) + 10}{-1 - 3}$$

   $$\dfrac{1 + 6 + 10}{-4}$$

   $$\dfrac{17}{-4}$$

   $$\dfrac{-17}{4}$$

2) Evaluate: $\dfrac{2x^2 - 4x + 3}{2x - 6}$, for

   a) x = −1
   b) x = 5
   c) x = 0

   Solutions
   a) x = −1

      Substitute x = −1 in $\dfrac{2x^2 - 4x + 3}{2x - 6}$

218

$$\frac{2(-1)^2 - 4(-1) + 3}{2(-1) - 6}$$

$$\frac{2(1) + 4 + 3}{-2 - 6}$$

$$\frac{2 + 7}{-8}$$

$$\frac{9}{-8}$$

$$= \frac{-9}{8}$$

b) x = 5

Substitute x = 5 in $\dfrac{2x^2 - 4x + 3}{2x - 6}$

$$\frac{2(5)^2 - 4(5) + 3}{2(5) - 6}$$

$$\frac{2(25) - 20 + 3}{10 - 6}$$

$$\frac{50 - 20 + 3}{4}$$

$$\frac{50 + 3 - 20}{4}$$

$$\frac{53 - 20}{4}$$

$$\frac{33}{4}$$

c) x = 0

Substitute x = 0 in $\dfrac{2x^2 - 4x + 3}{2x - 6}$

$$\frac{2(0)^2 - 4(0) + 3}{2(0) - 6}$$

$$\frac{2(0) - 4(0) + 3}{0 - 6}$$

$$\frac{0 - 0 + 3}{-6}$$

$$\frac{3}{-6}$$

$$\frac{-3}{6}$$

$$= -\frac{1}{2} \text{ (Divide both sides by 3)}$$

Note: -When substituting k for x in a rational expression $\frac{P(x)}{Q(x)}$ and if it yields the result

of the form $\frac{M}{0}$, where $M \neq 0$, we say that the expression $\frac{P(x)}{Q(x)}$ is undefined for x = k.

Examples: -

1) Evaluate: $\frac{3x - 1}{4x - 4}$, for x = 1

2) Evaluate: $\frac{2x - 3}{x - 7}$, for x = 7

Solution: -

1) $\frac{3x - 1}{4x - 4}$, for x = 1

   Substitute x = 1 in $\frac{3x - 1}{4x - 4}$

   $$\frac{3 - 1}{4 - 4}$$

   $$\frac{2}{0}$$

   Here, $\frac{2}{0}$ is undefined.

Solution: -

2) $\frac{2x - 3}{x - 7}$, for x = 7

220

Substitute $x = 7$ for $\dfrac{2x-3}{x-7}$

$$\dfrac{2(7)-3}{7-7}$$

$$\dfrac{14-3}{0}$$

$$\dfrac{11}{0}$$

Here, $\dfrac{11}{0}$ is undefined.

Note: - When substituting k for x in a rational expression $\dfrac{p(x)}{q(x)}$ and if it yields the result of the form $\dfrac{0}{0}$, we say that the expression $\dfrac{p(x)}{q(x)}$ is said to be indeterminate at $x = k$.

Examples: -

1) Evaluate: $\dfrac{x^2-6x+9}{x-3}$, for $x = 3$

Solution:

$$\dfrac{3^2-6(3)+9}{3-3}$$

$$\dfrac{9-18+9}{0}$$

$$\dfrac{18-18}{0}$$

$$\dfrac{0}{0}$$

Here, both the numerator and denominator are zero. Therefore, the expression $\dfrac{x^2-6x+9}{x-3}$ is indeterminate for $x = 3$.

2) Evaluate: $\dfrac{x^2-4}{x+2}$, for $x = -2$.

Solution:

$$\frac{(-2)^2 - 4}{-2 + 2}$$

$$\frac{4 - 4}{0}$$

$$\frac{0}{0}$$

Here, both the numerator and denominator are zero. Hence, the expression $\frac{x^2 - 4}{x + 2}$ is indeterminate when x = –2.

Note: $\frac{x}{0}$, for all x ≠ 0 is called undefined, whereas $\frac{x}{0}$, for x = 0 that is $\frac{0}{0}$ is called indeterminate. The terms undefined and indeterminate have different meaning and you will learn it more in detail in Algebra Two.

9.5. Solving Rational Equations

A rational equation is an equation containing one or more rational expressions. The best way to solve such equation is to eliminate all the denominators using the idea of LCM (Least Common Multiple). By doing this, the leftover equation to be solved with is usually either linear equation or quadratic equation.

Examples of rational equations

1) $\frac{1}{x} + \frac{1}{5} = \frac{2}{3} + \frac{x}{4}$

2) $3x + \frac{1}{3} = \frac{x + 5}{8}$

Solving rational equations

Examples

1) Solve: $\frac{2}{x + 3} = \frac{5}{4}$

Solution: The denominators are (x + 3) and 4. The least common multiple (LCM) for (x+3) and 4 is 4(x+3). We begin solving by multiplying both the left and the right sides of the equation by 4(x+3). Make sure that we will also write the restriction that x cannot be equal to –3 for the right side of the equation.

$\frac{2}{x + 3} = \frac{5}{4}$, x ≠ –3 ....Given equation

$4(x+3) \cdot \dfrac{2}{x+3} = \dfrac{5}{4} \cdot 4(x+3)$ ....Multiply both sides by 4(x+3).

$2 \times 4 = 5(x+3)$ ....(x+3) on the left side and 4 on the right side cancelled out.

$8 = 5x + 15$

$5x + 15 = 8$

$5x + 15 - 15 = 8 - 15$..................................Subtract 15 from both sides

$5x = -7$

$\dfrac{5x}{5} = \dfrac{-7}{5}$ ....Dividing both sides by 5.

$x = \dfrac{-7}{5}$

To make sure whether the answer is correct or not we can check our solution by substituting $\dfrac{-7}{5}$ into the original equation.

Note: - The original restriction in our equation is $x \neq -3$, thus, the solution set is $\left\{ \dfrac{-7}{5} \right\}$.

2) Solve: $\dfrac{x-2}{2x} + 1 = \dfrac{x+1}{x}$

Solution: We see that the denominator is zero when x = 0, we must avoid any values of the variable x that make the denominators zero. In this case we have to avoid x = 0, since it makes the denominators zero.

$\dfrac{x-2}{2x} + 1 = \dfrac{x+1}{x}$

We see that the denominators are 2x, 1 and x. The least common multiple (LCM) of 2x,1 and x is 2x. We multiply both sides of the equation by 2x. We will also write the restriction that x cannot be equal to 0 for both the right and the left sides of the equation.

$\dfrac{x-2}{2x} + 1 = \dfrac{x+1}{x}$, $x \neq 0$.................................Given equation

$2x \left(\dfrac{x-2}{2x} + 1 \right) = \left(\dfrac{x+1}{x} \right) \cdot 2x$ ....Multiply both sides by 2x.

$2x \left(\dfrac{x-2}{2x} \right) + 2x(1) = \left(\dfrac{x+1}{x} \right) \cdot 2x$ .... Use the distributive property.

$x - 2 + 2x = 2x + 2$...2x on the left side and x on the right side cancelled out.

$3x - 2 = 2x + 2$

$3x - 2 + 2 = 2x + 2 + 2$ .... Add 2 on both sides

$3x = 2x + 4$

$3x - 2x = 2x - 2x + 4$ .... Subtract 2x from both sides.

$x = 4$

Our solution set satisfies the restriction $x \neq 0$. Thus, the solution set is $\{4\}$.

Check the solution set by substituting $x = 4$ in the original equation.

$$\frac{x-2}{2x} + 1 = \frac{x+1}{x}$$

$$\frac{4-2}{2(4)} + 1 = \frac{4+1}{4}$$

$$\frac{2}{8} + 1 = \frac{5}{4}$$

$$\frac{1}{4} + 1 = \frac{5}{4}$$

$$\frac{1+4}{4} = \frac{5}{4}$$

$$\frac{5}{4} = \frac{5}{4} \checkmark$$

3) Solve: $\dfrac{1}{x-3} = \dfrac{-x}{2}$

Solution: The denominator on the left side is zero for $x = 3$, so the restriction is $x \neq 3$. The LCM for the denominators (x-3) and 2 is $2(x - 3)$

$\dfrac{1}{x-3} = \dfrac{-x}{2}$, $x \neq 3$ .... Given equation

$2(x - 3) \cdot \dfrac{1}{(x-3)} = \left(\dfrac{-x}{2}\right) \cdot 2(x - 3)$ .... Multiply both sides by 2(x – 3).

$2 = -x\,(x - 3)$ .... (x – 3) On the left side and 2 on the right side cancelled out.

$2 = -x^2 + 3x$ .... Distributive property

$x^2 - 3x + 2 = 0$ .... Collecting terms to one side.

$x^2 - 2x - x + 2 = 0$ .... –2x – x = –3x and (–2x) • (–x) = 2x²

$x\,(x - 2) - 1(x - 2) = 0$

$(x - 1)\,(x - 2) = 0$ .... Factoring

$x - 1 = 0$ or $x - 2 = 0$

$x = 1$ or $x = 2$

Since the restriction is $x \neq 3$, our solution set $x = 1$ or $x = 2$ is valid. Thus, the solution set is $\{1, 2\}$.

4) Solve: $\dfrac{x}{x-2} = \dfrac{2}{x-2} + 8$

Solution: - We see that x cannot be equal to 2, because it makes the denominators zero. Thus, we must avoid any values of the variable x that make the denominator zero. Hence, $x \neq 2$. The LCM of the denominator $(x - 2)$ and 1 is $(x - 2)$.

$$\frac{x}{x-2} = \frac{2}{x-2} + 8, x \neq 2 \quad\text{.................................Given equation}$$

$$(x-2)\left(\frac{x}{x-2}\right) = (x-2)\left(\frac{2}{x-2} + 8\right) \quad\text{.......Multiply both sides by } (x-2)$$

$$(x-2)\left(\frac{x}{x-2}\right) = (x-2)\left(\frac{2}{x-2}\right) + (x-2)\cdot 8 \quad\text{......Use distributive property.}$$

$$(x-2)\left(\frac{x}{x-2}\right) = (x-2)\left(\frac{2}{x-2}\right) + 8(x-2) \quad\text{......On both right and left sides (x-2) by}$$
$$\text{(x-2) are cancelled out.}$$

x = 2 + 8(x – 2)

The resulting equation doesn't have any fractions. We now solve for x.

x = 2 + 8x – 16 ....Use distributive property
x = 8x – 14 ....Combine like terms
x – 8x = 8x – 8x – 14 ....Subtract 8x from both sides
–7x = –14 ....Simplify

$$\frac{-7x}{-7} = \frac{-14}{-7} \quad\text{....................................Divide both sides by } -7$$

x = 2

The value x=2, is not a solution because of the restriction that x≠2.There is no solution set to this equation. Thus, the solution set is Ø, **(The empty set).**

5) Solve: $\frac{6x-7}{4} = \frac{5}{x}$

Solution

x=0, makes the denominator zero, so that our domain (value of x) for this equation must be {x:x≠0}.

$$\frac{6x-7}{4} = \frac{5}{x}, x \neq 0$$

We see that the LCM of the denominators 4 and x is $4x$.

$$(4x)\left(\frac{6x-7}{4}\right) = \left(\frac{5}{x}\right)(4x) \quad\text{...........................Multiply both sides by LCM, which } 4x$$

(x) (6x-7) = 5 × 4 ....4 on the left side and x on the right side are cancelled out.

$6x^2 - 7x = 20$ ....Use distributive property
$6x^2 - 7x - 20 = 20 - 20$ ....Subtract 20 from both sides

$6x^2 - 7x - 20 = 0$

$6x^2 + 8x - 15x - 20 = 0$ ......$(8x-15x) = -7x$ and$8x(-15x)=-120x^2$.....(Sum and product of roots respectively)

$2x(3x + 4) - 5(3x + 4) = 0$ ....Taking common factors

$(2x - 5)(3x + 4) = 0$ ....Taking the outside and the inside terms.

$2x - 5 = 0$ or $3x + 4 = 0$

$2x = 5$ or $3x = -4$

$\dfrac{2x}{2} = \dfrac{5}{2}$ or $\dfrac{3x}{3} = \dfrac{-4}{3}$

$x = \dfrac{5}{2}$ or $x = \dfrac{-4}{3}$

Thus, the solution set is: $\left\{\dfrac{-4}{3}, \dfrac{5}{2}\right\}$

6) Solve: $\dfrac{5}{2y} = \dfrac{2}{y-1}$

Solution:

The denominators become zero for $y = 0$ and $y = 1$; therefore, the restriction is $y \neq 0$, $y \neq 1$. The LCM of the denominators (2y) and (y-1) is $2y(y-1)$.

$\dfrac{5}{2y} = \dfrac{2}{y-1}$, $y \neq 0$, $y \neq 1$..................................Given equation

$2y(y-1)\left(\dfrac{5}{2y}\right) = \dfrac{2}{y-1}\left[2y(y-1)\right]$

..................Multiply both sides by the LCM of the denominators: $2y(y-1)$.

$5(y-1) = 2(2y)$ ....$2y$ on the left side and $y-1$ on the right side cancelled out.

$5y - 5 = 4y$ ....Use distributive property.

$5y - 4y = 5$ ....Collect like terms

$y = 5$

Thus, the solution set is {5}.

7) Solve: $\dfrac{12}{x^2 - 9} + \dfrac{5}{x+3} = \dfrac{8}{x-3}$

Solution:

We see that x cannot be equal to 3 or –3. Our restriction is $x \neq 3$, $x \neq -3$, The denominators are $x^2 - 9$, $x + 3$, and $x - 3$. Their Least Common Multiple (LCM) is $x^2 - 9 = (x + 3)(x - 3)$. We multiply both sides of the equation by $(x + 3)(x - 3)$.

$\dfrac{12}{x^2 - 9} + \dfrac{5}{x+3} = \dfrac{8}{x-3}$, $x \neq 3$, $x \neq -3$ ....Given equation

$(x+3)(x-3)\left(\dfrac{12}{x^2 - 9} + \dfrac{5}{x+3}\right) = \dfrac{8}{x-3}(x+3)(x-3)$...Multiply both sides by LCM

$\dfrac{12\,(\cancel{x^2 - 9})}{\cancel{x^2 - 9}} + \dfrac{5\,\cancel{(x+3)}(x-3)}{\cancel{x+3}} = \dfrac{8(x+3)\,\cancel{(x-3)}}{\cancel{x-3}}$ ....Use distributive property

$12 + 5(x - 3) = 8(x + 3)$ ....$x^2 - 9$ and $x + 3$ on the left side and $x - 3$ on the right side are cancelled out.

$12 + 5x - 15 = 8x + 24$ ....Use distributive property

$5x - 3 = 8x + 24$

$5x - 3 + 3 = 8x + 24 + 3$ ....Adding 3 on both sides.

$5x = 8x + 27$

$5x - 8x = 8x - 8x + 27$ ....Subtracting 8x from both sides.

$-3x = 27$

$\dfrac{-3x}{-3} = \dfrac{27}{-3}$ ....Dividing both sides by –3

$x = -9$

Thus, the solution set is {–9}.

8) Solve: $\dfrac{x^2 - 36}{x - 6} = 0$

Solution:

The denominator is zero when x = 6; thus, our restriction (the domain) is x ≠ 6.

$\dfrac{x^2 - 36}{x - 6} = 0$, x ≠ 6 ....Given equation

$\dfrac{(x + 6)\,(x - 6)}{(x - 6)} = 0$ ....$x^2 - 36$ is factored and (x – 6) is cancelled each other.

$x + 6 = 0$

$x + 6 - 6 = 0 - 6$............................Subtract 6 from both sides.

$x = -6$

Thus, the solution set is {–6}.

9) Solve: $\dfrac{4}{3x-1} = 2x$

Solution:

We see that the denominator (3x-1) is equal to zero when $x = \dfrac{1}{3}$, thus, our domain (value of x) must be {x: x ≠ $\dfrac{1}{3}$ }

Such type of equation is solved in different methods, the easiest method is to apply the rule of cross multiplication.

$\dfrac{4}{3x-1} = 2x$, x ≠ $\dfrac{1}{3}$ ....Given equation

$\dfrac{4}{3x-1} = \dfrac{2x}{1}$ ....Denominator of 2x is 1

$(3x - 1)\,(2x) = 4 \times 1$ ....Cross Multiplication

$6x^2 - 2x = 4$ .... Multiplying using distributive property

$6x^2 - 2x - 4 = 0$ .... Taking 4 to the left side

$\left(\dfrac{1}{2}\right)(6x^2 - 2x - 4) = 0 \left(\dfrac{1}{2}\right)$ .... Multiplying both sides by $\dfrac{1}{2}$ to simplify and make it easy to solve.

$3x^2 - x - 2 = 0$

$3x^2 - 3x + 2x - 2 = 0$ .... $(-3x+2x) = -x$ and $(-3x)(2x) = -6x^2$ (Sum and product of two numbers respectively)

$3x(x-1) +2(x-1) = 0$ .... Taking common factors

$(3x+2)(x-1) = 0$ .... collecting outside terms

$3x + 2 = 0$ or $x - 1 = 0$

$3x = -2$ or $x = 1$

$X = \dfrac{-2}{3}$ or $x = 1$

Thus, the solution set is $\left\{\dfrac{-2}{3}, 1\right\}$

10) Solve: $\dfrac{x-4}{x-8} = \dfrac{1}{x+1}$

Solution

The domain of the given equation is $x \neq 8$ and $x \neq -1$, since they make the denominators zero.

$\dfrac{x-4}{x-8} = \dfrac{1}{x+1}$, $x \neq 8$, $x \neq -1$

We can solve such equations in different methods. The easiest way is to use cross multiplication.

$\dfrac{x-4}{x-8} = \dfrac{1}{x+1}$, $x \neq 8$, $x \neq -1$, ... Given equation

$(x + 1)(x - 4) = (1)(x - 8)$ .... Cross multiplication

$x^2 - 4x + x - 4 = x - 8$ .... Multiplying

$x^2 - 3x - 4 = x - 8$ .... Adding

$x^2 - 3x - x - 4 + 8 = x - x - 8 + 8$ .... Subtracting x from and adding 8 on both sides.

$x^2 - 4x + 4 = 0$ .... Adding and subtracting

$x^2 - 2x - 2x + 4 = 0$ .... sum: $-2x - 2x = -4x$ and product: $(-2x)(-2x) = 4x^2$

$x(x - 2) - 2(x - 2) = 0$ .... Factoring

$(x - 2)(x - 2) = 0$ .... Collecting outside terms.

$x - 2 = 0$ or $x - 2 = 0$

$x = 2$ or $x = 2$

Thus, the solution set is {2}.

228

Exercise 9.1 - 9.5

1) Determine the domain of each of the following expressions.

a) $f(x) = \dfrac{2x^2 + 2}{x^2 - 1}$

b) $g(x) = \dfrac{x^2 - 1}{x^2 + 4}$

c) $f(x) = \dfrac{-3x}{3x - 9}$

d) $h(x) = \dfrac{12}{x^2 + 3}$

e) $g(x) = x^2 - 4$

f) $h(x) = \dfrac{x^2 - 8}{x + 4}$

g) $h(x) = \dfrac{x^2 + 2}{x^2 + 2}$

h) $f(x) = \dfrac{x^2 - 3x + 4}{x^2 - 4x}$

i) $f(x) = \sqrt{x^2}$

j) $f(x) = \sqrt{x - 1}$

k) $f(x) = \sqrt[3]{x}$

2) Solve each of the following rational equations algebraically.

a) $\dfrac{2}{x - 3} + \dfrac{1}{x - 2} = 1$

b) $\dfrac{1}{x} = \dfrac{1}{5} + \dfrac{3}{2x}$

c) $\dfrac{x}{x - 3} = \dfrac{3}{x - 3} + 9$

d) $\dfrac{x + 3}{6} = \dfrac{3x}{8} + \dfrac{x - 5}{4}$

e) $\dfrac{x}{3} = \dfrac{x}{2} - 2$

f) $\dfrac{2}{x - 2} = \dfrac{4}{x - 2} - 2$

g) $\dfrac{4}{x} + \dfrac{8}{x + 2} = 4$

h) $\dfrac{1}{x}+2=\dfrac{3}{x}$

3) Determine each of the following expressions, whether they are rational or not.

    a) $f(x) = x^2 + 3$

    b) $f(x) = \dfrac{x^2 - 4}{x - 2}$

    c) $h(x) = \dfrac{x^2 + 4}{x^2 + 2}$

    d) $g(x) = \dfrac{x^2 - 3x + 4}{\sqrt{x} + 6}$

    e) $f(x) = \pi x^2 - \sqrt{8}\, x + 2$

    f) $f(x) = 3x + 2$

    g) $f(x) = \dfrac{1}{2}x - \dfrac{1}{4}$

    h) $f(x) = \sqrt{2x + 20} + 3$

4) Simplify each of the following expressions regardless of their domains.

    a) $\dfrac{x^2 - 4}{x^2 - 2x}$

    b) $\dfrac{x^2 - 25}{x + 5}$

    c) $\dfrac{6x^2 + 7x - 5}{4x^2 - 1}$

    d) $\dfrac{7}{2x + 1} + \dfrac{2x + 5}{2x^2 - x - 1}$

    e) $\dfrac{5x^2 - 13x - 6}{3x^2 - 4x - 15}$

    f) $\left(\dfrac{x^2 - 7x + 10}{x - 2}\right)\left(\dfrac{1}{x^2 - 4}\right)$

    g) $\left(\dfrac{\dfrac{1}{4x - 8}}{x^2 - 4}\right)\left(\dfrac{2x - 4}{x - 2}\right)$

    h) $\dfrac{1}{\dfrac{1}{y} + \dfrac{1}{2y}} + \dfrac{2y}{y^2 - 4}$

    i) $\dfrac{\dfrac{1}{x} + \dfrac{1}{2x}}{\dfrac{x}{2} + \dfrac{x}{4}}$

5) Evaluate: $\dfrac{3x^2 - 4x + 1}{2x - 2}$, for

   a) $x = 3$
   b) $x = 2$
   c) $x = 1$
   d) $x = -4$

6) Evaluate: $\sqrt{2x - 8}$, for
   a) $x = 8$
   b) $x = 4$
   c) $x = 62$
   d) $x = 6$

7) If $A = x + 3$, $B = \dfrac{2x - 3}{2}$ and $C = \dfrac{2x - 1}{4}$, then find:

   a) $\dfrac{2A + 2B}{C}$

   b) $\dfrac{A - 2B}{2C}$

   c) $\dfrac{2A - B}{C + B}$

   d) $\dfrac{A - 3C}{2B}$

8) If $A + B = 3x - 1$ and $A - B = 8x - 4$, then evaluate for A in terms of x.

9) Solve for x: $\dfrac{3}{x + 2} - \dfrac{1}{2x - 4} = \dfrac{1}{4}$

10) What is the domain of $\dfrac{1}{x - 4} - \dfrac{1}{2x}$

11) Simplify: $\dfrac{x + \dfrac{1}{6}}{x - \dfrac{1}{3}}$

12) Simplify: $\dfrac{x^2 - y^2}{(x - y)^2} \cdot \dfrac{1}{x + y}$

13) Simplify: $\dfrac{\dfrac{r - s}{s}}{\dfrac{r^2 - s^2}{rs}}$

14) Simplify: $\dfrac{a - \dfrac{a}{b}}{b - \dfrac{b}{a}}$

CHAPTER 10

Evaluating Radical Expression

10.1. Radical Expression

An algebraic expression that contains radicals is called a radical expression. If n is a positive integer that is greater than 1 and x is a real number, then

$\sqrt[n]{x} = x^{1/n}$ where n is called the index, x is called the radicand, and the symbol: $\sqrt{}$ is the radical, Thus,

Parts of a Radical

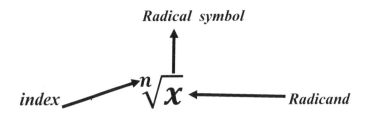

Radical symbol

index $\quad n\sqrt{x} \quad$ *Radicand*

The form $\sqrt[n]{x}$ is called the radical form and $(x)^{1/n}$ is called the exponent form. To evaluate a root expression for x = k, we replace x with k in the root-expression. just as we do for with polynomials and rational expressions.

Remark:

- If $\sqrt[n]{x}$, when n = 2, i.e., If the index is 2, it is customary to leave out the index and use only the radical symbol like $\sqrt{x}$. which is read as the square root of x.

- When n = 3, $\sqrt[3]{x}$ is called the cube root of x.
- When n = 4, $\sqrt[4]{x}$ is called the fourth root of x.
- When n = 5, $\sqrt[5]{x}$ is called the fifth root of x and so on.

Definitions
For all integer and n≥2

- If n is odd, then $\sqrt[n]{a^n} = \left(a^{1/n}\right)^n = a^{\frac{n}{n}} = a$

- If n is even then $\sqrt[n]{a^n} = |a|$, where $|a|$ is the absolute value of 'a'.

Examples:

a) $\sqrt[3]{x^3} = x$

b) $\sqrt[5]{x^5} = x$

c) $\sqrt[13]{y^{13}} = y$

d) $\sqrt[11]{m^{11}} = m$

e) $\sqrt{x^2} = |x|$

f) $\sqrt[4]{x^4} = |x|$

g) $\sqrt[8]{y^8} = |y|$

h) $\sqrt[4]{x^8} = |x^2|$

i) $\sqrt[6]{m^{36}} = |m^6|$

Examples:
1) Evaluate: $\sqrt{x-6}$, for
   a) $x = 6$
   b) $x = 12$
   c) $x = 9$
   d) $x = 4$
   e) $x = 15$

Solutions:
a) For $x = 6$
$$\sqrt{x-6}$$
$$\sqrt{6-6}$$
$$\sqrt{0}$$
$$0$$

b) For $x = 12$

$$\frac{\sqrt{x-6}}{\sqrt{12-6}}$$

$$\sqrt{6}$$

c) For x=9
$$\sqrt{x-6}$$

$$\sqrt{9-6}$$

$$\sqrt{3}$$

d) For x=4
$$\sqrt{x-6}$$

$$\sqrt{4-6}$$
$$\sqrt{-2}$$

Thus, $\sqrt{-2}$ doesn't exist in the set of real numbers.

e) For x=15
$$\sqrt{x-6}$$

$$\sqrt{15-6}$$

$$\sqrt{9}$$

$$\frac{\sqrt{3\times3}}{3}$$

2) Evaluate $\sqrt[3]{x}$, for
   a) x= 27
   b) x= 8
   c) x= –27
   d) x= 0
   e) x= 729

Solutions
   a) For x = 27
      $$\sqrt[3]{x}$$
      $$\sqrt[3]{27}$$
      3

b) For x = 8

$$\sqrt[3]{x}$$

$$\sqrt[3]{8}$$

2

c) For x = –27

$$\sqrt[3]{x}$$

$$\sqrt[3]{-27}$$

$$\sqrt[3]{(-3)\times(-3)\times(-3)}$$

-3

d) For x = 0

$$\sqrt[3]{x}$$

$$\sqrt[3]{0}$$

0

e) For x = 729

$$\sqrt[3]{x}$$

$$\sqrt[3]{729}$$

$$\sqrt[3]{9\times9\times9}$$

9

10.2. Simplifying Radical Terms

Simplifying radical equations requires applying the basic rules of exponents and following some basic algebraic principles. We state below the basic rules of radicals that we will use in the process of their simplification.

1. $\sqrt[n]{a^n} = |a|$, if n is even and the value of 'a' is not restricted.
   Why? because the absolute value accounts for the fact that if the value of 'a' is negative and raised to an even power, then the result is positive. Hence, we have to put the value of 'a' in an absolute value. If n is even and the value of 'a' is restricted: say when
   $a \geq 0$, $\sqrt[n]{a^n} = a$. This is because the positive value of 'a' to the power of an even number is always positive.

   Note: For $\sqrt[n]{a^n}$, where n is an even number, If the value of 'a' is restricted, as $a \geq 0$, the result is always 'a'. If not restricted, the result is |a|.

235

Examples if n is odd, $\sqrt[n]{a^n} = a$.

2. $\sqrt[n]{ab} = \sqrt[n]{a} \cdot \sqrt[n]{b}$

3. $\sqrt[n]{\dfrac{a}{b}} = \dfrac{\sqrt[n]{a}}{\sqrt[n]{b}}$, for $b \neq 0$

1) Simplify each of the following radical expressions.

a) $\sqrt{45}$

b) $\sqrt{9x^2}$

c) $\sqrt{9x^2}$, $x \geq 0$

d) $\sqrt{x^3 y^2}$

e) $\sqrt[3]{27x^3 y^6}$

f) $\sqrt[3]{\dfrac{81x^3}{64y^6}}$, $y \neq 0$

g) $\sqrt{8x^6 y^8}$

h) $\sqrt{8x^6 y^8}$, $x, y \geq 0$

i) $\sqrt[4]{x^8 y^4}$

j) $\sqrt{x^8 y^4}$, $x, y \geq 0$

k) $\sqrt{72}$

l) $\sqrt[3]{54x^4}$

m) $\sqrt[4]{81a^4}$

n) $\sqrt[5]{-64a^6 b^{10} x^8 y^{11}}$

Solutions

a) $\sqrt{45}$

First write 45 as a prime factor

236

$45 = 3 \times 3 \times 5 = 3^2 \times 5$

$\sqrt{45} = \sqrt{9 \times 5} = \sqrt{3^2 \times 5} = \sqrt{3^2} \cdot \sqrt{5} = 3\sqrt{5}$

b) $\sqrt{9x^2}$

$\sqrt{9x^2} = \sqrt{3^2 x^2} = 3|x|$

c) $\sqrt{9x^2}$, $x \geq 0$

$\sqrt{9x^2} = \sqrt{3^2 x^2} = 3x$

Recall: $\sqrt{a^n} = |a|$, for n is an even number and 'a' is not restricted.

$\sqrt{a^n}$, $a \geq 0 = a$, for n is an even number and 'a' is restricted.

d) $\sqrt{x^3 y^2}$

$\sqrt{x^3 y^2} = \sqrt{x(x^2)y^2} = |xy|\sqrt{x}$

e) $\sqrt[3]{27x^3 y^6}$

$\sqrt[3]{27x^3 y^6} = \sqrt[3]{3^3 x^3 (y^3)^2} = 3xy^2$

f) $\sqrt[3]{\dfrac{81x^3}{64y^6}}$, $y \neq 0$

$\sqrt[3]{\dfrac{81x^3}{64y^6}} = \sqrt[3]{\dfrac{3 \times (3^3) \times x^3}{4^3 \times (y^3)^2}} = \dfrac{3x}{4y^2}\sqrt[3]{3}$

g) $\sqrt[3]{\dfrac{81x^3}{64y^6}}$, $y \neq 0$

$\sqrt[3]{\dfrac{81x^3}{64y^6}} = \sqrt[3]{\dfrac{3 \times (3^3) \times x^3}{4^3 \times (y^3)^2}} = \dfrac{3x}{4y^2}\sqrt[3]{3}$

h) $\sqrt{8x^6 y^8}$, $x, y \geq 0$

$\sqrt{8x^6 y^8} = \sqrt{(2^2) \times 2(x^2)^3 \times (y^2)^4} = 2x^3 y^4 \sqrt{2}$

i) $\sqrt[4]{x^8 y^4}$

$\sqrt[4]{x^8 y^4} = \sqrt[4]{(x^4)^2 y^4} = x^2|y|$

j) $\sqrt[4]{x^8 y^4}$; $x, y \geq 0$

$$\sqrt[4]{x^8y^4} = \sqrt[4]{(x^4)^2 y^4} = x^2 y$$

k) $\sqrt{72}$

$$\sqrt{72} = \sqrt{9 \times 8} = \sqrt{3^2} \times \sqrt{2^2} \times \sqrt{2} = 3 \times 2\sqrt{2} = 6\sqrt{2}$$

l) $\sqrt[3]{54x^4} = \sqrt[3]{2 \times 3^3 \times x \times x^3} = 3x\sqrt[3]{2x}$

m) $\sqrt[4]{81a^4} = \sqrt[4]{3^4 a^4} = 3|a|$

n) $\sqrt[5]{-64a^6b^{10}x^8y^{11}}$

$$= \sqrt[5]{-2 \times 2^5 \times a \times a^5 \times (b^5)^2 \times x^3 \times x^5 \times y \times (y^5)^2}$$
$$= -2ab^2xy^2 \sqrt[5]{2ax^3y}$$

2) Simplify each of the following

a) $\sqrt[3]{\dfrac{81x^4y^6}{72a^6b^3}}$, a, b $\neq$ 0

b) $\sqrt[4]{16x^4y^8z^{16}}$

c) $\sqrt[5]{32x^5y^{10}}$

d) $\sqrt[3]{\dfrac{8y^7z^5w^3}{27a^3b^4}}$, a, b, $\neq$ 0

e) $\sqrt[5]{3125x^{10}y^{25}z^{15}}$

Solutions

a) $\sqrt[3]{\dfrac{81x^4y^6}{72a^6b^3}}$, a b, $\neq$ 0

$$= \sqrt[3]{\dfrac{3^3 \times 3 \times x^3 \times x \times (y^3)^2}{72 \times (a^3)^2 \times b^3}} = \dfrac{3xy^2}{a^2b}\sqrt[3]{\dfrac{3x}{72}} = \dfrac{3xy^2}{a^2b}\sqrt[3]{\dfrac{x}{24}} = \dfrac{3xy^2}{2a^2b}\sqrt[3]{\dfrac{x}{3}}$$

b) $\sqrt[4]{16x^4y^8z^4} = \sqrt[4]{2^4x^4(y^4)^2(z^4)^4} = 2y^2z^4|x|$

c) $\sqrt[5]{32x^5y^{10}} = \sqrt[5]{2^5 x^5 (y^5)^2} = 3xy^2$

d) $\sqrt[3]{\dfrac{8y^7z^5w^3}{27a^3b^4}} = \sqrt[3]{\dfrac{2^3(y^3)^2y\, z^3z^2w^3}{3^3a^3b^3b}} = \dfrac{2y^2zw}{3ab}\sqrt[3]{\dfrac{yz^2}{b}}$

e) $\sqrt[5]{3125x^{10}y^{25}z^{15}} = \sqrt[5]{5^5(x^5)^2(y^5)^5(z^5)^3} = 5x^2y^5z^3$

10.3. Rational Exponents

1. $a^{m/n} = \left(\sqrt[n]{a}\right)^m = \sqrt[n]{a^m}$

    Where m and n are integers with no common factors, $n > 1$ and $\sqrt[n]{a}$ is a real number.

2. $a^{-n} = \dfrac{1}{a^n}$, $a \neq 0$

3. $\dfrac{1}{a^n} = a^{-n}$, $a \neq 0$

4. $\left(\dfrac{a}{b}\right)^{-n} = \left(\dfrac{b}{a}\right)^n$, $a, b \neq 0$

- If x and y are rational numbers and a and b are real numbers, then

1. $a^x \bullet a^y = a^{x+y}$
2. $(a^x)^y = a^{xy}$
3. $(ab)^x = a^x b^x$

    Provided that all of the expressions that we used are defined.

Examples

1) Write each of the following in the radical symbol.

a) $x^{1/2}$

b) $x^{1/3}$

c) $x^{1/4}$

d) $x^{1/5}$

e) $x^{2/5}$

f) $x^{7/3}$

g) $x^{3/2}$

h) $x^{5/6}$

2) Evaluate each of the following expressions.

a) $9^{1/2}$

b) $27^{1/3}$

c) $16^{1/4}$

d) $243^{1/5}$

e) $32^{3/2}$

f) $81^{2/3}$

g) $625^{3/4}$

h) $\left(\dfrac{25}{196}\right)^{-\frac{1}{2}}$

3) Rewrite each of the following expressions in radical form and simplify if possible.

    a) $32^{2/5}$

    b) $81^{3/4}$

    c) $\left(27x^4y^5\right)^{1/3}$

    d) $\left(\dfrac{3ab}{4c}\right)^{-3}$

    e) $(81)^{5/4}$

    f) $\left(\dfrac{32a^3b^3}{24ab}\right)^{1/2}$, $a, b \neq 0$

Solutions

1) a) $x^{1/2}$

       $x^{1/2} = \sqrt{x}$

    b) $x^{1/3}$

       $x^{1/3} = \sqrt[3]{x}$

    c) $x^{1/4}$

       $x^{1/4} = \sqrt[4]{x}$

    d) $x^{1/5}$

       $x^{1/5} = \sqrt[5]{x}$

    e) $x^{2/5}$

       $x^{2/5} = \sqrt[5]{x^2}$

    f) $x^{7/3}$

       $x^{7/3} = \sqrt[3]{x^7}$

    g) $x^{3/2}$

       $x^{3/2} = \sqrt{x^3}$

h) $x^{5/6}$

$$x^{5/6} = \sqrt[6]{x^5}$$

2) a) $9^{1/2}$

$$9^{1/2} = \sqrt{9} = \sqrt{3^2} = 3$$

b) $27^{1/3}$

$$27^{1/3} = \sqrt[3]{27} = \sqrt[3]{3^3} = 3$$

c) $16^{1/4}$

$$16^{/} = \sqrt{16} = \sqrt{2} = 2$$

d) $243^{1/5}$

$$243^{1/5} = \sqrt[5]{243} = \sqrt[5]{3^5} = 3$$

e) $32^{3/2}$

$$(32)^{\frac{3}{2}} = \sqrt{(32)^3} = \sqrt{(16)^3 \times (2)^3}$$
$$= \sqrt{(16)^3 \times 2^2 \times 2^1}$$
$$= \quad 4^3 \times 2^1\sqrt{2}$$
$$= (2^2)^3 \times 2^1\sqrt{2}$$
$$= \quad 2^6 \times 2^1\sqrt{2}$$
$$= \quad 2^{6+1}\sqrt{2} \quad \ldots\ldots\ldots\ldots \text{ (Apply rule: } a^x a^y = \boldsymbol{a^{x+y}}\text{)}$$
$$= \boldsymbol{2^7\sqrt{2}}$$
$$= 128\sqrt{2}$$

Alternate Method

$$(32)^{\frac{3}{2}} = \sqrt{(32)^3} = \sqrt{(2^5)^3} = \sqrt{(2^{15})}$$
$$= \sqrt{2^{14} 2^1}$$
$$= \sqrt{2^{14}} \times \sqrt{2}$$
$$= 2^7\sqrt{2}$$
$$= 128\sqrt{2}$$

Recall: $a^x \cdot a^y = a^{x+y}$

f) $81^{2/3}$

$$(81^{\frac{2}{3}} = \sqrt[3]{(81)^2} = \sqrt[3]{(27)^2 \times (3^2)}$$
$$\sqrt[3]{(3^3)^2 \times 3^2}$$
$$\sqrt[3]{(3^3)^2} \times \sqrt[3]{3^2}$$
$$9 \sqrt[3]{9}$$

g) $625^{3/4}$

$$625^{3/4} = \sqrt[4]{(625)^3} = \sqrt[4]{(5^4)^3}$$
$$= 5^3$$
$$= 125$$

h) $\left(\dfrac{25}{196}\right)^{\frac{-1}{2}} = \left(\dfrac{196}{25}\right)^{1/2} = \sqrt{\dfrac{196}{25}} = \dfrac{14}{5}$

Recall: $\left(\dfrac{a}{b}\right)^{\frac{-1}{n}} = \left(\dfrac{b}{a}\right)^{\frac{1}{n}}$, a, b $\neq$ 0

3) a) $32^{2/5}$

$$32^{2/5} = \sqrt[5]{(32)^2} = \sqrt[5]{(2^5)^2} = 2^2 = 4$$

b) $81^{3/4} = \sqrt[4]{(81)^3} = \sqrt[4]{(3^4)^3} = 3^3 = 27$

c) $(27x^4y^5)^{\frac{1}{3}}$

$$(27x^4y^5)^{\frac{1}{3}} = \sqrt[3]{27x^4y^5}$$
$$= \sqrt[3]{3^3x^3x^1y^3y^2}$$
$$= 3xy\sqrt[3]{xy^2}$$

d) $\left(\dfrac{3ab}{4c}\right)^{-3} = \left(\dfrac{4c}{3ab}\right)^3$

Recall that: $\left(\dfrac{a}{b}\right)^{-1} = \left(\dfrac{b}{a}\right)$; a, b $\neq$ 0

e) $(81)^{5/4} = \sqrt[4]{(81)^5} = \sqrt[4]{(3^4)^5} = 3^5 = 243$

f) $\left(\dfrac{32a^3b^3}{24ab}\right)^{1/2} = \left(\dfrac{32a^3b^3}{24a^1b^1}\right)^{1/2} = \left(\dfrac{8 \times 4\, a^{3-1}b^{3-1}}{8 \times 3}\right)^{1/2} = \left(\dfrac{4a^2b^2}{3}\right)^{1/2} = \sqrt{\dfrac{4a^2b^2}{3}} = 2|ab|\sqrt{\dfrac{1}{3}}$

Recall that: $\left(\dfrac{a^x}{b^y}\right) = a^{x-y}$, b $\neq$ 0

4) Write each of the following expressions without radical symbol.

a) $\sqrt{y^3}$

b) $\sqrt[3]{x^4 y^5} + \sqrt{3 x^5 y^7}$

c) $\dfrac{\sqrt[5]{x^4 y^3 z}}{\sqrt{x y^3 z}}$

d) $\dfrac{\sqrt{x^3 y^5 z^9}}{\sqrt[4]{2 x^2 y^3 z^5}}$

e) $\dfrac{\sqrt[3]{2xyz}}{\sqrt[7]{3xyz}}$

f) $\sqrt{2xyz}$

Solutions

a) $\sqrt{y^3} = y^{3/2}$

b) $\sqrt[3]{x^4 y^5} + \sqrt{3 x^5 y^7}$

$= x^{4/3} y^{5/3} + 3^{1/2} x^{5/2} y^{7/2}$

c) $\dfrac{\sqrt[5]{x^4 y^3 z}}{\sqrt{x y^3 z}} = \dfrac{x^{4/5} y^{3/5} z^{1/5}}{x^{1/2} y^{3/2} z^{1/2}}$

d) $\dfrac{\sqrt{x^3 y^5 z^9}}{\sqrt[4]{2 x^2 y^3 z^5}} = \dfrac{x^{3/2} y^{5/2} z^{9/2}}{2^{1/4} x^{2/4} y^{3/4} z^{5/4}}$

e) $\dfrac{\sqrt[3]{2xyz}}{\sqrt[7]{3xyz}} = \dfrac{2^{1/3} x^{1/3} y^{1/3} z^{1/3}}{3^{1/7} x^{1/7} y^{1/7} z^{1/7}}$

f) $\sqrt{2xyz} = 2^{1/2} x^{1/2} y^{1/2} z^{1/2}$

Examples
Answer each of the following questions

a) $\left(3\sqrt{2} + 3\sqrt{2}\right)\left(3\sqrt{2} - 3\sqrt{2}\right)$

b) $\left(\sqrt{x} + \sqrt{y}\right)\left(\sqrt{x} - \sqrt{y}\right)$

c) $\left(\sqrt{x+1} + \sqrt{x}\right)\left(\sqrt{x+1} - \sqrt{x}\right)$

d) $\left(\sqrt{x+y}+\sqrt{y}\right)\left(\sqrt{x+y}-\sqrt{y}\right)$

e) $\left(\sqrt{x}+\sqrt{y}\right)\left(\sqrt{x}+\sqrt{y}\right)$

a) $\left(3\sqrt{2}+3\sqrt{2}\right)\left(3\sqrt{2}-3\sqrt{2}\right)$

Solution:

$$=3\sqrt{2}\cdot 3\sqrt{2}-3\sqrt{2}\cdot 3\sqrt{2}+3\sqrt{2}\cdot 3\sqrt{2}-3\sqrt{2}\cdot 3\sqrt{2}$$
$$=9\sqrt{4}-9\sqrt{4}+9\sqrt{4}-9\sqrt{4}$$
$$9\sqrt{4}+9\sqrt{4}-9\sqrt{4}-9\sqrt{4}$$
$$18\sqrt{4}-18\sqrt{4}$$
$$18(2)-18(2)$$
$$36-36$$
$$0$$

$$\left(3\sqrt{2}+3\sqrt{2}\right)\left(3\sqrt{2}-3\sqrt{2}\right)$$
$$\left(3\sqrt{2}\right)^2-\left(3\sqrt{2}\right)^2$$
$$(9\times2)-(9\times2)$$
$$18-18$$

Recall that: (a + b) (a – b) = a² – b²

Note: a² – b² = (a + b) (a – b) is called the difference of two squares.

b) $\left(\sqrt{x}+\sqrt{y}\right)\left(\sqrt{x}-\sqrt{y}\right)$

Solution:

$$\sqrt{x}\cdot\sqrt{x}-\sqrt{y}\cdot\sqrt{x}+\sqrt{y}\cdot\sqrt{x}-\sqrt{y}\cdot\sqrt{y}$$

$$\sqrt{x^2}-\sqrt{y}\cdot\sqrt{x}+\sqrt{y}\cdot\sqrt{x}-\sqrt{y^2}$$
$$=x-\sqrt{xy}+\sqrt{xy}-y$$
$$=x+0-y$$
$$=x-y$$

Assuming that $\sqrt{x^2}$ and $\sqrt{y^2}$ are positive real numbers.

c) $\left(\sqrt{x+1}+\sqrt{x}\right)\left(\sqrt{x+1}-\sqrt{x}\right)$

Solution:

$$\sqrt{x+1} \cdot \sqrt{x+1} - (\sqrt{x+1})(\sqrt{x}) + \sqrt{x}(\sqrt{x+1}) - \sqrt{x} \cdot \sqrt{x}$$

$$\sqrt{(x+1)^2} - \sqrt{x}(\sqrt{x+1}) + \sqrt{x}(\sqrt{x+1}) - \sqrt{x^2}$$

$$\sqrt{(x+1)^2} - \sqrt{x}(\sqrt{x+1}) + \sqrt{x}(\sqrt{x+1}) - \sqrt{x^2}$$

$x+1+0-\sqrt{x^2}$

$x+1-x$

$x-x+1$

$0+1$

1

Assuming that $\sqrt{x^2}$ and $\sqrt{(x+1)^2}$ are positive real numbers.

d) $\left(\sqrt{x+y}+\sqrt{y}\right)\left(\sqrt{x+y}-\sqrt{y}\right)$

Solution:

$$\left(\sqrt{x+y}+\sqrt{y}\right)\left(\sqrt{x+y}-\sqrt{y}\right)$$

$$\sqrt{x+y} \cdot \sqrt{x+y} - \sqrt{y}(\sqrt{x+y}) + \sqrt{y}(\sqrt{x+y}) - \sqrt{y} \cdot \sqrt{y}$$

$$\sqrt{(x+y)^2} - \sqrt{y}(\sqrt{x+y}) + \sqrt{y}(\sqrt{x+y}) - \sqrt{y^2}$$

$(x+y) +0-y$

$x+y-y$

$= x$

Assuming that $\sqrt{(x+y)^2}$ and $\sqrt{y^2}$ are positive real numbers.

Using the difference of two squares method

Recall that: $(a + b)(a - b) = a^2 - b^2$

$$\left(\sqrt{x+y}+\sqrt{y}\right)\left(\sqrt{x+y}-\sqrt{y}\right)$$

$$\left(\sqrt{x+y}\right)^2 - \left(\sqrt{y}\right)^2$$

$x+y-y$

$x+0$

x

e) $\left(\sqrt{x}+\sqrt{y}\right)\left(\sqrt{x}+\sqrt{y}\right)$

$$\left(\sqrt{x} + \sqrt{y}\right)\left(\sqrt{x} + \sqrt{y}\right)$$
$$= \sqrt{x} \cdot \sqrt{x} + \sqrt{x} \cdot \sqrt{y} + \sqrt{y} \cdot \sqrt{x} + \sqrt{y}\sqrt{y}$$
$$= \sqrt{x^2} + 2\sqrt{xy} + \sqrt{y^2}$$
$$= x + 2\sqrt{xy} + y$$

Assuming that $\sqrt{x^2}$ and $\sqrt{y^2}$ are positive real numbers.

Note: (a + b) (a + b) = a² + 2ab + b²

More Related Examples

Solve each of the following equations.

a) $\sqrt{x-2} = 2 - x$

b) $\sqrt{x+2} = x$

c) $\sqrt{2x+1} - x + 1 = 0$

d) $\sqrt{x+6} = -x$

e) $\sqrt{3x-5} + x - 1 = 0$

f) $(x+3)^{2/3} = 4$

g) $\sqrt{3x-5} + \sqrt{x-1} = 2$

h) $\sqrt{x+5} + \sqrt{20-x} = 7$

i) $\sqrt{\sqrt{x+3}} = 4$

j) $\sqrt{x-1} = 4$

k) $\sqrt{\sqrt{\sqrt{x-2}}} = 3$

Solutions

a) $\sqrt{x-2} = 2 - x$

$\left(\sqrt{x-2}\right)^2 = (2-x)^2$ ....Squaring both sides

$x - 2 = (2-x)(2-x)$

$x - 2 = 4 - 2x - 2x + x^2$

$x - 2 = 4 - 4x + x^2$

$x^2 - 4x - x + 4 + 2 = 0$ ....Collecting like terms

$x^2 - 5x + 6 = 0$

$x^2 - 3x - 2x + 6 = 0$ ....Factoring

$x(x-3) - 2(x-3) = 0$

$(x-2)(x-3) = 0$

$x-2=0$ or $x-3=0$

$x=2$ or $x=3$

Checking when x=2

$\sqrt{x-2} = 2-x$

$\sqrt{2-2} = 2-2$

$\sqrt{0} = 0$

$0 = 0$ .... Valid

Checking when x=3

$\sqrt{x-2} = 2-x$

$\sqrt{3-2} = 2-3$

$\sqrt{1} \neq -1$

$1 \neq -1$ .... Invalid

Therefore, x=3 is extraneous solution

Thus, the solution set is {2}

b) $\sqrt{x+2} = x$

$(\sqrt{x+2})^2 = x^2$ ....Squaring both sides to eliminate the square root sign.

$x+2 = x^2$

$x^2 - x - 2 = 0$ ....Bring x and 2 to the right sides.

$x^2 + x - 2x - 2 = 0$ ....Factoring

$x(x+1) - 2(x+1)$...............Taking common factors.

$(x-2)(x+1) = 0$ ....Taking outside terms and (x+1).

$x-2=0$ or $x+1=0$

$x=2$ or $x=-1$

Checking when x=2 and Checking when x= −1

$\sqrt{x+2} = x$	$\sqrt{x+2} = x$
$x = 2$	$x = -1$
$\sqrt{2+2} = 2$	$\sqrt{-1+2} = -1$
$\sqrt{4} = 2$	$\sqrt{1} = -1$
$2 = 2$ ......Valid	$1 \neq -1$ ......Invalid

Recall that x= −1 is called **extraneous solution.**

The square root of any number can never be negative number. Thus, the solution set is {2}.

Note: - When solving radical equations, make sure to check first the calculated solution whether it is valid or not before making decision.

c) $\sqrt{2x+1} - x + 1 = 0$

$\sqrt{2x+1} = x - 1$ .... Bring –x and 1 to the right side. Make sure to change the signs.

Or $(2x+1)^{1/2} - x + 1 + x - 1 = 0 + x - 1$

$\left(\sqrt{2x+1}\right)^2 = (x-1)^2$ .... Square both sides.

$2x + 1 = (x - 1)(x - 1)$

$2x + 1 = x^2 - 2x + 1$

$x^2 - 2x + 1 - 2x - 1 = 0$ .... Bring 2x and 1 to the right side and change the signs.

$x^2 - 4x + 1 - 1 = 0$ .... Adding

$x^2 - 4x + 0 = 0$ .... Subtracting

$x^2 - 4x = 0$

$x(x - 4)$ .... Taking x as common factor.

$x = 0$ or $x - 4 = 0$

$x = 0$ or $x = 4$

When we check the solution sets, the only solution set is {4}, since 0 doesn't satisfy the equation.

Recall that '0' is an **extraneous solution**.

d) $\sqrt{x+6} = -x$

$\left(\sqrt{x+6}\right) = (-x)$ .... Squaring both sides.

$x + 6 = x^2$

$x^2 - x - 6 = 0$ .... Bring x and 6 to the right side and change the signs.

$x^2 + 2x - 3x - 6 = 0$ .... sum of $2x + (-3x) = -x$ and product of $(2x)$ $(-3x) = -6x^2$.

$x(x + 2) - 3(x + 2) = 0$ .... Taking common factors.

$(x - 3)(x + 2) = 0$ .... Collecting outside terms.

$x - 3 = 0$ or $x + 2 = 0$

$x = 3$ or $x = -2$

When we check the solution sets, the correct one is $x = -2$, $x = 3$ does not satisfy the equation. Thus, the solution set is {–2}.

Recall that x=3 is an **extraneous solution.**

e) $\sqrt{3x-5} + x - 1 = 0$

$\sqrt{3x-5} = -x + 1$ .... To make it easy for solving, bring x – 1 to the right side and change the signs.

$\left(\sqrt{3x-5}\right)^2 = (1-x)^2$ .... Square both sides to get rid of the radical sign.

$3x - 5 = (1 - x)(1 - x)$

$3x - 5 = 1 - x - x + x^2$ .... Multiplying the right side

$3x - 5 = 1 - 2x + x^2$

$x^2 - 2x + 1 - 3x + 5 = 0$ ....Bring the left side expression to the right side and change the signs.

$x^2 - 5x + 6 = 0$ ....Collecting like terms.

$x^2 - 3x - 2x + 6 = 0$..................................$-3x + (-2x) = -5x$ and $(-3x) \cdot (-2x) = 6x^2$

$x(x - 3) - 2(x - 3) = 0$ ....Taking common factors.

$(x - 2)(x - 3) = 0$ ....Collecting outside terms.

$x - 2 = 0$ or $x - 3 = 0$

$x = 2$ or $x = 3$

When we substitute and check $x = 2$ or $x = 3$ in $\sqrt{3x - 5} + x - 1 = 0$, both $x = 2$ and $x = 3$, **do not** satisfy the equation; thus, the solution set is empty set. S.S = { }.

Recall that both x=2 and x=3 are **extraneous solutions.**

f) $(x + 3)^{\frac{2}{3}} = 4$

$(x + 3)^{\frac{2}{3} \times \frac{3}{2}} = (4)^{\frac{3}{2}}$ ....Multiplying both side by reciprocal of power of the left side. (i.e., By the reciprocal of $\frac{2}{3}$)

$(x+3) = (4^3)^{\frac{1}{2}}$

$(x+3) = (4 \times 4 \times 4)^{\frac{1}{2}}$

$x+3 = (64)^{\frac{1}{2}}$

$x+3 = \sqrt{64}$

$x+3 = \pm 8$

$x+3 = 8$ or $x+3 = -8$

$x = 8-3$ or $x = -3-8$

$x = 5$ or $x = -11$

Thus, the value of x is -11 or 5

Solution set = {-11,5}

g) $\sqrt{3x - 5} + \sqrt{x - 1} = 2$

$\left(\sqrt{3x - 5} + \sqrt{x - 1}\right)^2 = 2^2$ ....Squaring both sides to get rid of the radical signs.

To make it easy for solving the equation, it is customized to let each radical by the variable. Thus,

Let $a = \sqrt{3x - 5}$ and $b = \sqrt{x - 1}$

$(a + b)^2 = a^2 + 2ab + b^2$

$\left(\sqrt{3x - 5} + \sqrt{x - 1}\right)^2 = \left(\sqrt{3x - 5}\right)^2 + 2\left(\sqrt{3x - 5} \times \sqrt{x - 1}\right) + \left(\sqrt{x - 1}\right)^2$

$\left(\sqrt{3x - 5}\right)^2 + 2\left(\sqrt{3x - 5} \times \sqrt{x - 1}\right) + \left(\sqrt{x - 1}\right)^2 = 2^2$

$3x - 5 + 2\left(\sqrt{(3x - 5)(x - 1)}\right) + x - 1 = 4$ ...Squaring both sides

$3x+x-5-1+2(\sqrt{3x^2-3x-5x+5})=4$ ..... collecting like terms.

$4x-6+2(\sqrt{3x^2-3x-5x+5})=4$

$2\left(\sqrt{3x^2-8x+5}\right)=4-4x+6$ .... Moving **4x – 6** to the right side by changing the signs.

$2\left(\sqrt{3x^2-8x+5}\right)=-4x+10$

$\sqrt{3x^2-8x+5}=-2x+5$ .... Dividing both sides by 2.

$\left(\sqrt{3x^2-8x+5}\right)^2=(-2x+5)^2$ .... Squaring both sides to get rid of the left side radical sign.

$3x^2-8x+5=4x^2-20x+25$
$4x^2-20x+25-3x^2+8x-5=0$ .... Taking the left side expression to the right side
$4x^2-3x^2-20x+8x+25-5=0$ .... Collecting like terms.
$x^2-12x+20=0$
$x^2-2x-10x+20=0$ .... Sum of –2x and –10x is –12x and product of –2x and –10x is 20x², (factorizing)
$x(x-2)-10(x-2)=0$ .... Taking common factor.
$(x-10)(x-2)=0$ .... Collecting outside terms.
$x-10=0$ or $x-2=0$
$x=10$ or $x=2$

When we substitute and check x=10 or x=2, in the original equation $\sqrt{3x-5}+\sqrt{x-1}=2$, **x = 10 doesn't satisfy** the equation and it is called **extraneous solution.** Therefore, the solution set = {2}.

h) $\sqrt{x+5}+\sqrt{20-x}=7$

$\left(\sqrt{x+5}+\sqrt{20-x}\right)^2=7^2$ .... Squaring both sides to get rid of the radical signs

Recall that: .... $(a+b)^2=a^2+2ab+b^2$

Let $a=\sqrt{x+5}$ and $b=\sqrt{20-x}$

$\left(\sqrt{x+5}+\sqrt{20-x}\right)^2=7^2$ .... Squaring both sides.

$\left(\sqrt{x+5}\right)^2+2\left(\sqrt{x+5}\cdot\sqrt{20-x}\right)+\left(\sqrt{20-x}\right)^2=7^2$

$x+5+2\left(\sqrt{(x+5)(20-x)}\right)+20-x=49$

Note that: $\sqrt{a}\sqrt{b}=\sqrt{ab}$

$x+5+20-x+2\left(\sqrt{(x+5)(20-x)}\right)=49$ ......Collecting like terms.

$25+2\left(\sqrt{(5+x)(20-x)}\right)=49$

$25-25+2\left(\sqrt{(5+x)(20-x)}\right)=49-25$ ......Subtracting 25 from both sides.

$$2\left(\sqrt{(x+5)(20-x)}\right) = 24$$

$$\frac{2\sqrt{(x+5)(20-x)}}{2} = \frac{24}{2} \quad \text{.........................Dividing both sides by 2.}$$

$$\sqrt{(x+5)(20-x)} = 12$$

$$\sqrt{-x^2 + 15x + 100} = 12 \quad \text{...................Multiplying the inside terms in the radical.}$$

$$\left(\sqrt{-x^2 + 15x + 100}\right)^2 = (12)^2 \quad \text{....Squaring both sides to eliminate the radical sign}$$

$-x^2 + 15x + 100 = 144$

$-x^2 + 15x + 100 - 144 = 0$.........................Bring 144 to the left side by changing the sign.

$-x^2 + 15x - 44 = 0$

$(-1)(-x^2 + 15x - 44) = 0(-1)$.....................Multiply both sides by −1 to make the coefficient of x^2 positive.

$x^2 - 15x + 44 = 0$

$x^2 - 4x - 11x + 44 = 0$...............................Sum of −4x and −11x is −15x and product of −4x and −11x is $44x^2$.

$x(x - 4) - 11(x - 4) = 0$.............................Taking common factors.

$(x - 11)(x - 4) = 0$....................................Collecting outside terms.

$x - 11 = 0$ or $x - 4 = 0$

$x = 11$ or $x = 4$

When you plug and check x = 11 or x = 4, in the original equation $\sqrt{x+5} + \sqrt{20-x} = 7$, both numbers satisfy the equation. Thus, the solution set is: {4, 11}.

i) $\sqrt{\sqrt{x+3}} = 4$

Observe that in this equation we see that there are two radical signs, to find the value of x, we need to get rid of the radical signs step by step by squaring both the left side and the right side of the equation twice. Thus,

$$\sqrt{\sqrt{x+3}} = 4$$

$$\left(\sqrt{\sqrt{x+3}}\right)^2 = 4^2 \quad \text{...Squaring both sides and get rid of the upper radical sign first}$$

$$\sqrt{x+3} = 16$$

$$\left(\sqrt{x+3}\right)^2 = (16)^2 \quad \text{.................................Squaring both sides and get rid of the lower radical sign.}$$

$x + 3 = 256$

$x + 3 - 3 = 256 - 3$.................................Subtracting 3 from both sides.

$x = 253$

Thus, the solution set is: {253}

j) $\sqrt{x-1} = 4$

$\left(\sqrt{x-1}\right)^2 = 4^2$ ....Squaring both sides and get rid of the radical sign.

$x - 1 = 16$

$x - 1 + 1 = 16 + 1$ ....Adding 1 on both sides.

$x = 17$

The solution set is: {17}.

k) $\sqrt{\sqrt{\sqrt{x-2}}} = 3$

Observe that there are three radical signs in this equation. To solve such kinds of equation, square both sides of the equation three times to make sure that all the radical signs are omitted from the equation. Thus,

$\sqrt{\sqrt{\sqrt{x-2}}} = 3$

$\left(\sqrt{\sqrt{\sqrt{x-2}}}\right)^2 = 3^2$ ....Squaring both sides to omit the first radical sign

$\sqrt{\sqrt{x-2}} = 9$

$\left(\sqrt{\sqrt{x-2}}\right)^2 = 9^2$ ....Squaring both sides to omit the second radical sign.

$\sqrt{x-2} = 81$

$\left(\sqrt{x-2}\right)^2 = (81)^2$ ....Squaring both sides to omit the third radical sign.

$x - 2 = 6,561$

$x = 6,561 + 2$

$x = 6,563$

T.S. = {6,563}

Exercise 10.1-10.3

1. Simplify each of the following questions

   a) $\sqrt{20}$

   b) $\sqrt{90}$

   c) $\sqrt{84x^2}$

   d) $\sqrt{x^3 y^4 z^3}$

   e) $\sqrt[3]{27x^6 y^9}$

f) $\sqrt{256x^4y^6}$

g) $\sqrt[3]{\dfrac{162x^{18}y^{16}}{81}}$

h) $\sqrt{0.25x^4y^4z^6}$

i) $\sqrt[5]{x^{10}y^{25}z^{30}}$

j) $\sqrt[5]{x^{10}y^{15}z^{25}}$

k) $\sqrt{x^4y^4}$

l) $\sqrt[6]{x^6y^{12}z^{18}}$

m) $\sqrt[5]{x^5y^6z^{10}}$

2. Rewrite each of the following expressions in radical form and simplify.

a) $64^{5/6}$

b) $\left(49x^3y^2\right)^{1/2}$

c) $\left(x^2y^2z^2\right)^{-1/2}$

d) $\left(x^4y^3z^6\right)^{1/3}$

e) $\left(a^2b^2c^2\right)^{1/4}$

f) $(xyz)^{1/7}$

3. Rewrite each expression without radical symbol.

a) $\sqrt{m^3}$

b) $\sqrt{16a^2b^2c^2}$

c) $\dfrac{\sqrt[5]{x^3y^4}}{\sqrt[3]{a^2b^3}}$

d) $\sqrt[5]{32x^3y^4z^9}$

e) $\dfrac{\sqrt[4]{7a^5b^3c^2}}{\sqrt[5]{27x^3y^4z^5}}$

f) $\sqrt[3]{x^3y^2z^6}$

g) $\sqrt{2xy}$

h) $\sqrt[7]{xyz}$

i) $\sqrt[3]{2x^4y^5}$

4. Multiply each of the following expressions.

   a) $\left(\sqrt{3}-2\sqrt{2}\right)\left(\sqrt{3}+2\sqrt{2}\right)$

   b) $\left(\sqrt{x}+\sqrt{2y}\right)\left(\sqrt{x}-\sqrt{2y}\right)$

   c) $\left(\sqrt{2x}-\sqrt{y}\right)\left(\sqrt{2x}+\sqrt{y}\right)$

   d) $\left(\sqrt{a}-b\right)\left(\sqrt{a}+b\right)$

   e) $\left(\sqrt{a}+b\right)\left(\sqrt{a}+b\right)$

   f) $\left(\sqrt{2}+\sqrt{3}\right)\left(\sqrt{2}-\sqrt{3}\right)$

   g) $\left(\sqrt{2}+5\right)\left(\sqrt{2}-5\right)$

   h) $\left(3\sqrt{3}+1\right)\left(3\sqrt{3}-1\right)$

5. Solve each of the following radical equations.

   a) $\sqrt{2x-1}+2=4$
   b) $\sqrt{3x+1}-\sqrt{x+4}=1$

   c) $\sqrt{x+5}-\sqrt{x-3}=2$

   d) $\sqrt{x+5}=6$

   e) $\sqrt{x+3}=-4$

   f) $\sqrt{x+10}=x-2$

   g) $\sqrt{3x+18}=x$

   h) $\sqrt{\sqrt{2x-1}}=6$

   i) $\sqrt{\sqrt{\sqrt{x+1}}}=9$

   j) $\sqrt{x}+10=2\sqrt{x}+9$

   k) $\left(3\sqrt{x}+1\right)=\sqrt{x}$

   l) $\sqrt{x+3}=x-3$

m) $\sqrt{6x+1} = x-1$

n) $\sqrt{2x+15} - 6 = x$

o) $\sqrt{3x-5} = \sqrt{4x-1}$

6. Solve each of the following expressions

a) $(x+5)^{2/3} = 4$

b) $(x-4)^{2/3} = 16$

c) $x^{3/2} = 64$

d) $\dfrac{x^{1926}}{x^{1925}} = 32$

CHAPTER 11

Quadratic Equations and Functions

11.1. Quadratic Equation

A quadratic equation is an equation that can be written as a standard (general) form:

$f(x) = ax^2 + bx + c = 0$; where the numbers a, b and c are real numbers and $a \neq 0$. Such numbers are called the coefficients of the terms. Moreover, a is called the leading coefficient of the equation whereas c is the constant term. This function is a second-degree polynomial function.

11.2. 11.2 Solving Quadratic Equations.

There are some different methods of solving quadratic equations. These are: -

- Factoring method.
- Quadratic formula
- Square root property
- Completing the square and
- Graphic method.

Before we are going to examine how to solve quadratic equations using these methods, let us first learnt some of the general properties of these equations.

11.3. Properties of Quadratic Equations

- A quadratic equation is a polynomial equation of degree 2.
- The quadratic equation has the graph of "U" or inverted "U" shape and it is called a Parabola.
- A quadratic equation has a maximum of two solutions. The solution sets may be two distinct real solutions, one solution or two imaginary solutions.

11.4. How to Solve Quadratic Equations by Factoring Method.

Factoring is the easiest method of solving the quadratic equation. Factoring a quadratic equation is defined as the process of breaking the equation into product of its factors. A quadratic equation may or may not be factored. If it can be factored, it is written as a product of two factors. Solving a quadratic equation by factoring method depends on the zero-product property, which states that: -

If the product of two numbers or two expressions equal to zero, then at least one of the numbers or one of the expressions must be equal to zero. Because zero multiplied by any number equals zero. Thus,

If a x b = 0, then either a = 0 or b = 0, or both a and b = 0 where a and b are real numbers of algebraic expressions.

Examples: -Solving factored quadratic equations.

1) Solve the quadratic equation:

$(x + 1)(x - 2) = 0$

Solution: -Before you are going to solve such equation, make sure that you have to answer the question, why $(x + 1)(x - 2) = 0$ is a quadratic equation since it does not have anything raise to the second power? The reason is that, it has two expressions whose product contains the variables of degree two. In other words, if we expand these products, we will get the quadratic equation. To show this fact,

$(x + 1)(x - 2) = 0$

$x(x-2) +1(x-2) =0$.........................(use distributive property)

$x^2 - 2x + x - 2 = 0$......................(Adding like terms)

$x^2 - x - 2 = 0$

Now $(x + 1)(x - 2) = 0$ is already factored. To find the solution set, we use zero product property. Thus,

To make this expression true, either $x + 1 = 0$ or $x - 2 = 0$

$(x + 1)(x - 2) = 0$

$x + 1 = 0$ or $x - 2 = 0$

$x = -1$ or $x = 2$

The solution set is: {-1, 2}

2) Solve: $(2x + 1)(x - 4) = 0$

Solution: $(2x + 1)(x - 4) = 0$

$2x + 1 = 0$ or $x - 4 = 0$

$2x = -1$ or $x = 4$

$x = \dfrac{-1}{2}$ or $x = 4$

Solution set $= \left\{\dfrac{-1}{2}, 4\right\}$

3) Solve: $(x - 4)(x - 4) = 0$

Solution: $x - 4 = 0$ or $x - 4 = 0$

$x = 4$ or $x = 4$

The solution set is {4}

4) Solve: $(x - 6)(x + 5) = 0$

Solution: $x - 6 = 0$ or $x + 5 = 0$

$x = 6$ or $x = -5$

The solution set is: $\{-5, 6\}$

5) Solve: $x(x - 8) = 0$

Solution: $x(x - 8) = 0$

$x = 0$ or $x - 8 = 0$

$x = 0$ or $x = 8$

The solution set is $\{0, 8\}$.

6) Solve: $(x - 6)^2 = 0$

Solution: $(x - 6)^2 = (x - 6)(x - 6)$

$(x - 6)(x - 6) = 0$

$x - 6 = 0$ or $x - 6 = 0$

$x = 6$ or $x = 6$

Solution set $= \{6\}$

7) Solve: $2y(y - 3) = 0$

Solution: $2y(y - 3) = 0$

$2y = 0$ or $y - 3 = 0$

$y = 0$ or $y = 3$

Solution set: $\{0, 3\}$

8) Solve: $(2x - 1)(4x - 3) = 0$

Solution: $(2x - 1)(4x - 3) = 0$

$2x - 1 = 0$ or $4x - 3 = 0$

$$x = \frac{1}{2} \text{ or } x = \frac{3}{4}$$

Solution set: $\left\{\dfrac{1}{2}, \dfrac{3}{4}\right\}$

9) Solve: $x^2 = 81$

Solution: $x^2 = 81$

$x^2 - 81 = 0$ ………. (Bring 81 to the left side of equal sign and make it

negative)

$(x + 9)(x - 9) = 0$ ……… (Factoring rule)

$x + 9 = 0$ or $x - 9 = 0$

$x = -9$ or $x = 9$

Solution set: $\{-9, 9\}$

10) Solve: $y^2 = -9$

Solution: The square of any number can never be negative number and $y^2 = -9$ does not have solution set.

Solution set: $\{\ \ \}$

Exercise 11.1-11.4

Solve each of the following factored quadratic equations

1) $(x - 7)(x - 4) = 0$
2) $(2x - 1)(3x - 2) = 0$
3) $(x - 5)(2x + 1) = 0$
4) $2y(y - 1) = 0$
5) $(a - 4)(2a + 3) = 0$
6) $(x - 8)^2 = 0$
7) $(6x - 4)(x - 1) = 0$
8) $(2x - 7)(x - 6) = 0$
9) $(4 - x)(2 - x) = 0$
10) $(x - 7)(6 - 2x) = 0$

11.5. Solving Quadratic Equations by Factoring

Suppose we want to solve the quadratic equation $x^2 + 5x + 6 = 0$, then we will see the steps how to deal with it.

Example: Find the solution set of $x^2 + 5x + 6 = 0$.

To get its answer, first factor $x^2 + 5x + 6 = 0$ by sum-product pattern. To do this, find two numbers whose sum is 5x and whose product is 6 (x^2). These two numbers are 2x and 3x. Thus,

$x^2 + 5x + 6 = 0$
$x^2 + 2x + 3x + 6 = 0$
$x(x + 2) + 3(x + 2) = 0$.................................(Taking common factor)
$(x + 3)(x + 2) = 0$...(Collecting the outside expression and taking one of the bracket numbers)

$x + 3 = 0$ or $x + 2 = 0$
$x = -3$ or $x = -2$
Thus, the solution set is: $\{-3, -2\}$

Examples:

1) Solve: $x^2 + 6x + 9 = 0$
Solution: $x^2 + 6x + 9 = 0$
Find two numbers whose sum is 6x and whose product is $(9)(x^2)$, these two numbers are 3x and 3x; that is, $3x + 3x = 6x$ and $(3x)(3x) = 9x^2$.
$x^2 + 6x + 9 = 0$
$x^2 + 3x + 3x + 9 = 0$
$x(x + 3) + 3(x + 3) = 0$.................................(Taking common factors)
$(x + 3)(x + 3) = 0$...(Taking the outside terms and one of the expressions in bracket)

$x + 3 = 0$ or $x + 3 = 0$
$x = -3$ or $x = -3$

S.S = {–3}

2) Solve: $2x^2 – 6x + 4 = 0$

Solution: $2x^2 – 6x + 4 = 0$

Find two numbers whose sum is –6x and whose product is $(2x^2)(4) = 8x^2$, these numbers are –2x and –4x. Thus, $(–2x) + (–4x) = –6x$ and $(–2x)(–4x) = 8x^2$.

$2x^2 – 2x – 4x + 4 = 0$

$2x(x – 1) – 4(x – 1) = 0$.................................(Taking common factor)

$(2x – 4)(x – 1) = 0$

$2x – 4 = 0$ or $x – 1 = 0$

$2x = 4$ or $x = 1$

$x = \dfrac{4}{2}$ or $x = 1$

$x = 2$ or $x = 1$

S.S. {1, 2}

3) Solve: $5x^2 – x – 4 = 0$

Solution: $5x^2 – x – 4 = 0$

Find two numbers whose sum is –x and whose product is $(5x^2)(– 4) = –20x^2$, these two numbers are –5x and 4x.

$5x^2 – x – 4 = 0$

$5x^2 – 5x + 4x – 4 = 0$

$5x(x – 1) + 4(x – 1) = 0$.................................(Taking common factors)

$(5x + 4)(x – 1) = 0$

$5x + 4 = 0$ or $x – 1 = 0$

$5x = –4$ or $x = 1$

$x = \dfrac{–4}{5}$ or $x = 1$

S.S. $= \left\{ \dfrac{–4}{5}, 1 \right\}$

4) Solve: $3x^2 + 11x – 4 = 0$

Solution: Find two numbers whose sum is 11x and whose product is $3x^2(–4) = –12x^2$, these numbers are 12x and –x.

$3x^2 + 11x – 4 = 0$

$3x^2 + 12x – x – 4 = 0$

$3x(x + 4) – 1(x + 4) = 0$

$(3x – 1)(x + 4) = 0$ .... (Taking common factors)

$3x – 1 = 0$ or $x + 4 = 0$

$3x – 1 = 0$ or $x + 4 = 0$

$x = \dfrac{1}{3}$ or $x = –4$

S.S. $= \left\{ –4, \dfrac{1}{3} \right\}$

5) Solve: $-2x^2 + 8x - 6 = 0$

Solution: - Multiply both sides of the equation by -1 in order to make the coefficient of x^2 positive. We can also solve this equation without multiplying both sides by -1. The reason why we are multiplying by -1 is to make it easy while we are solving it.

$-2x^2 + 8x - 6 = 0$

$(-1)(-2x^2 + 8x - 6) = (0)(-1)$

$2x^2 - 8x + 6 = 0$

$$\frac{2x^2}{2} - \frac{8x}{2} + \frac{6}{2} = \frac{0}{2} \quad ....(\text{Divide both sides by 2})$$

$x^2 - 4x + 3 = 0$

Find two numbers whose sum is $-4x$ and whose product is $3(x^2) = 3x^2$, these numbers are $-3x$ and $-x$

$x^2 - 3x - x + 3 = 0$

$x(x - 3) - 1(x - 3) = 0$

$(x - 1)(x - 3) = 0$

$x - 1 = 0$ or $x - 3 = 0$

$x = 1$ or $x = 3$

S.S. = $\{1, 3\}$

6) Solve: $3x^2 - x - 4 = 0$

Solution: $3x^2 - x - 4 = 0$

Find two numbers whose sum is the middle term $-x$, and whose product is the first term and the last term $(3x^2)(-4) = -12x^2$. These two numbers are $3x$ and $-4x$.

$3x^2 - x - 4 = 0$

$3x^2 + 3x - 4x - 4 = 0$

$3x(x + 1) - 4(x + 1) = 0$

$(3x - 4)(x + 1) = 0 \quad ....(\text{Taking common factors and one of the brackets})$

$3x - 4 = 0$ or $x + 1 = 0$

$3x = 4$ or $x = -1$

$x = \dfrac{4}{3}$ or $x = -1$

S.S. = $\left\{-1, \dfrac{4}{3}\right\}$

7) Solve: $x^2 + 49 = 0$

Solution: The equation $x^2 + 49 = 0$ can't be factored, so that the solution set does not exist.
Solution set: $\{\ \ \}$.

Exercise 11.5

Solve each of the following quadratic equations by factoring method.
1) $x^2 - 3x + 2 = 0$
2) $5x^2 - 11x + 6 = 0$
3) $x^2 - 2x + 1 = 0$
4) $-3x^2 + 9x - 6 = 0$
5) $x^2 + 10x + 25 = 0$
6) $x^2 - x - 20 = 0$
7) $2y^2 + 7y + 5 = 0$
8) $x^2 - 81 = 0$
9) $4x^2 - 16x + 16 = 0$
10) $x^2 + 25 = 0$

11.6. Solving Quadratic Equation by the Quadratic Formula

The quadratic formula is a formula that helps to provide the solution(s) to a given quadratic equation.

The general quadratic equation is given by the form: -

$$ax^2 + bx + c = 0$$

where a, b and c represent constants with a $\neq$ 0. To solve such equations, the general formula is given by: -

$$x = \frac{-b \pm \sqrt{b^2 - 4ac}}{2a}$$

i.e., 'x' equals negative 'b' plus or minus the square root of $b^2 - 4ac$ all divided by 2a.

Examples: -

1) Solve: $2x^2 - 8x + 6 = 0$

Solution: The general form of a quadratic equation is $ax^2 + bx + c = 0$. The given equation is: $2x^2 - 8x + 6 = 0$. Comparing this general quadratic equation form and the given equation yields a = 2, b = -8 and c = 6.

Substituting these values into the quadratic formula and simplifying gives the solution of the equation. Hence,

$$x = \frac{-b \pm \sqrt{b^2 - 4ac}}{2a}$$

$$x = \frac{-(-8) \pm \sqrt{(-8)^2 - 4(2)(6)}}{2(2)}$$

$$x = \frac{8 \pm \sqrt{64 - 48}}{4}$$

$$x = \frac{8 \pm \sqrt{16}}{4}$$

$$x = \frac{8 \pm 4}{4}$$

$$x = \frac{8+4}{4} \text{ or } x = \frac{8-4}{4}$$

$$x = \frac{12}{4} \text{ or } x = \frac{4}{4}$$

$$x = 3 \text{ or } x = 1$$

$$S.S = \{1, 3\}$$

2) Solve: $2x^2 - 7x - 9 = 0$

Solution: Comparing the equation $2x^2 - 7x - 9 = 0$ with quadratic equation form $ax^2 + bx + c = 0$

Shows: a = 2, b = −7 and c = −9

$$x = \frac{-b \pm \sqrt{b^2 - 4ac}}{2a}$$

$$x = \frac{-(-7) \pm \sqrt{(-7)^2 - 4(2)(-9)}}{2(2)}$$

$$x = \frac{7 \pm \sqrt{49 + 72}}{4}$$

$$x = \frac{7 \pm \sqrt{121}}{4}$$

$$x = \frac{7 \pm 11}{4}$$

$$x = \frac{7+11}{4} \text{ or } \frac{7-11}{4}$$

$$x = \frac{18}{4} \text{ or } x = \frac{-4}{4}$$

$$x = \frac{9}{2} \text{ or } x = -1$$

$$S.S = \left\{-1, \frac{9}{2}\right\}$$

3) Solve: $x^2 + 6x + 8 = 0$

Solution: Similarly, from the equation $x^2 + 6x + 8 = 0$, we can conclude that $a = 1$, $b = 6$ and $c = 8$. Substituting these values in the general formula yields:

$$x = \frac{-b \pm \sqrt{b^2 - 4ac}}{2a}$$

$$x = \frac{-6 \pm \sqrt{(6)^2 - 4(1)(8)}}{2(1)}$$

$$x = \frac{-6 \pm \sqrt{36 - 32}}{2}$$

$$x = \frac{-6 \pm \sqrt{4}}{2}$$

$$x = \frac{-6 \pm 2}{2}$$

$$x = \frac{-6 + 2}{2} \text{ or } x = \frac{-6 - 2}{2}$$

$$x = \frac{-4}{2} \text{ or } x = \frac{-8}{2}$$

$$x = -2 \quad \text{or } x = -4$$

S.S = {−4, −2}

4) Solve: $3x^2 - 8x + 2 = 0$

Solution: From $3x^2 - 8x + 2 = 0$, we can say that : $a = 3$, $b = -8$ and $c = 2$

$$x = \frac{-b \pm \sqrt{b^2 - 4ac}}{2a}$$

$$x = \frac{-(-8) \pm \sqrt{(-8)^2 - 4(3)(2)}}{2(3)}$$

$$x = \frac{8 \pm \sqrt{64 - 24}}{6}$$

$$x = \frac{8 \pm \sqrt{40}}{6}$$

$$x = \frac{8 \pm \sqrt{4 \times 10}}{6}$$

$$x = \frac{8 \pm 2\sqrt{10}}{6}$$

$$x = \frac{8 + 2\sqrt{10}}{6} \text{ or } x = \frac{8 - 2\sqrt{10}}{6} \quad ....(\text{Multiplying both 'x' values by } \frac{2}{2})$$

$$x = \frac{4 + \sqrt{10}}{3} \text{ or } x = \frac{4 - \sqrt{10}}{3}$$

$$\text{S.S.} = \text{ or } \left\{ \frac{4 \pm \sqrt{10}}{3} \right\} \text{ or } \left\{ \frac{4 - \sqrt{10}}{3}, \frac{4 + \sqrt{10}}{3} \right\}$$

11.7. Discriminant of Quadratic Equation

Note: - In the general quadratic formula:

$$x = \frac{-b \pm \sqrt{b^2 - 4ac}}{2a}$$

- The quantity $b^2 - 4ac$, which appears under the radical sign, is called the Discriminant.
- This discriminant determines the number and type of solutions.
- In the quadratic equation $ax^2 + bx + c = 0$, if the discriminant $b^2 - 4ac > 0$, then the quadratic equation has two distinct solutions.

Example:

How many solutions has $x^2 - 6x + 8 = 0$?
Answer: $x^2 - 6x + 8 = 0$
    $a = 1, b = -6$ and $c = 8$
The discriminant: $b^2 - 4ac$
    $= (-6)^2 - 4(1)(8)$
    $= 36 - 32$
    $= 4$
Here, 4 is greater than 0. Thus, the quadratic equation has two distinct solution sets.

- In the quadratic equation $ax^2 + bx + c = 0$, if the discriminant $b^2 - 4ac < 0$, then the quadratic equation **does not** have solution set. (No real solution).

Example: How many solutions has $2x^2 + 3x + 8 = 0$?
    Answer: $2x^2 + 3x + 8 = 0$ implies $a = 2, b = 3$ and $c = 8$.
The discriminant $b^2 - 4ac$:
    $= 3^2 - 4(2)(8)$
    $= 9 - 64$
    $= -55$
It is true that -55 is less than 0. Thus, the quadratic equation $2x^2 + 3x + 8 = 0$ does not have solution. Therefore, the solution set is empty.

- In the quadratic equation $ax^2 + bx + c = 0$, if the discriminant $b^2 - 4ac = 0$, then the quadratic equation has only one solution set.

Example:

How many solution(s) has/have the quadratic equation: $x^2 - 4x + 4 = 0$?

    Answer: For equation $x^2 - 4x + 4 = 0$, $a = 1$, $b = -4$ and $c = 4$.

    Its discriminant $b^2 - 4ac$:

$$= (-4)^2 - 4(1)(4)$$
$$= 16 - 16$$
$$= 0$$

    Thus, the quadratic equation $x^2 - 4x + 4 = 0$ has only one solution set.

Exercise 11.6-11.7

A) Solve each of the following quadratic equations using quadratic formula.
1) $x^2 - x - 3 = 0$
2) $3x^2 + 4x - 2 = 0$
3) $5x^2 + 3x - 3 = 0$
4) $x^2 + 2x - 8 = 0$
5) $x^2 - 12x + 36 = 0$
6) $3x^2 + 5x + 9 = 0$
7) $3x^2 - 2x + 4 = 0$
8) $x^2 - 25 = 0$
9) $-x^2 - x + 2 = 0$
10) $x^2 + 5x + 6 = 0$

B) Determine the number of solutions for each of the following quadratic equations using the discriminant $b^2 - 4ac$.
1) $-4x^2 - 7x + 12 = 0$
2) $x^2 + 6x + 9 = 0$
3) $2x^2 + 5x + 6 = 0$
4) $x^2 + 6x + 4 = 0$
5) $3x^2 + 4x + 2 = 0$
6) $3x^2 - 9x - 12 = 0$
7) $x^2 + 10x + 25 = 0$
8) $x^2 - x - 1 = 0$
9) $6x^2 + 3x - 2 = 0$
10) $4x^2 + 16x + 12 = 0$

11.8. Solving Quadratic Equation by Square Root Property

    This method of solving quadratic equation is usually used if the quadratic equation is written in the form of $x^2 - c = 0$. In contrast, if it is written in the form of $x^2 + c = 0$, then the roots are imaginaries or complex numbers.

Definition: The square root property states that, if m is an algebraic expression and n is a non-zero real number, then $m^2 = n$ has exactly two solutions. In short, if the quadratic equation $m^2 = n$ is given, then its solution set is both $m = \sqrt{n}$ and $m = -\sqrt{n}$; abbreviated as, $m = \pm\sqrt{n}$.

Examples: -

Solve each of the following quadratic equations using square root property.
a) $x^2 - 25 = 0$
b) $5x^2 - 45 = 0$
c) $2x^2 - 36 = 0$
d) $x^2 + 81 = 0$
e) $(x - 9)^2 = 7$
f) $3x^2 + 27 = 0$

Solutions:

a) $x^2 - 25 = 0$
    Soln.
    $x^2 - 25 + 25 = 0 + 25$
    $x^2 = 25$
    $(x^2)^{1/2} = (25)^{1/2}$
    $x = \pm\sqrt{25}$
    $x = \pm 5$
    $x = 5$ or $x = -5$
    S.S = {−5, 5} or {±5}

b) $5x^2 - 45 = 0$
    Solution: $5x^2 - 45 = 0$
    $\quad\quad 5x^2 - 45 + 45 = 0 + 45$
    $\quad\quad 5x^2 = 45$
    $\quad\quad \dfrac{5x^2}{5} = \dfrac{45}{5}$
    $\quad\quad x^2 = 9$
    $\quad\quad (x^2)^{1/2} = (9)^{1/2}$
    $\quad\quad x = \pm\sqrt{9}$
    $\quad\quad x = \pm 3$
    $\quad\quad x = 3$ or $x = -3$
    $\quad\quad$ S.S = {±3} or {−3, 3}

c) $2x^2 - 36 = 0$
    Solution: $2x^2 - 36 = 0$
    $\quad\quad 2x^2 - 36 = 0$

$$\frac{2x^2}{2} - \frac{36}{2} = \frac{0}{2}$$

$$x^2 - 18 = 0$$

$$(x^2)^{1/2} = (18)^{1/2}$$

$$x = \pm\sqrt{18}$$

$$x = \pm\sqrt{9 \times 2}$$

$$x = \pm 3\sqrt{2}$$

$$x = 3\sqrt{2} \text{ or } -3\sqrt{2}$$

$$\textbf{S.S} = \left\{-3\sqrt{2}, 3\sqrt{2}\right\} \text{ or } \left\{\pm 3\sqrt{2}\right\}$$

d) $x^2 + 81 = 0$

Solution: $x^2 + 81 = 0$

$$x^2 = -81$$

$$x = \pm\sqrt{-81} \quad\text{.............................(Apply the square root property)}$$

$$x = \pm 9i \quad\text{........................(Solution sets expressed in terms of i)}$$

$$x = 9i \text{ or } x = -9i$$

The solution set is: $\{-9i, 9i\}$ or $\{\pm 9i\}$

Note: An imaginary number is a complex number that can be written as a real number multiplied by the imaginary unit i.

Which is defined by its property $\mathbf{i^2 = -1}$

For example, if 2i is an imaginary number, then its square (i.e., 2i x 2i = $4i^2$ = 4 x -1) is –4.

e) $(x - 9)^2 = 7$

Solution: $(x - 9)^2 = 7$

$$((x - 9)^2)^{1/2} = (7)^{1/2}$$

$$x - 9 = \pm\sqrt{7}$$

$$x - 9 + 9 = 9 \pm\sqrt{7}$$

$$x = 9 \pm\sqrt{7}$$

$$x = 9 + \sqrt{7} \text{ or } x = 9 - \sqrt{7}$$

Thus, S.S $= \{9 \pm\sqrt{7}\}$ or $\{9 - \sqrt{7}, 9 + \sqrt{7}\}$

f) $3x^2 + 27 = 0$

Solution: $3x^2 + 27 = 0$

$$3x^2 = -27$$

$$\frac{3x^2}{3} = \frac{-27}{3} \quad\text{..............................(Divide both sides by 3)}$$

$$x^2 = -9$$

$$(x^2)^{1/2} = (-9)^{1/2}$$

$x = \pm\sqrt{-9}$ ....(Apply the square root property)

$x = \pm 3i$..(Solution sets expressed in terms of i)

S.S = {–3i, 3i} or {±3i}

Exercise 11.8

Find the solution set of each of the following quadratic equations using square root property.

1) $(x – 1)^2 = 4$
2) $x^2 – 16 = 0$
3) $2x^2 – 6 = 44$
4) $x^2 + 16 = 0$
5) $x^2 + 25 = 0$
6) $3x^2 – 54 = 0$
7) $x^2 – 48 = 0$
8) $(x – 4)^2 = 10$
9) $x^2 – 288 = 0$
10) $x^2 + 45 = 0$

11.9. Solving a Quadratic Equation by Completing the Square Method.

One of the methods used for solving quadratic equation is completing the square method. In the quadratic equation $ax^2 + bx + c = 0$, if the trinomial $ax^2 + bx + c$ cannot be factored, using completing the square method technique, we can convert the given equation into an equivalent equation so as to solve it using the square root property.

11.9.1. Completing the Square Method

If $x^2 + bx$ is a binomial with coefficient of x^2 equal to 1, then by adding $\left(\dfrac{b}{2}\right)^2$, which is the square of half the coefficient of x, a perfect square trinomial will result.

Thus, $x^2 + bx + \left(\dfrac{b}{2}\right)^2 = \left(x + \dfrac{b}{2}\right)^2$

To solve $ax^2 + bx + c = 0$ by the completing the square technique,

1) Rewrite the equation $ax^2 + bx + c = 0$ by moving the constant term 'c' to the right side of an equal sign.

   i.e., $ax^2 + bx + = -c$

2) If the coefficient of x^2 i.e., 'a' is not equal to 1, divide both the right and left sides by 'a' .

   i.e., $\dfrac{ax^2}{a} + \dfrac{bx}{a} + \dfrac{c}{a} = \dfrac{0}{a}$, which is equal to:

$$x^2 + \frac{bx}{a} + \frac{c}{a} = 0$$

3) Move the constant term $\frac{c}{a}$ to the right-hand side.

i.e., $x^2 + \frac{bx}{a} = -\frac{c}{a}$

4) Add the square of half of the coefficient of the x-term, $\left(\frac{b}{2a}\right)^2$ to both the right and left sides of the equation.

i.e., $x^2 + \frac{b}{a}x + \left(\frac{b}{2a}\right)^2 = \frac{-c}{a} + \left(\frac{b}{2a}\right)^2$ or

$x^2 + \frac{b}{a}x + \left(\frac{b}{2a}\right)^2 = \left(\frac{b}{2a}\right)^2 - \frac{c}{a}$

5) Factor the left side as the square of a binomial.

i.e., $\left(x + \frac{b}{2a}\right)^2 = \frac{b^2}{4a^2} - \frac{c}{a}$

$= \left(x + \frac{b}{2a}\right)^2 = \frac{b^2 - 4ac}{4a^2}$

6) Take the square root of both sides.

$x + \frac{b}{2a} = \frac{\pm\sqrt{b^2 - 4ac}}{2a}$

7) Then solve for the variable x

$x = \frac{-b \pm \sqrt{b^2 - 4ac}}{2a}$

Note: - When you add a constant term to one side of the equation to complete the square, make sure to add the same constant to the other side of the equation.

- Completing the square method is mostly used for quadratic equation, when it is difficult to find two numbers which gives the sum of the roots and the product of the roots of the given quadratic equation.
- Completing the square formula is used to drive the 'General quadratic formula':

$$X = \frac{-b \pm \sqrt{b^2 - 4ac}}{2a}$$

Examples: -

Solve each of the following quadratic equations by using the completing the square method.
1) $x^2 - 4x - 8 = 0$
2) $3x^2 + 6x + 8 = 0$
3) $4x^2 + 24x - 48 = 0$
4) $x^2 + 6x + 4 = 0$
5) $-5x^2 - 10x - 25 = 0$

Solutions: -
1) $x^2 - 4x - 8 = 0$

Step i)   we cannot solve this equation using factoring method because we can't find two numbers whose sum is **-4x** and whose product is **-8x²**. So, for such kind of equation, it is easy to use and find the solution set using completing the square method, in this given equation, the coefficient of x^2 is 1, rewrite it by taking the constant term to the right side of the equal sign.
$x^2 - 4x = 8$

Step (ii) Complete the square by adding the square of half of the coefficient of x to both sides of the equation to make it a perfect square. Thus,
$$\left(\frac{-4}{2}\right)^2 = 4$$
$x^2 - 4x + 4 = 8 + 4$
$x^2 - 4x + 4 = 12$

Step (iii) Write the left side (perfect square) as a square.
$(x - 2)^2 = 12$

Step(iv) Apply square root property and solve for x.
$((x - 2)^2)^{1/2} = (12)^{1/2}$
$x - 2 = \pm\sqrt{12}$
$x - 2 + 2 = 2 \pm \sqrt{12}$ ....(Add 2 on both sides)
$$x - 2 + 2 = 2 \pm \left(\sqrt{2 \times 2 \times 3}\right)^{\frac{1}{2}}$$

271

$$x = 2 \pm 2\sqrt{3} \quad \text{................................} \quad \left(\sqrt{12} = 2\sqrt{3}\right)$$

$$\text{S.S} = \left\{2 \pm 2\sqrt{3}\right\} \text{ or } \left\{2 - 2\sqrt{3}, \ 2 + 2\sqrt{3}\right\}$$

2) $3x^2 + 6x + 8 = 0$

Step (i)   Since the coefficient of x^2 is 3, divide both sides by 3.

$$\frac{3x^2 + 6x + 8}{3} = \frac{0}{3}$$

$$x^2 + 2x + \frac{8}{3} = 0$$

Step (ii) Rewrite the equation by taking the constant term to the right side of equal sign.

$$x^2 + 2x = \frac{-8}{3}$$

Step (iii) Complete the square by adding the square of half of the coefficient of x to both sides of the equation to make it the perfect square.

$$\text{Thus, } \left(\frac{2}{2}\right)^2 = 1$$

$$x^2 + 2x + 1 = \frac{-8}{3} + 1$$

Step (iv) Write the left side (perfect square) as a square.

$$(x + 1)^2 = \frac{-5}{3}$$

Step (v) Apply square root property and solve for x.

$$(x + 1) = \pm\sqrt{\frac{-5}{3}}$$

$$x = -1 \pm \sqrt{\frac{-5}{3}}$$

$$x = -1 \pm i\sqrt{\frac{5}{3}}$$

$$x = -1 \pm i\sqrt{\frac{5}{3}} \text{ or } x = -1 - i\sqrt{\frac{5}{3}}$$

$$S.S = \left\{ -1 \pm i\sqrt{\frac{5}{3}} \right\} \text{ or } \left\{ -1 - i\sqrt{\frac{5}{3}} \text{ or } -1 + i\sqrt{\frac{5}{3}} \right\}$$

3) $4x^2 + 24x - 48 = 0$

(i) Since the coefficient of x^2 is 4, divide both sides by 4.

$$\frac{4x^2 + 24x - 48}{4} = \frac{0}{4}$$

$$x^2 + 6x - 12 = 0$$

(ii) Rewrite the equation by taking the constant term to the right side of the equal sign.

$x^2 + 6x - 12 + 12 = 0 + 12$

$x^2 + 6x = 12$

(iii) Complete the square by adding the square of half of the coefficient of x to both sides of the equation to make the perfect square. Thus,

$$\left(\frac{6}{2} \right)^2 = 9$$

$x^2 + 6x + 9 = 12 + 9$

$x^2 + 6x + 9 = 21$

(iv) Write the left side (perfect square) as a square.

$(x + 3)^2 = 21$

(v) Apply square root property and solve for x.

$(x + 3)^2 = 21$

$x + 3 = \pm\sqrt{21}$

$x + 3 - 3 = -3 \pm \sqrt{21}$

$x = -3 \pm \sqrt{21}$

$x = -3 - \sqrt{21}$ or $x = -3 + \sqrt{21}$

$S.S = \left\{ -3 \pm \sqrt{21} \right\}$ or $\left\{ -3 - \sqrt{21}, -3 + \sqrt{21} \right\}$

4) $x^2 + 6x + 4 = 0$

$x^2 + 6x = -4$

$$x^2 + 6x + \left(\frac{6}{2} \right)^2 = -4 + \left(\frac{6}{2} \right)^2$$

$x^2 + 6x + 9 = -4 + 9$

$x^2 + 6x + 9 = 5$

$(x + 3)^2 = 5$

$$x + 3 = \pm\sqrt{5}$$

$$x = -3 \pm \sqrt{5}$$

$$x = -3 - \sqrt{5} \text{ or } -3 + \sqrt{5}$$

$$S.S = \left\{-3 \pm \sqrt{5}\right\} \text{ or } \left\{-3 - \sqrt{5}, -3 + \sqrt{5}\right\}$$

5) $-5x^2 - 10x - 25 = 0$

   (i) Since the coefficient of x^2 is –5, divide both sides by –5.

   $$\frac{-5x^2 - 10x - 25}{-5} = \frac{0}{-5}$$

   $$x^2 + 2x + 5 = 0$$

   (ii) Rewrite the equation by taking the constant term to the right side of the equal sign.

   $$x^2 + 2x = -5$$

   (iii) Complete the square by adding the square of half of the coefficient of x to both sides of the equation to make the perfect square. Thus,

   $$\left(\frac{2}{2}\right)^2 = 1$$

   $$x^2 + 2x + 1 = -5 + 1$$

   (iv) Write the left side (perfect square) as a square.

   $$x^2 + 2x + 1 = -4$$

   $$(x + 1)^2 = -4$$

   (v) Apply the square root property and solve for x.

   $$(x + 1) = \pm\sqrt{-4}$$

   $$x + 1 - 1 = -1 \pm \sqrt{-4}$$

   $$x = -1 \pm \sqrt{-4}$$

   $$x = -1 \pm 2i$$

   $$x = -1 - 2i \text{ or } -1 + 2i$$

   $$S.S = \{-1 - 2i, -1 + 2i\} \text{ or } \{-1 \pm 2i\}$$

Exercise 11.9 - 11.9.1

1) Solve each of the following quadratic equations using completing the square method.

   a) $x^2 - 6x - 11 = 0$

   b) $x^2 - 4x - 5 = 0$

   c) $3x^2 - 6x + 9 = 0$

d) $x^2 + 8x + 15 = 0$

e) $x^2 + 5x + 2 = 0$

f) $2x^2 + 9x + 16 = 0$

g) $3x^2 + 2x + 9 = 0$

h) $2x^2 - 4x + 4 = 0$

i) $-3x^2 + 9x + 15 = 0$

j) $x^2 + x + 2 = 0$

2) Solve each of the following equations.

a) $x^2 - 4x = 0$

b) $2x^2 - 8x - 16 = 0$

c) $x^2 - \dfrac{1}{64} = 0$

d) $x^3 - 5x = 0$

e) $x^4 - 9x^3 + 20x^2 = 0$

f) $a^2 - 9 = 0$

11.10. Sum and Product of Roots (or zero) of Quadratic Equation:

As you learned before, a standard form of a quadratic equation is written in the form of $ax^2 + bx + c = 0$, where a, b, and c are real numbers and $a \neq 0$. You are familiar with any quadratic equations that have two roots even though one may be either repeated root or the roots may not even real number. The roots of a quadratic equation have different natures. One of its natures is the sum and product of the roots. If the quadratic equation is given in standard form, $ax^2 + bx + c = 0$, where a, b and c are real numbers and $a \neq 0$. we can find the sum and product of the roots using coefficients of x^2, x and constant term c.

Note: - As mentioned in the previous chapter, to find the roots of a quadratic equation, the general formula is given by:

$$x = \frac{-b \pm \sqrt{b^2 - 4ac}}{2a}$$

Let us see below how to get the sum and product of the roots of a quadratic equation using this general quadratic formula.

Let r_1 and r_2 be the two roots of a quadratic equation, $ax^2 + bx + c = 0$, where a, b and c are real numbers and $a \neq 0$. Then, to get the sum and product of these roots of a quadratic equation, we use the following formula: -

11.10.1. Sum of the Roots

$$r_1 + r_2 = \frac{-b + \sqrt{b^2 - 4ac}}{2a} + \frac{-b - \sqrt{b^2 - 4ac}}{2a} \quad \ldots\ldots\ldots \text{(sum of roots)}$$

$$= \frac{-2b+0}{2a}$$

$$r_1 + r_2 = \frac{-b}{a}$$

Thus, the sum of the roots of a quadratic equation is given by $\frac{-b}{a}$.

$$r_1 + r_2 = \frac{-b}{a}$$

11.10.2. Product of the Roots

We can multiply r_1 and r_2 as follows. But to obtain the product of the roots, first recall that the simplified form of the difference of the two squares.

$(a + b)(a - b) = a^2 - b^2$ ....(Difference of two squares)

$$r_1 r_2 = \left(\frac{-b+\sqrt{b^2-4ac}}{2a}\right)\left(\frac{-b-\sqrt{b^2-4ac}}{2a}\right) \ldots\ldots \text{(product of roots)}$$

$$= \frac{(-b)^2-\left(\sqrt{b^2 4ac}\right)^2}{(2a)^2}$$

$$= \frac{b^2-\left(b^2-4ac\right)}{4a^2}$$

$$r_1 r_2 = \frac{4ac}{4a^2}$$

$$r_1 r_2 = \frac{c}{a}$$

Thus, the product of the roots of a quadratic equation is given by $\frac{c}{a}$.

$$r_1 r_2 = \frac{c}{a}$$

Examples: -

1. Find the sum and product of the roots of the following quadratic equations.
    a) $x^2 - 14x + 48 = 0$
    b) $3x^2 - 1 = 0$
    c) $x^2 - 4x + 4 = 0$
    d) $x^2 - 18x + 72 = 0$
    e) $3y^2 - 12y + 9 = 0$
    f) $-2x^2 - 8x + 17 = 0$

g) $5x^2 + 11x - 12 = 0$

h) $3x^2 - 6x + 2 = 0$

Solutions

a) $x^2 - 14x + 48 = 0$

Comparing this quadratic equation with the general form $ax^2 + bx + c = 0$, implies, a = 1, b = –14 and c = 48

As explained above, sum of its roots $= \dfrac{-b}{a}$

Substituting 1 for a and -14 for b yields:

$$\dfrac{-(-14)}{1}$$

$$= 14$$

Thus, the sum of the roots is 14.

Product of roots $= \dfrac{c}{a}$

Substituting 1 for a and 48 for c yields:

$$= \dfrac{48}{1}$$

$$= 48$$

Thus, the sum of the roots is 14 while their product is 48.

b) $3x^2 - 1 = 0$

a = 3, b = 0 and c = –1

Note: $3x^2 - 1 = 3x^2 + 0x - 1 = 0$

Substituting gives the sum of the roots $= \dfrac{-b}{a}$

$$= \dfrac{0}{3}$$

$$= 0$$

The product of its roots $= \dfrac{c}{a}$

Again, substituting results $\dfrac{c}{a} = \dfrac{-1}{3}$

Thus, the sum of the roots is 0 while their product is $\dfrac{-1}{3}$.

c) $x^2 - 4x + 4 = 0$

$a = 1$, $b = -4$, and $c = 4$

Likewise, the sum of the roots $= \dfrac{-b}{a}$

Substituting:

$$= \dfrac{-(-4)}{1}$$

$$= 4$$

Product of roots $= \dfrac{c}{a}$

$$= \dfrac{4}{1}$$

$$= 4$$

Thus, the sum of the roots is 4 whereas their product is also 4.

d) $x^2 - 18x + 72 = 0$

$a = 1$, $b = -18$ and $c = 72$.

Sum of the roots $= \dfrac{-b}{a}$

$$= \dfrac{-(-18)}{1}$$

$$= 18$$

The product of the roots $= \dfrac{c}{a}$.

As, c= 72 and a = 1, substituting and yields:

$$= \dfrac{72}{1}$$

$$= 72$$

Thus, the sum of the roots is 18 and their product is 72.

e) $3y^2 - 12y + 9 = 0$

$a = 3$, $b = -12$, and $c = 9$

The sum of the roots $= \dfrac{-b}{a}$

(Substituting)

$$= \frac{-(-12)}{3}$$

$$= \frac{12}{3}$$

$$= 4$$

The product of the roots $= \frac{c}{a}$ (Substituting)

$$= \frac{9}{3}$$

$$= 3$$

Thus, the sum of the roots is 4 and their product is 3.

f) $-2x^2 - 8x + 7 = 0$

$a = -2, b = -8$ and $c = 7$

The sum of the roots $= \frac{-b}{a}$ (Substituting)

$$= \frac{-(-8)}{-(2)}$$

$$= \frac{-8}{2}$$

$$= -4$$

The product of the roots $= \frac{c}{a}$ (Substituting)

$$= \frac{7}{-2}$$

$$= \frac{-7}{2}$$

Thus, the sum of the roots is -4 and their product is -7/2.

g) $5x^2 + 11x - 12 = 0$

$a = 5, b = 11,$ and $c = -12$

The sum of the roots $= \frac{-b}{a}$ (Substituting)

$$= \frac{-11}{5}$$

The product of the roots $= \frac{c}{a}$ (Substituting)

$$= \frac{-12}{5}$$

Product of roots $= \dfrac{-12}{5}$

Thus, the sum of the roots is -11/5 while their product is -12/5.

h) $3x^2 - 6x + 2 = 0$
    $a = 3, b = -6$ and $c = 2$

The sum of the roots $= \dfrac{-b}{a}$ (Substituting)

$$= \frac{-(-6)}{3}$$

$$= \frac{6}{3}$$

$$= 2$$

The product of the roots $= \dfrac{c}{a}$ (Substituting)

$$= \frac{2}{3}$$

Thus, the sum of the roots is 2 whereas their product is 2/3.

2) Find the value of k such that the quadratic equation $x^2 - kx + 18 = 0$ has the sum of roots equal to 9.
    Solution: - Compare $x^2 - kx + 18 = 0$ with $ax^2 + bx + c = 0$
    $a = 1, b = -k$ and $c = 18$

    The given sum of the roots = 9. This is also equals to the sum of the roots $= \dfrac{-b}{a}$

    $$9 = \frac{-(-k)}{1}$$

    $$K = 9$$

    Thus, the value of k is 9.

3) If the roots of the quadratic equation $5x^2 + (k + 1) x - 7 = 0$ are 1 and $\dfrac{-7}{5}$, then what is the value of k?
    Solution: - Compare: $5x^2 + (k + 1) x - 7 = 0$ with $ax^2 + bx + c = 0$
    $a = 5, b = (k + 1)$, and $c = -7$

    Let $r_1 = 1$ and $r_2 = \dfrac{-7}{5}$

The sum of the roots $= \dfrac{-b}{a}$

$$r_1 + r_2 = \dfrac{-b}{a}$$

$$1 + \left(\dfrac{-7}{5} \right) = \dfrac{-(k+1)}{5}$$

$$\dfrac{5 + (-7)}{5} = \dfrac{-(k+1)}{5}$$

$$\dfrac{-2}{5} = \dfrac{-(k+1)}{5}$$

$-2 = -(k + 1)$ ....(5 is cancelled each other)

$k + 1 = 2$ ....(Dividing both sides by -1)

$k + 1 - 1 = 2 - 1$

$k = 1$

Thus, the value of k is equal to 1.

Therefore, the quadratic equation is

$5x^2 + (k + 1) x - 7 = 0$

$5x^2 + (1 + 1) x - 7 = 0$

$5x^2 + 2x - 7 = 0$

4) If the two roots of a quadratic equation are 3 and 5, then what is their parent quadratic equation?

Solution: $(x - r_1) (x - r_2) = 0$

$r_1 = 3$ and $r_2 = 5$

$(x - r_1) (x - r_2) = 0$

$(x - 3) (x - 5) = 0$

$x(x - 5) + -3(x - 5)$

$x^2 - 5x - 3x + 15 = 0$

$x^2 - 8x + 15 = 0$

Thus, the parent quadratic equation is $x^2 - 8x + 15 = 0$

5) If the roots of $3x^2 + mx - n = 0$ are -3 and 4, what are the values of m and n? What is the original quadratic equation?

Solution: -

Given: $r_1 = -3$, $r_2 = 4$, $a = 3$, $b = m$ and $c = -n$

$$r_1 + r_2 = \frac{-b}{a}$$

$$-3 + 4 = \frac{-m}{3}$$

$$1 = \frac{-m}{3}$$

$$m = -3$$

$$r_1 r_2 = \frac{c}{a}$$

$$(-3)(4) = \frac{-n}{3}$$

$$-12 = \frac{-n}{3}$$

$$n = 36$$

Thus, m = –3 and n = 36

To find the original quadratic equation, substitute the values of m and n, in

$3x^2 + mx - n = 0,$

$3x^2 + (-3x) - (36) = 0$

$3x^2 - 3x - 36 = 0$

Thus, the original quadratic equation is: $3x^2 - 3x - 36 = 0$

Note: We can divide the equation $3x^2 - 3x - 36 = 0$ by their common factor 3 and obtain the same equation. i.e., $\frac{1}{3}(3x^2 - 3x - 36) = 0$

$x^2 - x - 12 = 0$

6) One of the roots of the equation $2x^2 + 5x - (k + 4) = 0$ is –3. Then, what is the value of k?

Solution: -

Given: $r_1 = -3$, a = 2, b = 5, and c = – (k + 4)

The value of k can be found by two methods: by direct substitution of the given root for x or by product of the root's method.

(i) By direct substitution of the given root for x:

Thus, $2x^2 + 5x - (k + 4) = 0$, $r_1 = -3$

$2(-3)^2 + 5(-3) - (k + 4) = 0$ ....(Substitute x = –3)

$18 - 15 - k - 4 = 0$

$18 - 19 - k = 0$

$-1 - k = 0$

$-1 + 1 - k = 0 + 1$

$-k = 1$

$k = -1$

Thus, the value of k is –1.

(ii) By product of the root's method:

Let one of the given root $r_1 = -3$ and the other non-given root be r_2.

$$r_1 + r_2 = \frac{-b}{a}$$

$$-3 + r_2 = \frac{-5}{2}$$

$$r_2 = \frac{-5}{2} + 3$$

$$r_2 = \frac{-5}{2} + \frac{6}{2}$$

$$r_2 = \frac{1}{2}$$

The product of the two roots i.e., $r_1 (r_2) = \frac{c}{a}$

$$-3 (r_2) = \frac{c}{a}$$

$$-3 (r_2) = \frac{c}{a} = \frac{-(k+4)}{2}$$

$$-3 \left(\frac{1}{2}\right) = \frac{-(k+4)}{2} \quad \ldots\ldots\ldots\ldots\ldots\ldots\ldots \text{Substituting the value of } r_2 = \frac{1}{2}$$

$$\frac{-3}{2} = \frac{-(k+4)}{2} \quad \ldots\ldots\ldots\ldots\ldots\ldots\ldots \text{Multiplying both sides by 2}$$

$$-3 = -(k+4)$$

$$3 = k+4 \ldots\ldots\ldots\ldots\ldots\ldots\ldots \text{Multiplying both sides by -1}$$

$$k = 3 - 4$$

Thus, the value of $k = -1$.

7) For the quadratic equation $kx^2 - 4x + 7 = 0$, if the sum of the roots is $\frac{-4}{3}$, then find the value of k.

Solution: - Recall that the sum of the roots

$$r_1 + r_2 = \frac{-b}{a}, \text{ where } a = k, b = -4 \text{ and } c = 7$$

$$r_1 + r_2 = \frac{-4}{3}$$

$$\frac{-4}{3} = \frac{-(-4)}{k}$$

$$\frac{-4}{3} = \frac{4}{k} \quad \ldots\ldots\ldots\ldots\ldots\ldots\ldots\ldots \text{substituting } r_1 + r_2 \text{ by -4/3}$$

$$-4k = 12 \ldots\ldots\ldots\ldots\ldots\ldots\ldots\ldots\ldots \text{(cross multiplication)}$$

$$k = \frac{12}{-4}$$

$$k = -3$$

Thus, the value of k is –3.

8) Find the value of k, if the product of the roots of the equation $(k + 1) x^2 + (k + 4) x + 3 = 0$ is $\frac{3}{5}$.

Solution: - Recall that the product of the root is: $r_1 r_2 = \frac{c}{a}$.

In the given equation $(k + 1) x^2 + (k + 4) x + 3 = 0$, $a = (k + 1)$, $b = (k + 4)$ and $c = 3$.

Given that: $r_1 r_2 = \dfrac{3}{5}$ .... Product of roots.

$r_1 r_2 = \dfrac{c}{a}$ (substitution of 3/5 by $r_1 r_2$, 3 for c and k + 1 for a)

$\dfrac{3}{5} = \dfrac{3}{(k+1)}$

$3(k + 1) = 3 \times 5$ ....(Cross multiplication)
$k + 1 = 5$ ....(Divide both sides by 3)
$k + 1 - 1 = 5 - 1$
$k = 4$

Thus, the value of k is 4.

Exercise 11.10-11.10.2

Answer each of the following questions.

1) Find the value of k such that $3x^2 - 4kx + 9 = 0$ has the sum of their roots equal to 4.
2) If the quadratic equation has the form $2kx^2 + 5x + 60 = 0$ has one of the roots -3, then find the value of the other root and k.
3) Find the value of 'm' and 'n' such that $mx^2 + 18x - n = 0$ has one of the roots is equal to 2 and sum of roots 6.
4) If the roots of a quadratic equation are 2 and -3, then what is the parent quadratic equation?
5) If $(2k - 1)x^2 + 6x - 1 = 0$ has one of its roots equal to $- 4$, then what is the value of k?
6) If one of the roots of a quadratic equation $x^2 + kx + 20 = 0$ is 4, then what is the value of k?
7) If one of the roots of the equation $x^2 - 8x + k = 0$ exceeds the other by 4, then what is the value of k?
8) If $(k - 1)x^2 - 8x + 4 = 0$ has sum of roots equal to $\dfrac{8}{3}$, what is the value of k?
9) $x^2 - 9x + k + 6 = 0$ has the difference between its roots equal to 1, what is the value of k?
10) Find the value of k such that $3x^2 + (k+1) x + 54 = 0$ has one root equal to to twice the other.

11.11. Graphing Quadratic Functions

A quadratic function is a function that can be written in the standard form $y = ax^2 + bx + c$, where a, b and c are real numbers with $a \neq 0$. Every quadratic function has a U-shaped graph called a Parabola. If the leading coefficient 'a' is positive the parabola opens upward. In contrast, if the leading coefficient 'a' is negative, the parabola opens downward. The highest or the lowest point at which the graph turns is called the vertex. When compared to the other methods, solving a quadratic equation using graphical method gives an estimate to the solution(s). The solution of the graph of a quadratic function is the point at which the graph crosses the x-axis. If the graph of a quadratic function crosses the x-axis at two points, then the quadratic function has two solutions. If the graph touches the x-axis at one point, then the graph of a quadratic function has one solution. If the graph does not intersect or touches the x-axis, then the equation has no real solution.

Using the discriminant of the quadratic equation: $ax^2 + bx + c = 0$, where $a \neq 0$, we can determine the number and type of a solution.

Note: - As you learned before, the quadratic formula says that:

$$x = \frac{-b \pm \sqrt{b^2 - 4ac}}{2a}$$

for any quadratic equation as in the form of $ax^2 + bx + c = 0$, where $a \neq 0$

- The discriminant is the part of the quadratic formula under the square root.
- In the above quadratic formula, the discriminant is: $b^2 - 4ac$
- The discriminant can be positive, negative or zero and this determines how many solutions there are to the given quadratic equation.

The following graphs show the types of solutions that a quadratic equation can have, that is; two solutions, one solution and no real solution.

In the quadratic equation $ax^2 + bx + c = 0$, where $a \neq 0$.

- If the discriminant $b^2 - 4ac > 0$, then the graph crosses the x-axis at two points; thus, the equation has two distinct solutions.

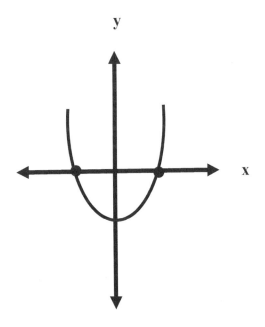

Fig. 11.1

Two solutions
Two x-intercepts

- The x-intercepts of the graph are the solutions of the equation.
- If the discriminant $b^2 - 4ac = 0$, then the quadratic equation has exactly one solution. (The graph touches the x-axis only at one point).

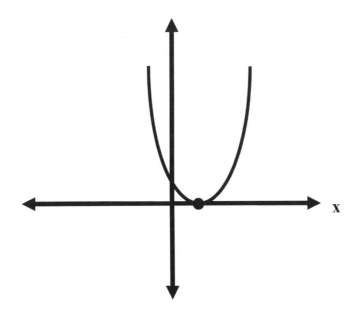

Fig. 11.2

One solution
One x-intercept

- If the discriminant $b^2 - 4ac < 0$, then the quadratic does not have solution. i.e., the graph doesn't cross or touch the x-axis.

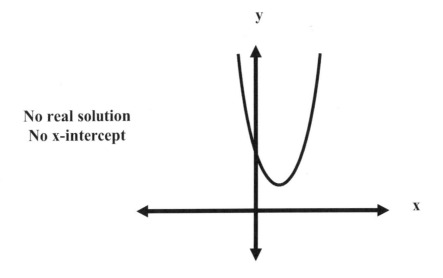

No real solution
No x-intercept

Fig. 11.3

The graph does not cross or touch the x-axis

Examples

1) Use the graph of $y = x^2 - 2x - 3$ to solve $x^2 - 2x - 3 = 0$
   Solution: - Make the table of values for x and y.

$y = x^2 - 2x - 3$	x	−2	−1	0	1	2	3	4
	y	5	0	−3	−4	−3	0	5

y $y = x^2 - 2x - 3$

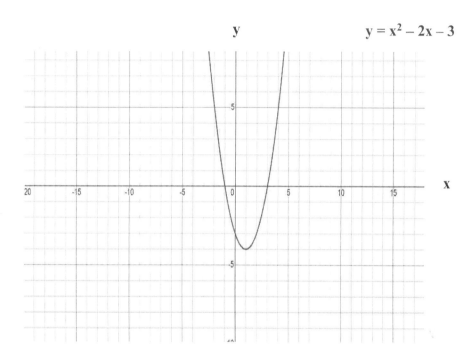

Fig. 11.4

The graph of the given parabola intersects the x-axis at two distinct points −1 and 3; thus, the solution set is: S.S = {−1, 3}

2) Use the graph of $y = x^2 - x - 2$ to solve $x^2 - x - 2 = 0$

Solution: - Make table of values for x and y.

$y = x^2 - x - 2$	x	−2	−1	0	1	2	3
	y	4	0	−2	−2	0	4

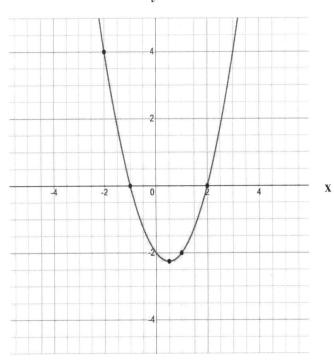

Fig. 11.5

The graph of the given parabola intersects the x-axis at two distinct points −1 and 2; thus, the solution set is: S.S = {−1, 2}.

3) Use the graph of $y = -x^2 + 1$ to solve $-x^2 + 1 = 0$

Solution: - Make table of values for x and y.

$y = -x^2 + 1$	x	–2	–1	0	1	2
	y	–3	0	1	0	–3

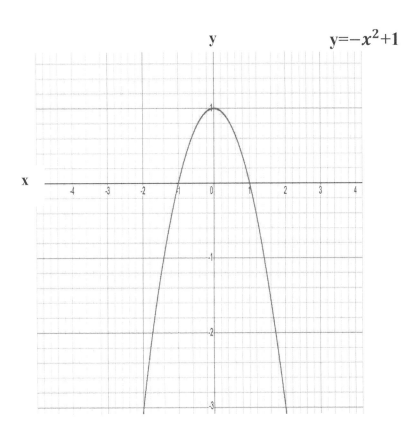

Fig. 11.6

The graph of the given parabola intersects the x-axis at two distinct points –1 and 1; thus, the solution set is: S.S = {–1, 1}.

4) Use the graph of $y = x^2$ to solve $x^2 = 0$

Solution: - Make table of values of x and y.

$y = x^2$	x	–2	–1	0	1	2
	y	4	1	0	1	4

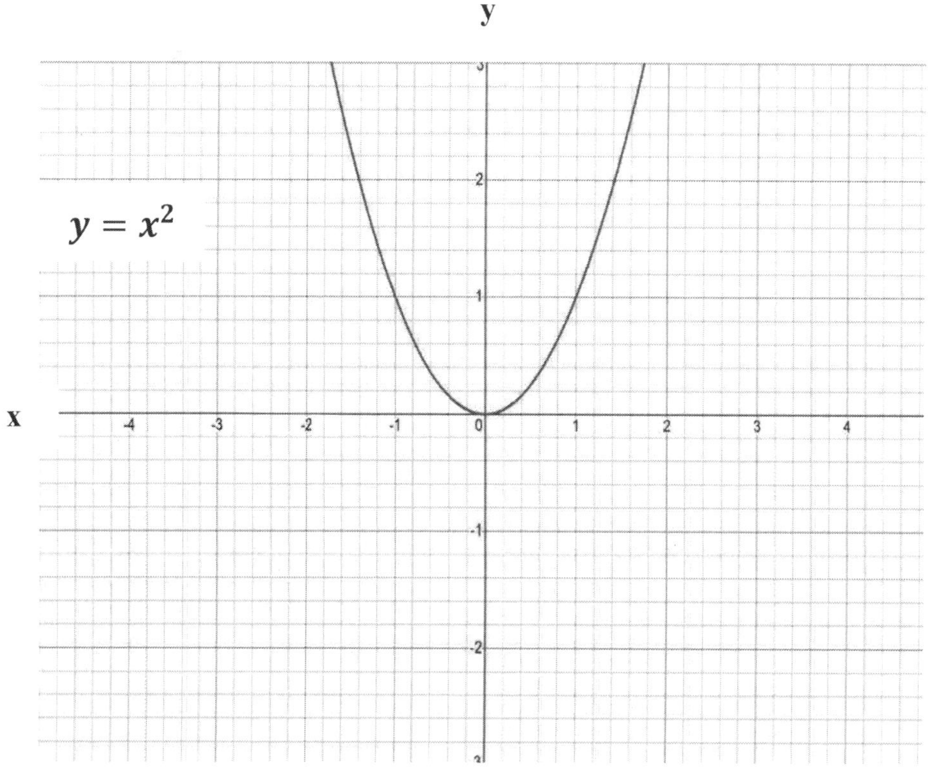

y

$y = x^2$

x

Fig. 11.7

The graph of the given parabola touches the x-axis at one point, 0. Thus, the solution set of the given equation is {0}.

5) Use the graph of $y = x^2 - 4x + 4$ to solve $x^2 - 4x + 4 = 0$

Solution: - Make a table of values for x and y.

$y = x^2 - 4x + 4$	x	−2	−1	0	1	2	3	4	5
	y	16	9	4	1	0	1	4	9

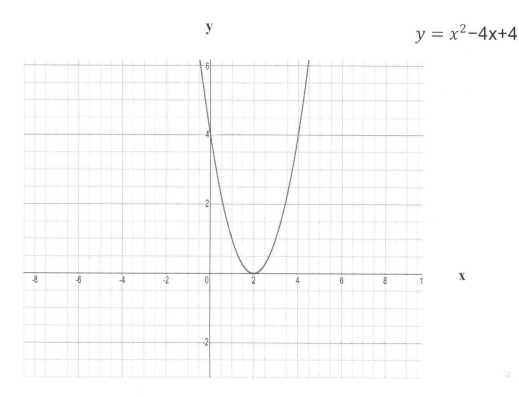

$$y = x^2 - 4x + 4$$

Fig. 11.8

The graph of the given parabola touches the x-axis at one point, 2. Thus, the solution set of the equation is {2}

6) Use the graph of $y = x^2 + 2$ to solve the equation: $x^2 + 2 = 0$

Solution: - Make table of values for x and y.

$y = x^2 + 2$	x	−2	−1	0	1	2
	y	6	3	2	3	6

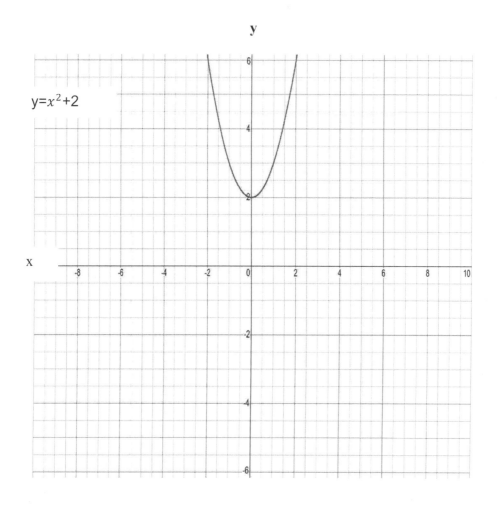

Fig. 11.9

The graph of the given parabola does not cross or touch the x-axis. Therefore, the equation does not have solution set. Meaning the solution set is empty set, **{   }.**

Exercise 11.11

Draw the graph of each of the following quadratic functions and find their solution set.

1. $y = x^2 + 3x + 2$
2. $y = x^2 - 2x + 1$
3. $y = x^2 + 3$
4. $y = x^2 - 2x - 3$
5. $y = -x^2 + 2x - 1$

11.12. Important Features of a Parabola

In the previous section, we learned that the way how to solve the quadratic equations using different methods. Now, let us discuss important features of the graphs of the quadratic function. We are familiar with a quadratic function. The graph of a quadratic functions are called parabolas. All parabolas are roughly, 'U' shaped. If a > 0, the parabola opens upward and if a < 0, the parabola opens downward. The highest or lowest point at which the graph turns is called the Vertex. The vertical line passing through the vertex that divides the parabola into two symmetric parts is called the axis of symmetry. The two symmetric parts are the mirror images of each other.

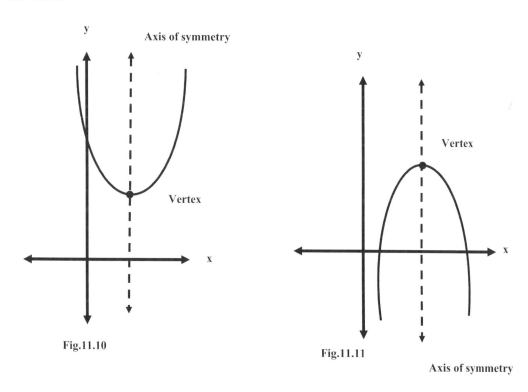

Fig.11.10

Fig.11.11

Note:- The graph is symmetrical about its axis.

The standard equation of a quadratic function: $y = ax^2 + bx + c$, where a, b and c are real numbers with $a \neq 0$ can also be rewritten in vertex form as:

$y = a (x - h)^2 + k$

In the above equation, the vertex of a parabola is the point (h, k). To find the formula that gives h and k, we are going to use the process of completing the square method.

For the function f(x) = ax^2 + bx + c, factoring out 'a' from the first two terms gives,

$$ax^2 + bx + c = a\left[x^2 + \frac{b}{a}x\right] + c$$

$$= a\left[\left(x + \frac{b}{2a}\right)^2 - \frac{b^2}{4a^2}\right] + c \quad\ldots\ldots\ldots\ldots\ldots\ldots\text{(Completing the square)}$$

$$= a\left(x + \frac{b}{2a}\right)^2 - \frac{b^2}{4a} + c$$

$$= a\left(x + \frac{b}{2a}\right)^2 + \frac{4ac - b^2}{4a}$$

$$== a\left(x - \left(-\frac{b}{2a}\right)\right)^2 + \frac{4ac - b^2}{4a}$$

$$\downarrow \qquad\qquad \downarrow$$

$$h \qquad\qquad k$$

Comparing with y = a(x − h)^2 + k, we get, h = $\frac{-b}{2a}$ and k = $\frac{4ac-b^2}{4a}$.

Thus, the coordinate of the vertex of the parabola y = ax^2 + bx + c is $\left[\frac{-b}{2a}, \frac{4ac-b^2}{4a}\right]$ and the axis of symmetry is x = $\frac{-b}{2a}$

Examples: -

1) Find the axis of symmetry of the graph y= x^2 + 6x + 8, using the formula.
    Solution: For a quadratic equation in standard form, y = ax^2 + bx + c, the axis of symmetry is a vertical line x = $\frac{-b}{2a}$, in y= x^2 + 6x + 8, Here, a=1, b=6 and c=8
    Substituting a and b in x = $\frac{-b}{2a}$ gives:-

$$x = \frac{-b}{2a}$$

$$x = \frac{-(6)}{2(1)}$$

$$x = \frac{-6}{2}$$

$$x = -3$$

Thus, the axis of symmetry is x= −3

2) Find the axis of symmetry and the vertex of the parabola y = 3x^2 − 9x + 6
    Solution: Compare y = 3x^2 − 9x + 6 with y = ax^2 + bx + c, we get a = 3, b = −9 and c = 6

Now substituting a, b and c in $\left(-\dfrac{b}{2a}, \dfrac{4ac - b^2}{4a}\right)$ we get:

$$\left(\dfrac{-(-9)}{2(3)}, \dfrac{4(3)(6) - (-9)^2}{4(3)}\right)$$

$$\left(\dfrac{9}{6}, \dfrac{72 - 81}{12}\right)$$

$$\left(\dfrac{9}{6}, \dfrac{-9}{12}\right)$$

$$\left(\dfrac{3}{2}, \dfrac{-3}{4}\right)$$

Thus, the vertex V of the parabola is $\left(\dfrac{3}{2}, \dfrac{-3}{4}\right)$ and the axis of symmetry is $x = \dfrac{3}{2}$, $\left(x = \dfrac{-b}{2a}\right)$

3) If $y = 3x^2 - 2x + 4$, then find the value of h and k, comparing with $y = a(x - h)^2 + k$.

Solution: $h = -\dfrac{b}{2a}$ and $\dfrac{4ac - b^2}{4a}$

In the quadratic equation $y = 3x^2 - 2x + 4$, a = 3, b = –2 and c = 4. Substitute:

$$h = -\dfrac{b}{2a}$$

$$h = \dfrac{-(-2)}{2(3)}$$

$$h = \dfrac{2}{6}$$

$$h = \dfrac{1}{3}$$

$$k = \dfrac{4ac - b^2}{4a}$$

$$k = \dfrac{4(3)(4) - (-2)^2}{4(3)}$$

$$k = \dfrac{48 - 4}{12}$$

$$k = \dfrac{44}{12}$$

$$k = \dfrac{11}{3}$$

Thus, $h = \dfrac{1}{3}$ and $k = \dfrac{11}{3}$

Note: - h is the same as $x = -\dfrac{b}{2a}$ and k is the same as $\dfrac{4ac - b^2}{4a}$

Remark: If once you get the value of $x = -\dfrac{b}{2a}$, you can find the value of $\dfrac{4ac - b^2}{4a}$ by direct substituting $\dfrac{-b}{2a}$ for x in the original equation $y = ax^2 + bx + c$.

Example: - Find the vertex 'V' of the parabola $y = 4x^2 - 8x + 3$.
Solution: - Given a = 4, b = –8 and c = 3.

$$x = -\frac{b}{2a}$$

$$= \frac{-(-8)}{2(4)}$$

$$x = \frac{8}{8}$$

$$x = 1$$

Thus, the vertex V occurs at f (1). Substituting x = 1 in $y = 4x^2 - 8x + 3$ gives,

$y = 4(1)^2 - 8(1) + 3$

$y = 4 - 8 + 3$

$y = -1$

Therefore, the vertex of the parabola is (1, –1).

4) Find the axis of symmetry and the vertex of the parabola for $f(x) = 5x^2 + 6x + 1$.
Solution: for $f(x) = 5x^2 + 6x + 1$, a = 5, b = 6 and c = 1.

Axis of symmetry: $x = -\dfrac{b}{2a}$

$$x = -\frac{(6)}{2(5)}$$

$$x = -\frac{6}{10}$$

$$x = -\frac{3}{5}$$

Substituting $x = -\dfrac{3}{5}$ in $y = 5x^2 + 6x + 1$ so that you can find $k = \dfrac{4ac - b^2}{4a}$

$$y = 5x^2 + 6x + 1$$

$$k = 5\left(\frac{-3}{5}\right)^2 + 6\left(\frac{-3}{5}\right) + 1$$

$$= 5\left(\frac{9}{25}\right) + \left(\frac{-18}{5}\right) + 1$$

$$= \frac{9}{5} - \frac{18}{5} + 1$$

$$= \frac{9 - 18 + 5}{5} \qquad \text{............Using LCM}$$

$$k = \frac{-4}{5}$$

Thus, the vertex of the parabola is $\left(\frac{-3}{5}, \frac{-4}{5}\right)$

Alternate method to find 'K'

$$k = \frac{4ac - b^2}{4a}$$

$$k = \frac{4(5)(1) - 6^2}{4(5)}$$

$$k = \frac{20 - 36}{20}$$

$$k = \frac{-16}{20}$$

$$k = \frac{-4}{5}$$

5) Find the axis of symmetry and the vertex for a parabola: $y = 2x^2 - 5x + 3$

Solution: a=2, b= −5 and c=3

The axis of symmetry: $x = -\dfrac{b}{2a}$

$$x = \frac{-(-5)}{2(2)}$$

$$x = \frac{5}{4}$$

The vertex V is $\left(\dfrac{-b}{2a}, \dfrac{4ac - b^2}{4a}\right)$

$$\left(\frac{5}{4}, \frac{4(2)(3) - (-5)^2}{4(2)}\right)$$

$$\left(\frac{5}{4}, \frac{24 - 25}{8}\right)$$

$$\left(\frac{5}{4}, \frac{-1}{8}\right)$$

The vertex is $\left(\dfrac{5}{4}, \dfrac{-1}{8}\right)$

Exercise 11.12

Find the coordinates of the vertex and the axis of symmetry for each of the following quadratic function.

1. $y = 2x^2 - 5x + 1$
2. $y = -3x^2 + 8x + 11$
3. $y = -2x^2 + 6x + 3$
4. $y = x^2 + 6x + 9$
5. $y = 3x^2 + 6x + 36$
6. $y = (x - 2)(x - 4)$

11.13. Maximum and Minimum Values of a Quadratic Function.

Recall that every quadratic function has a parabolic graph which is either open upward or downward depending on the leading coefficient, x^2. It has a maximum or a minimum value, but not both at the same time. It has a minimum value, if the leading coefficient is greater than zero and the maximum value if the leading coefficient is negative.

Note: - The leading coefficient in a quadratic equation is the coefficient of x^2, which is 'a'.

- There is no maximum value for a parabola which opens upward; instead, there is minimum value.
- There is no minimum value for a parabola which opens downward; rather, there is maximum value.
- The maximum and minimum values of a function are respectively the largest and smallest values that the function occurs at a given point. Together, they are known as the Extrema. These extreme values of the function occur at the vertex of the parabola.

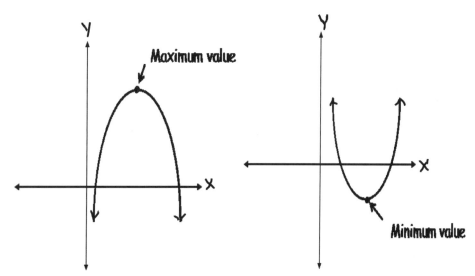

Note; The extreme value of f(x) is $\dfrac{4ac - b^2}{4a}$ and it occurs when $X = \dfrac{-b}{2a}$.

Examples:

Find the extreme values (maximum or minimum values) of the following quadratic functions.

1) $y = -2x^2 + 6x + 3$
2) $y = 3x^2 - 2x + 6$
3) $y = -x^2 - 2x + 3$
4) $y = 4x^2 + 8x - 9$
5) $y = x^2 - 6x + 9$

Solutions

1) The extreme value of f(x) is given by

$$\frac{4ac - b^2}{4a}$$

For $y = -2x^2 + 6x + 3$, $a = -2$, $b = 6$ and $c = 3$ since $a < 0$, it has a maximum value.

$$\frac{4(-2)(3) - (6)^2}{4(-2)}$$

$$\frac{-24 - 36}{-8}$$

$$\frac{-(24 + 36)}{-8}$$

$$\frac{60}{8}$$

$$\frac{15}{2}$$

Thus, the maximum value is $\frac{15}{2}$.

2) $y = 3x^2 - 2x + 6$

$a = 3$, $b = -2$ and $c = 6$

Since $a > 0$, it has minimum value.

$$\frac{4(3)(6) - (-2)^2}{4(3)}$$

$$\frac{72 - 4}{12}$$

$$\frac{68}{12}$$

$$\frac{17}{3}$$

Thus, the minimum value is $\dfrac{17}{3}$

3) $y = -x^2 - 2x + 3$

a = −1, b = −2 and c = 3

Since a < 0, it has maximum value.

$$\dfrac{4ac - b^2}{4a}$$

$$\dfrac{4(-1)(3) - (-2)^2}{4(-1)}$$

$$\dfrac{-12 - (4)}{-4}$$

$$\dfrac{-(12 + 4)}{-4}$$

$$\dfrac{16}{4}$$

4

Thus, the maximum value is 4.

4) $y = 4x^2 + 8x - 9$

a = 4, b = 8 and c = −9

Since a > 0, it has minimum value.

$$\dfrac{4ac \quad b}{4a}$$

$$\dfrac{4(4)(-9) - (8)}{4(4)}$$

$$\dfrac{-144 - 64}{16}$$

$$\dfrac{208}{16}$$

13

Thus, the minimum value is −13

5) $y = x^2 - 6x + 9$

a = 1, b = −6 and c = 9

Since a > 0, it has minimum value.

$$\frac{4ac - b^2}{4a}$$

$$\frac{4(1)(9) - (-6)^2}{4(1)}$$

$$\frac{36 - 36}{4}$$

$$\frac{0}{4}$$

Thus, the minimum value is 0.

Exercise 11.13

Find the extreme (maximum or minimum) value of each of the following graphs of quadratic functions.

1) $f(x) = -3x^2 + 2x + 1$
2) $y = 2x^2 - 4x + 8$
3) $f(x) = -x^2 - x - 1$
4) $y = 5x^2 - 2x + 1$
5) $y = (x - 3)^2$
6) $y = 4x^2 - 3x - 6$
7) $y = 3x^2$
8) $y = -2x^2 - 4x + 8$
9) $y = 2x - 7x^2 + 4$
10) $f(x) = 7x^2 - 8x + 6$

11.14. Application of Quadratic Functions

A quadratic function is applicable in many real-life situations that involves in geometry like the ways of maximizing areas of a fenced enclosure, in area when both dimensions are expressed in terms of some variables, in projectile motion problems and in shooting cannons. In many of these situations, the vertex of the parabola is used to find the maximum or minimum values, and roots to find solution sets of the given quadratic functions.

Examples:

1) If one leg of a right-angled triangle is one unit longer than the other leg, and the hypotenuse is 2 units longer than the smaller leg. Find the dimensions of the right-angled triangle.

Solution: Let x be the length of the smaller leg, (x + 1) be the leg one unit longer than the smaller leg and (x + 2) be the length of the hypotenuse.

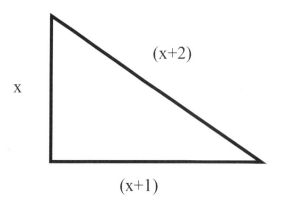

Fig. 11.14

You are familiar with Pythagoras formula $a^2 + b^2 = c^2$, thus, a = x, b = (x + 1) and c = (x + 2)

$x^2 + (x + 1)^2 = (x + 2)^2$

$x^2 + x^2 + 2x + 1 = x^2 + 4x + 4$

$2x^2 + 2x + 1 = x^2 + 4x + 4$

$2x^2 - x^2 + 2x - 4x + 1 - 4 = 0$ ....Collecting like terms

$x^2 - 2x - 3 = 0$

$x^2 + x - 3x - 3 = 0$...(Find two numbers whose sum is $-2x$ and whose product is $-3x^2$)

$x (x + 1) - 3(x + 1) = 0$

$(x - 3) (x + 1) = 0$

$x - 3 = 0$ or $x + 1 = 0$

$x = 3$ or $x = -1$

$x = -1$ cannot be the answer because no negative dimension exists for measuring length. Thus, the smaller leg is 3 units, the other leg is 4 and the hypotenuse is 5 units. i.e., x = 3 units, x + 1 = 3 + 1 = 4 units and the hypotenuse = x + 2 = 3 + 2 = 5 units.

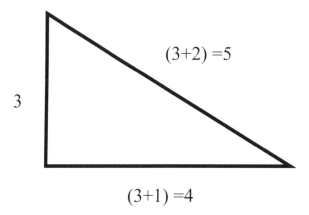

Fig. 11.15

2) As shown in the figure below a farmer with 400 ft of fence wants to enclose a rectangular field next to a river. He determined not to use a fence along the river. Suppose he uses x ft fence on the side perpendicular to the river bank.

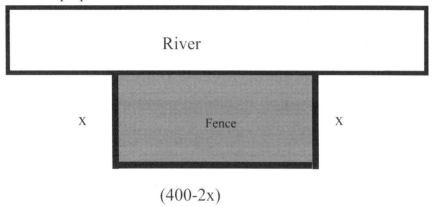

(400-2x)

Fig. 11.16

a) Find x that will give the maximum enclosed area.
b) What is the maximum enclosed area?

Solutions: - a) In the figure above, the area (A) of the rectangular field is a function of the variable x and it is given that the length of the fence is 400 ft so that the length of the side parallel to the river is (400 – 2x). Thus, the area (A) of the rectangular field is given by (A = LW).

$\quad$ A(x) = x • (400 – 2x)
$\quad$ A(x) = 400x – 2x²
$\quad$ A(x) = –2x² + 400x, which is a quadratic function.

The quadratic function A(x) = –2x² + 400x has a = –2, b = 400 and c = 0; Hence, its maximum value occurs at $x = \dfrac{-b}{2a}$

$$x = \dfrac{-b}{2a} = \dfrac{-400}{2(-2)} = \dfrac{-400}{-4}$$

$\quad$ x = 100 ft

Thus, the other side is 400 – 2x
$\quad$ 400 – 2(100)
$\quad$ 400 – 200
$\quad$ 200
$\quad$ 400 – 2x = 400 – 2(100)= 200ft

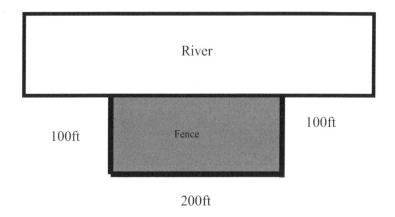

River

100ft

100ft

Fence

200ft

Fig. 11.17

Therefore, to get the maximum enclosed area, the farmer should enclose a rectangular field of dimensions having a width of 100 ft and a length of 200 ft.

c) The maximum enclosed area of a rectangular field is given by:

A = LW

A = x (400 – 2x)

A = 100 (400 – 2x100)

= (100) (200)

A = 20,000 ft²

Note: - The maximum enclosed area (A) can also be found by: $\dfrac{4ac - b^2}{4a}$

Thus, maximum area enclosed $= \dfrac{4ac - b^2}{4a}$

$= \dfrac{4(-2)(0) - (400)^2}{4(-2)}$

$= \dfrac{0 - (160,000)}{-8}$

Maximum area enclosed = 20,000 ft²

3) Sofia wants to clean a rectangular rug, which has an area of 400ft² and perimeter 82ft. What are the dimensions of the rug?

Solution: Let 'L' be the length and 'W' be the width of a rug.

Given the area, A=400ft² and perimeter, P=82ft

Area of a rectangle= LW

Perimeter of a rectangle= 2L+2W

W

L

Fig. 11.18

A=LW

LW=400

P=2L+2W

2L+2W=82

$\begin{cases} LW = 400 \\ 2L + 2W = 82 \end{cases}$ ....combine the two equations and solve.

$\begin{cases} LW = 400 \\ L + W = 41 \end{cases}$ ....Dividing both sides of the second equation by 2.

L+W=41

L=41-W ....Taking W to the right side

Substitute L=41-W in the first equation, LW=40

(41-W) W=400

41W-W^2=400

W^2-41W+400=0 ....Bring the left side of the equation to the right side.

W^2-16W-25W+400=0

W(W-16)-25(W-16) =0 ....Taking common factors

(W-25) (W-16) =0

W-25=0 or W-16=0

W=25 or W=16

To find the value of 'L', substitute the value of 'W' in either of the above equations.

Thus, when W=25

LW=400

25L= 400

L=$\frac{400}{25}$

L=16

(L, W) = (16,25)

When W=16

16L=400

$L=\dfrac{400}{16}$

L=25

(L, W) = (25,16)

Therefore, the dimensions of the rectangular rug are L=16ft and W=25ft or L=25ft and W=16ft.

4) The equation $h(t) = -16t^2 + 32t + 80$ provides the height h above the ground in feet for an object thrown at $t = 0$, straight upward from the top of 80 feet building.
   a) What is the maximum height reached by the object?
   b) How long will it take the object to reach its maximum point?
   c) When does the object hit the ground?

   Solutions
   a) The height h given above is a quadratic function. Its graph as a function of time, t provides a parabolic shape opened downward and its maximum height h occurs at the vertex of a parabola. For a quadratic function of the form:

   $h = at^2 + bt + c$, the vertex is located at $t = -\dfrac{b}{2a}$. For $h(t) = -16t^2 + 32t + 80$, where

   $a = -16$, $b = 32$ and $c = 80$, this maximum point (the vertex) occurs at:
   $$t = \dfrac{-(32)}{2(-16)} = \dfrac{-32}{-32} = 1 second$$

   Thus, the height (maximum value of h) is given by:
   $h(t) = -16t^2 + 32t + 80$

   $h = -16(1)^2 + 32(1) + 80$
   $h = -16 + 32 + 80$
   $h = 96$ feet
   Thus, the maximum height reached by an object is 96 feet.

   b) The object reached its maximum point at: $t = -\dfrac{b}{2a}$.

   $$t = \dfrac{-32}{2(-16)} = \dfrac{-32}{-32} = 1 second$$

   Thus, the object reached its maximum height at t=1second.

   c) At the ground, when h=0, its solution provides the time 't' at which the object hits the ground.

   $h(t) = -16t^2 + 32t + 80$

$-16t^2 + 32t + 80 = 0$

$-t^2 + 2t + 5 = 0$...Dividing both sides by 16.

$$t = \frac{-b \pm \sqrt{b^2 - 4ac}}{2a}$$

$$t = \frac{-2 \pm \sqrt{2^2 - 4(-1)(5)}}{2(-1)}$$

$$= \frac{-2 \pm \sqrt{4 + 20}}{-2}$$

$$= \frac{-2 \pm \sqrt{24}}{-2}$$

$$= \frac{-2 + \sqrt{24}}{-2} \text{ or } \frac{-2 - \sqrt{24}}{-2}$$

$$= \frac{-2 + \sqrt{4 \times 6}}{-2} \text{ or } \frac{-2 - \sqrt{4 \times 6}}{-2}$$

$$= \frac{-2 + 2\sqrt{6}}{-2} \text{ or } \frac{-2 - 2\sqrt{6}}{-2}$$

$$= \frac{-2(1 + \sqrt{6})}{-2} \text{ or } \frac{-2(1 - \sqrt{6})}{-2}$$

$1 + \sqrt{6}$ or $1 - \sqrt{6}$ ....-2 is cancelled out.

$1 + 2.45$ or $1 - 2.45$

t = 3.45 sec or t = −1.45 sec

The correct answer is t = 3.45 sec. Thus, the object hits the ground at 3.45 seconds.

Note: - The maximum point reached by an object can also be found by $\dfrac{4ac - b^2}{4a}$.

5) A bullet is fired upwards with an initial velocity of 1000 ft/sec from a point 10 ft above the ground, and its height above the ground at time t is given by $h(t) = -16t^2 + 1000t + 10$.
   a) What is the maximum height of the bullet?
   b) How long does it take for the bullet to reach the highest point?

Solutions: The figure below shows the height h(t) of the bullet at any time t. This height is given by a quadratic equation: $h(t) = -16t^2 + 1000t + 10$

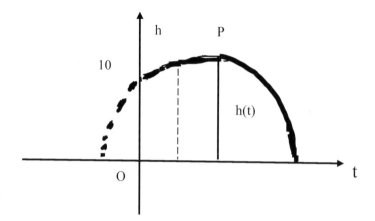

Fig. 11.19

a) The coordinate of the highest point (vertex) P is given by

$\dfrac{4ac - b^2}{4a}$, where, a = −16,

b = 1000

c = 10

$$\dfrac{4(-16)(10) - (1000)^2}{4(-16)}$$

$$= \dfrac{-640 - 1000000}{-64}$$

$$= \dfrac{-1,000,640}{-64}$$

$$= 15,635 \text{ ft.}$$

Thus, the bullet reaches a maximum height of 15,635 ft.

(In the figure above, the bullet reaches its maximum height at a point 'p')

b) The bullet reaches its maximum height at, $t = \dfrac{-b}{2a}$

$$t = \dfrac{-b}{2a} = \dfrac{-1000}{2(-16)} = \dfrac{-1000}{-32} = 31.25 \text{ seconds}$$

6) An object is launched directly upward at 39.2 m/s from 245-meter-tall building. The equation for the object's height (h) at time t seconds after launch is: h(t) = −4.9t^2 + 39.5t + 245, where h is in meter. At what time does the object reach the ground?

Solution: - When the object reaches on the ground, its height i.e h(t) is zero.

h(t) = −4.9t^2 + 39.5t + 245

−4.9t^2 + 39.5t + 245 = 0

where a = −4.9, b = 39.5 and c = 245

$$t = \frac{-b \pm \sqrt{b^2 - 4ac}}{2a}$$

$$t = \frac{-39.5 \pm \sqrt{(39.5)^2 - 4(-4.9)(245)}}{2(-4.9)}$$

$$t = \frac{-39.5 \pm \sqrt{1,560.25 + 4,802}}{-9.8}$$

$$t = \frac{-39.5 \pm 79.8}{-9.8}$$

$$t = \frac{-39.5 + 79.8}{-9.8} \quad \text{or} \quad t = \frac{-39.5 - 79.8}{-9.8}$$

t= −4.11 seconds or t=12.174 seconds

Thus, the object strikes the ground at 12.17 seconds.

Note: - The negative time $t = -4.11$ seconds is invalid.

Exercise 11.14

Answer each of the following questions.

1) An object is dropped from the top of a tower, which is 195m high.
   a) How long does it take for the object to reach the ground?
   b) How far will an object fall in 3 sec if it is thrown downward from the tower with an initial velocity of 16 m/s?

   (Hint: $h = \frac{1}{2} gt^2 + v_0 t = 4.9t^2 + v_0 t$)

2) A bullet is fired upward with an initial velocity of 100 ft/sec from a point 15 ft above the ground and its height above the ground at a time t is given by $h(t) = -16t^2 + 100t + 15$.
   a) What is the maximum height of the bullet?
   b) How long does it take for the object to reach the highest point?

3) A bullet is fired into air from the top of a tower. Its height (h) above the ground in feet is given by: $h(t) = -16t^2 + 10t + 20$
   a) What was the initial height of the bullet?
   b) When did the bullet reach its maximum height?

4) A farmer with 600 meter of fence wants to enclose a rectangular field next to a river. He determined not to use a fence along the river. Assume that he uses x meter on the side perpendicular to the river bank. (See the fig. below).

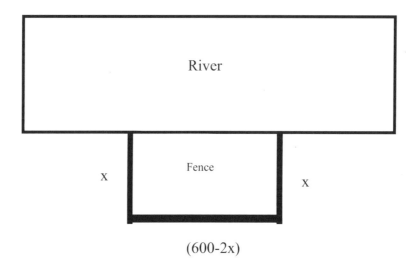

(600-2x)

Fig. 11.20

a) Find x that will give the maximum enclosed area.
b) What is the maximum enclosed area?

5) A farmer wants to fence in his garden in two equal rectangular sections. If he has 72 meter of fence and the area of the entire garden is 120 m². Find the dimension of the garden.

11.15. Quadratic Inequality

In a quadratic equation written in the form of $ax^2 + bx + c = 0$, where a, b and c are real numbers and $a \neq 0$, if the equal sign (=) is replaced by <, >, ≤ or ≥, then such type of equation is called Quadratic Inequality. Quadratic inequalities can be written in the form of:

$$ax^2 + bx + c < 0$$
$$ax^2 + bx + c \leq 0$$
$$ax^2 + bx + c > 0$$
$$ax^2 + bx + c \geq 0$$

There are different methods to solve quadratic inequalities. These are algebraically using sign chart method and graphically using graphs.

11.16. Steps to Solve Quadratic Inequality Using Sign Chart Method.

1) Set up quadratic inequality in the form of $ax^2 + bx + c < 0$, $ax^2 + bx + c \leq 0$, $ax^2 + bx + c > 0$, or $ax^2 + bx + c \geq 0$.

2) Factorize the quadratic inequality, if necessary.
3) Determine the value of x. (Determine the value of the given variable.)
   Note: - In this case the values of x are called the critical values.
4) Create sign chart by dividing into boundaries using critical values.
5) Complete the sign chart so as to determine which factor of the inequality is positive, negative, or zero on the number line.
   Note: - The left side from the reference point (zero) on the number line is negative and the right side from zero is positive. Thus,

On the sign chart, our reference points are the critical values and these number line rules hold true while completing the sign chart.

6) Write your final answer in inequality sign, on the number line, or in interval notation form.

Examples: -

1) Solve for x: $x^2 - 4x + 4 > 0$
   Solution
   Step 1: $x^2 - 4x + 4 = 0$
   Step 2: $x^2 - 4x + 4 = 0$
   $$x^2 - 2x - 2x + 4 = 0$$
   $$x (x - 2) - 2(x - 2) = 0$$
   $$(x - 2) (x - 2) = 0$$
   Step 3: $(x - 2) (x - 2) = 0$
   $$x - 2 = 0 \text{ or } x - 2 = 0$$
   $$x = 2 \text{ or } x = 2$$
Thus, critical value is 2.

Step 4:

	A	2	B
(x − 2)	−		+
(x − 2)	−		+
(x − 2) (x − 2)	+		+

Take any number to the left side of 2, boundary A, and substitute for x to determine whether (x − 2) is positive or negative. In this case, the number you substitute for x is known as the test number.

i.e.,

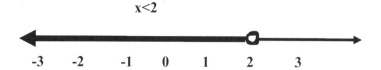

Now let us take 0 and substitute for x.

x = 0

(x – 2) =(0 – 2) = –2

Since the result is negative number, any number to the left side of the number 2 substituted for x gives negative numbers, so that we write negative sign (–) in the table.

•   Take any number to the right side of 2 and substitute in to (x – 2). In this case, let us take 3.

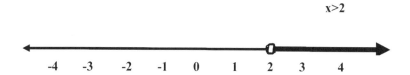

x = 3

(x – 2) = (3 – 2) = 1

The result is positive number. Generally, if we take any test number on the right side of 2, the result is positive number. Therefore, we write positive sign (+) in the table.

In this case the numbers 0 and 3 are called test numbers.

For (x – 2) (x – 2) = (–) (–) = (+). So, we write positive sign in the third column to the left of 2 for (x – 2) (x – 2).

(x – 2) (x – 2) = (+) (+) = (+), so we write positive sign in the third column to the right side of 2.

Note:- In the inequality (x – 2) (x – 2) > 0, the sign is not ≥; therefore, 2 is not a member of the solution. For this reason, we mark open circle (o) at 2 on the sign chart to show that 2 is exclusive.

Step 5)   See the completed table above on step 4.

Step 6)   As you see, the final result of the product (x – 2) (x – 2) to the left side of A is positive showing that all numbers to the left side of 2 are members of the solution set. Besides, the final result of (x – 2) (x – 2) to the right side of 2 is positive; showing that all numbers to the right side of 2 are members of the solution set.

To sum up, the solution set is all real numbers less than or greater than 2. We can write this solution set in different ways.

In set builder notation,

S.S = {x: x < 2 or x > 2}

In number line form,

Interval notation,

S.S = $(-\infty, 2) \cup (2, \infty)$

Remark: While solving inequality, it is not necessary to write all these three forms of expressing the solution set unless you are told to do so by your instructor. You can show your answer with any one of these three methods.

2) Solve the inequality: $x^2 - 5x + 6 \geq 0$

Solution:

Write $x^2 - 5x + 6 \geq 0$ in the form of $x^2 - 5x + 6 = 0$, then factorize: $x^2 - 5x + 6 = 0$.

$x^2 - 3x - 2x + 6 = 0$.................................factorizing

$x(x - 3) - 2(x - 3) = 0$

$(x - 2)(x - 3) = 0$

Find the solution sets (critical values).

$(x - 2)(x - 3) = 0$

$x - 2 = 0$ or $x - 3 = 0$

$x = 2$ or $x = 3$

Thus, the critical values are 2 and 3.

Create a sign chart.

	A	2	B	3	C
$(x - 2)$	$-$		$+$		$+$
$(x - 3)$	$-$		$-$		$+$
$(x - 2)(x - 3)$	$+$		$-$		$+$

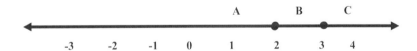

Boundary A: $x < 2$,

Take any number less than 2, say 1 and test for $(x - 2)$.

$= (1 - 2) = -1$ Negative

For $(x - 3)$,

$= (1 - 3) = -2$ Negative

Their product $(x - 2)(x - 3) = (-)(-) = (+)$ Positive

Boundary B: $2 < x < 3$

Take any number between 2 and 3, say 2.5 and test for $(x - 2)$

$= (2.5 - 2) = 0.5$ Positive

For $(x - 3)$,

$= (2.5 – 3) = –0.5$ Negative
Their product $(x – 2)(x – 3) = (+)(–) = (–)$ Negative

Boundary C: $x > 3$
Take 4 and test
For $(x – 2)$,
$= (4 – 2) = 2$ Positive
For $(x – 3)$,
$= (4 – 3) = 1$ Positive
Their product $(x – 2)(x – 3) = (+)(+) = (+)$ Positive

Since the inequality is $\geq$, the critical values 2 and 3 are members of the solution set.
As you have seen in the third row on the sign chart,
$(x – 2)(x – 3)$:   Boundary 'A' Positive✓
                    Boundary 'B' Negative ✗
                    Boundary 'C' Positive✓
Since, the inequality is ≥ 0, we take only positive boundaries.
Therefore, the solution set of $x^2 – 5x + 6 \geq 0$ is: $\{x: x \leq 2 \text{ or } x \geq 3\}$

3) Solve the inequality: $x^2 – 3x – 10 < 0$
    $x^2 – 3x – 10 = 0$
    $x^2 – 5x + 2x – 10 = 0$
    $x(x – 5) + 2(x – 5) = 0$
    $(x + 2)(x – 5) = 0$
    $x + 2 = 0 \text{ or } x – 5 = 0$
    $x = –2 \text{ or } x = 5$
    The critical values are –2 and 5. Creating and filling the sign chart below in similar ways as we done before results in:

	–2	5	
$(x + 2)$	–	+	+
$(x – 5)$	–	–	+
$(x + 2)(x – 5)$	+	–	+

    Observe that in the third row the negative numbers are found between –2 and 5, that is $–2 < x < 5$, thus, the solution set is: $\{x: –2 < x < 5\}$

4) Solve the inequality: $x^2 + 9 < 0$
    Solution:
    $x^2 + 9$ is always positive. That is, it is always greater than zero. So, $x^2 + 9 < 0$ has no solution set.
    The solution set is $\{\ \ \}$.

5) Solve for x: $x^2 + 8 > 0$.

Solution

$x^2 + 8$ is non-perfect square and it is not factorized. If you substitute any number for x, the equation $x^2 + 8$ is always positive number. Thus, the solution set is all real numbers.

S.S = {x: x ∈ |R}.

6) Solve the inequality: $x^2 - 7x + 12 > 0$

Solution:

$x^2 - 7x + 12 = 0$

$x^2 - 3x - 4x + 12 = 0$

$x(x - 3) - 4(x - 3) = 0$

$(x - 4)(x - 3) = 0$

$x - 4 = 0$ or $x - 3 = 0$

$x = 4$ or $x = 3$

		3		4	
(x − 3)	−		+		+
(x − 4)	−		−		+
(x − 3)(x − 4)	+		−		+

Observe that for x < 3, the third row is positive and for x > 4, the third row is also positive. Since the given inequality is greater than, the solution set is {x: x < 3 or x > 4}

Exercise 11.15-11.16

Solve for x and write your answer in set builder notation, number line form and in interval notation form.

1) $x^2 - 9 > 0$
2) $x^2 + 7x + 12 \geq 0$
3) $x^2 - x < 0$
4) $x^2 - 25 \leq 0$
5) $x^2 + 11x + 20 > -8$
6) $x^2 + 2 < 0$
7) $-x^2 + 49 \leq 0$
8) $(x - 3)^2 \geq 0$
9) $x^2 - 16x \leq 0$
10) $x^2 - 8x + 16 > 0$

11.17. Finding Solution Set of Quadratic Inequalities Using Graphs

As I mentioned before, the quadratic inequalities are written in one of the following form:

$$ax^2 + bx + c < 0 \qquad\qquad ax^2 + bx + c \leq 0$$
$$ax^2 + bx + c > 0 \qquad\qquad ax^2 + bx + c \geq 0$$

Such quadratic functions have parabolic graphs open either upward when $a > 0$ and downward when $a < 0$.

The graph of a quadratic inequality consists of the graph of all ordered pairs (x, y) that are solutions of the inequality. To graph a quadratic inequality, first draw the parabola and shade in the region either above or below it based on the given inequality. If the inequality symbol is strict inequality ($<$ or $>$), then the parabola should be drawn with broken line to show that the region does not include its boundary. While, if the inequality symbol is non-strict inequality ($\leq$ or $\geq$), then the region includes the boundary line of the parabola; so, it should be drawn by a solid line.

Examples

1) Graph the quadratic inequality: $y \geq x^2 - 4x + 4$

Solution

Write the equation in the form of $x^2 - 4x + 4 = 0$ and find x and y-intercepts.

$$x^2 - 4x + 4 = 0$$
$$x^2 - 2x - 2x + 4 = 0$$
$$x(x - 2) - 2(x - 2) = 0$$
$$(x - 2)(x - 2) = 0$$
$$x - 2 = 0 \text{ or } x - 2 = 0$$
$$x = 2 \text{ or } x = 2$$

Thus, the x-intercept is $(2, 0)$.

The y-intercept is where $x = 0$

$$y = x^2 - 4x + 4$$
$$y = 0^2 - 4(0) + 4$$
$$y = 0 - 0 + 4$$
$$y = 4$$

Thus, the y-intercept is $(0, 4)$.

The parabola has x-intercept at 2 and y-intercept at 4.

To find the vertex of the given parabola plug in the x-value: $x = 2$ in $y = x^2 - 4x + 4$ and we get: -

$$y = 2^2 - 4(2) + 4$$
$$y = 4 - 8 + 4$$
$$y = 0$$

Thus, the vertex is at $(2, 0)$.

Now we have enough information to draw the graph. Since the given inequality is $\geq$, we need to draw with a solid line.

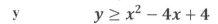

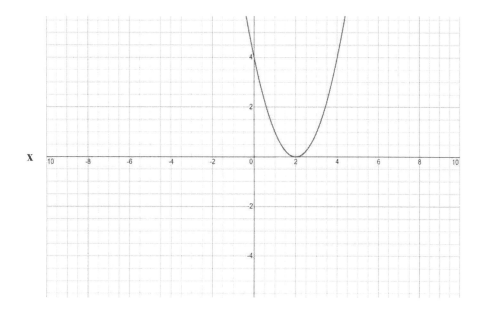

Fig. 11.21

To determine the solution set of the given inequality, that is, to determine the region to be shaded, take any ordered pair numbers and test it whether it is true or not. The easiest test number is taking an ordered pair (0, 0). Plug it in $y \geq x^2 - 4x + 4$.

$y \geq x^2 - 4x + 4$

$0 \geq 0^2 - 4(0) + 4$

$0 \geq 0 - 0 + 4$

$0 \geq 4$ is false assumption. Since a point (0, 0) or the origin is outside the parabola, we need to shade the part that does not include this point. This means, we have to shade the inside region of the parabola.

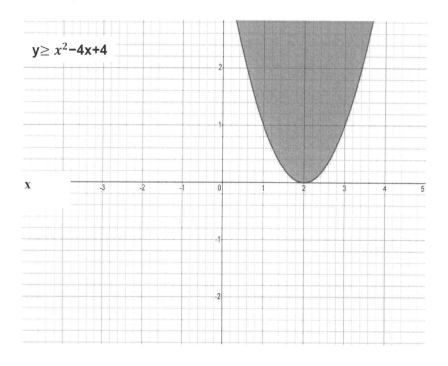

Fig. 11.22

Note: - The shaded region means, the solution set of the inequality.

Thus, the solution set of $y \geq x^2 - 4x + 4$ is the inside region of the parabola, including the solid line of the graph.

2) Sketch the graph of $y > x^2$.

Solution

Write the equation in the form of $y = x^2$ and find the x and y intercepts.

Both the x and y intercepts are 0.

To find the vertex of the parabola, plug in the value $x = 0$ in $y = x^2$

$\quad y = x^2$

$\quad y = 0^2$

$\quad y = 0$

Thus, the vertex is at $(0, 0)$.

Now, we have enough information to draw the graph. Since the given inequality is $>$, we have to draw with a broken line.

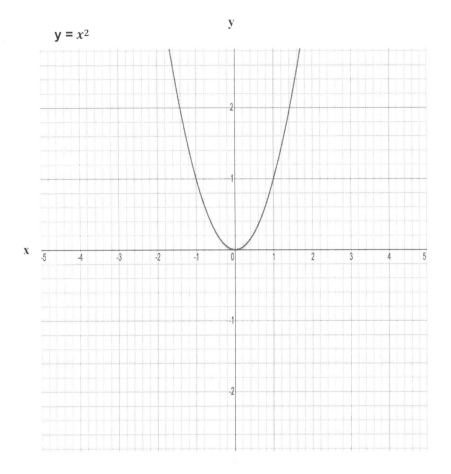

$y = x^2$

Fig. 11.23

To determine the solution set of the given inequality, that is, to determine the region to be shaded, take any ordered pair numbers and test it whether it is true or not, the easiest testing ordered pair number is to take (0, 0). Plug it in $y > x^2$.

$y > x^2$

$0 > 0^2$

$0 > 0$, this is false, since the given inequality is $>$ the point (0,0), the origin, is not in the solution set of the inequality. So, the graph of the parabola must be drawn by broken line and should be shaded its inside region(portion) as its answer.

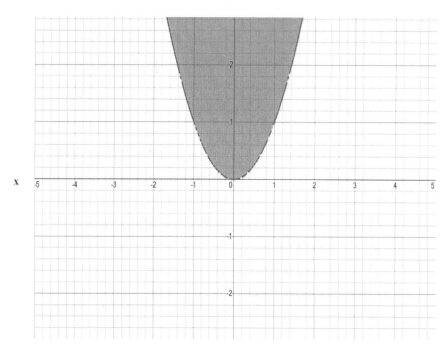

Fig. 11.24

Note: The line of the graph is not a member of the solution set. Because of this, we have used a broken line to draw the graph. Moreover, the origin (0, 0), is not the member of the solution set, too.

3) Sketch the graph of $y \leq x^2$

Solution:

Write the equation in the form of $y = x^2$ and find the x and y-intercepts. Fortunately, both the x and y-intercepts are 0.

To find the vertex of the parabola, plug in the x-value, x = 0 in $y = x^2$.

$y = x^2$

$y = 0^2$

$y = 0$

Therefore, the vertex is (0, 0).

Thus, we have enough information to draw the graph of the inequality. Since the given inequality is $\leq$, we have to draw the graph with a solid line.

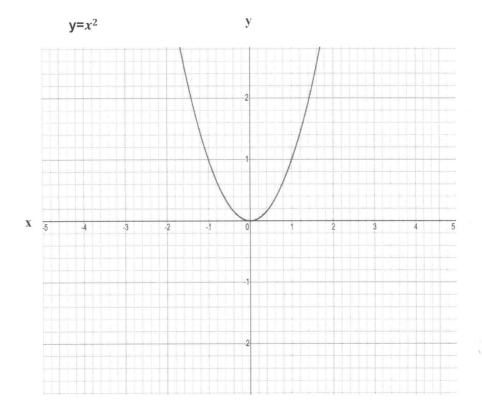

Fig. 11.25

To determine the solution set of the given inequality, that is; to determine the region which is supposed to be shaded, take any ordered pair numbers and test it whether it is in the solution set or not. The easiest testing ordered pair numbers is taking a point $(0, 0)$. Plug this value in $y \le x^2$.

$y \le x^2$

$0 \le 0^2$

$0 \le 0$; this is true. But it is not telling exactly the direction where to be shaded. Instead, it is better to take another pint outside the graph. For example, take $(-2, -2)$. Substituting this value in x and y gives $-2 < (-2)^2 = -2 \le 4$. This is fact. So, the point $(-2, -2)$ is definitely an element of the solution set. Since the given inequality is $\le$. the graph must be drawn with a solid line and shade the outside region of the parabola.

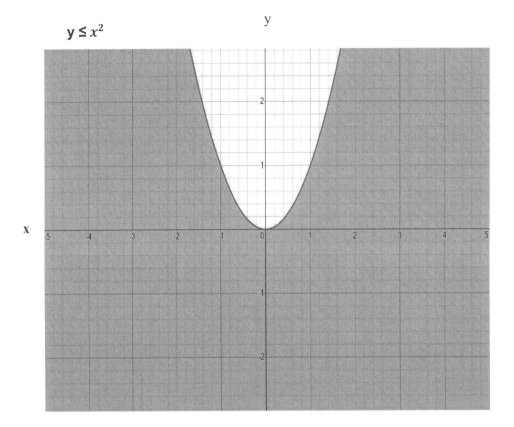

y ≤ x²

Fig. 11.26

4) Graph the quadratic inequality: $y \le x^2 - 2x - 3$

Solution: - First let us rewrite the given inequality in the form of: $x^2 - 2x - 3 = 0$, factorize and find the x and y intercepts of the inequality.

$$x^2 - 2x - 3 = 0$$
$$x^2 + x - 3x - 3 = 0$$
$$x(x + 1) - 3(x + 1) = 0$$
$$(x - 3)(x + 1) = 0$$
$$x - 3 = 0 \text{ or } x + 1 = 0$$
$$x = 3 \text{ or } x = -1$$

So, the parabola has x-intercepts at –1 and 3. The vertex of a parabola must lie midway between these points. The x-coordinate of the vertex is found at $\mathbf{x = \dfrac{-b}{2a}}$

$x^2 - 2x - 3 = 0$ where a = 1, b = -2 and c = -3.

Substituting this value in $x = \dfrac{-b}{2a}$ results in: -

$$x = \frac{-(-2)}{2(1)}$$

$$x = \frac{2}{2}$$

x = 1

The y-coordinate of the vertex is at f (1):

$y = x^2 - 2x - 3$

$y = 1^2 - 2(1) - 3$

$y = 1 - 2 - 3$

$y = -4$

Thus, the vertex of the parabola is at the point (1, –4).

Note: - To find the y-coordinate of the vertex of the parabola, we can also use $\dfrac{4ac - b^2}{4a}$.

Now we have enough information to draw the parabola. Make sure to graph it with a solid line, since the given inequality is "less than or equal to". That is, "≤".

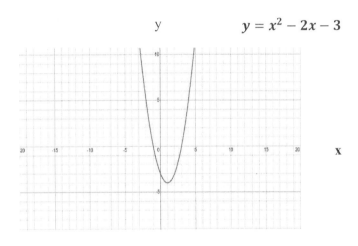

y $y = x^2 - 2x - 3$

Fig. 11.27

To determine the solution set of the given quadratic inequality, that is, the region to be shaded, take any ordered pair numbers and test it whether it satisfies or not. Thus, the easiest number for testing is the origin, (0, 0) and plug it in $y \le x^2 - 2x - 3$.

$0 \le 0^2 - 2(0) - 3$

$0 \le 0 - 0 - 3$

$0 \le -3$, this is not true, showing that the origin, (0, 0) is not a member of the solution set of the quadratic inequality. Thus, we have to shade the region which does not include the point (0, 0).

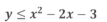

$$y \leq x^2 - 2x - 3$$

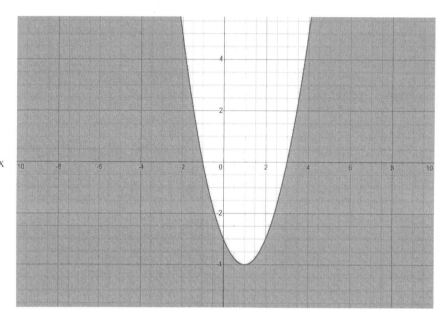

Fig. 11.28

5) Draw the graph of $y < -x^2 - 2x + 3$

Solution: Write the inequality as $-x^2 - 2x + 3 = 0$

Factorize it and find the x-intercepts.

$-x^2 - 2x + 3 = 0$

$-(x^2 + 2x - 3) = 0$

$-(x^2 + 3x - x - 3) = 0$

$-[x(x + 3) - 1(x + 3)] = 0$

$-(x - 1)(x + 3) = 0$

Thus, the parabola has x-intercepts at -3 and 1.

To find y-intercept, plug in $x = 0$ in $y = -x^2 - 2x + 3$

$y = -(0)^2 - 2(0) + 3$

$y = 3$

The parabola has y-intercept at 3.

Vertex of the parabola: -

$$x = \frac{-b}{2a} = \frac{-(-2)}{2(-1)} = \frac{-2}{2} = -1$$

$y = f(-1) = -(-1)^2 - 2(-1) + 3$, substitute -1 in $y = -x^2 - 2x + 3$

$= -1 + 2 + 3$

$= -1 + 5$

$y = 4$

Thus, the turning point of the parabola, i.e., the vertex is $(-1, 4)$.

Now, we have enough information to draw the graph of a parabola. Remember to use a broken line for drawing the graph as the given inequality is not inclusive. That is, "<".

Note: Since the leading coefficient is negative, the parabola is open downward.

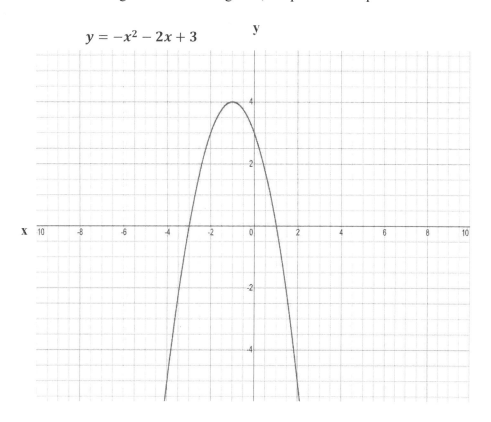

Fig. 11.29

To determine the solution set of the given quadratic inequality, that is, the region to be shaded, take any ordered pair numbers and test it whether it satisfies or not.

The easiest number for testing is plugging the point $(0, 0)$ in $y < -x^2 - 2x + 3$.

$y < -x^2 - 2x + 3$

$0 < -(0)^2 - 2(0) + 3$

$0 < 0 - 0 + 3$

$0 < 3$, this is true, showing that the origin $(0, 0)$ is a member of the solution set. Thus, we have to shade the inside region of the parabola.

$$y < -x^2 - 2x + 3$$

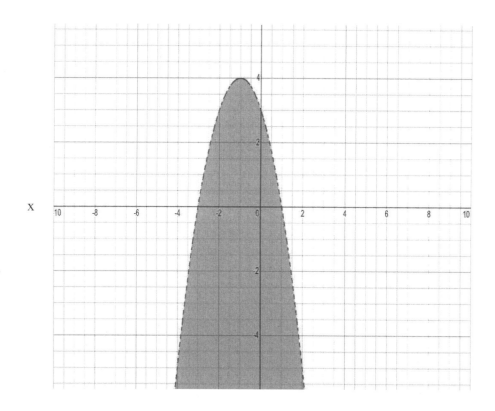

Fig. 11.30

Exercise 11.17

Sketch the graphs of each of the following inequalities.

1) $y \leq 2x^2$
2) $y < 2x^2$
3) $y > x^2 - 2x + 1$
4) $y \leq x^2 + 3x + 2$
5) $y < x^2 - 7x + 6$
6) $y \leq -x^2 + 2$

CHAPTER 12

Ratio, Proportion, Rates and Percentage

12.1. Ratio

Definition: A ratio is a comparison of two or more quantities of the same or different items or kinds, in the same measurement units that indicates their size in relation to each other.

- E.g., weight to length, boys to girls, apples to bananas and so on.

Notation: The ratio of a to b is written as: a to b, a: b or $\frac{a}{b}$.

In the above notation, the dividend or the number being divided 'a' is called antecedent whereas the divisor or the number by which the dividend is divided. 'b' is called Consequent.

Note: - The quantities that we supposed to compare should be expressed in the same units of measurement.

- It is desirable to write a ratio in its lowest terms.
- In the ratio $\frac{a}{b}$, b ≠ 0, a is called numerator and b is called denominator. The numerator represents how many parts you have and the denominator shows the total amount of parts that make up a whole.

Note: - The denominator of a fraction can never be zero.

i.e., in the ratio, $\frac{a}{b}$, b ≠ 0.

Example 1

1) $\frac{25}{100}$ can be written as its lowest term as $\frac{1}{4}$.

2) $\frac{6}{9}$ can be written as its lowest term as $\frac{2}{3}$.

3) $\frac{6}{14}$ can be written as its lowest term as $\frac{3}{7}$.

4) $\dfrac{10}{20}$ can be written as its lowest term as $\dfrac{1}{2}$.

Example 2

Find the value of x in each of the following ratio.

a) x:2 = 6:4

b) $\dfrac{x}{9} = \dfrac{6}{8}$

c) $\dfrac{16}{x} = \dfrac{x}{9}$

d) $\dfrac{24}{x} = \dfrac{6}{7}$

e) $\dfrac{2x-4}{x-3} = \dfrac{6}{5}$

f) $\dfrac{x-2}{6} = 2$

Solutions

a) x:2 = 6:4

x:2 = 6:4, can also be written as $\dfrac{x}{2} = \dfrac{6}{4}$

(2)(6) = (x)(4)...Use cross multiplication and equate them.

12 = 4x

$\dfrac{12}{4} = \dfrac{4x}{4}$ ....Divide both sides by 4.

3 = x

Thus, the value of x is 3.

b) $\dfrac{x}{9} = \dfrac{6}{8}$

(x)(8) = (6)(9) ....Cross multiplication and equate them.

8x = 54

$\dfrac{8x}{8} = \dfrac{54}{8}$ ....Divide both sides by 8.

$x = \dfrac{54}{8}$

$x = \dfrac{27}{4}$

Thus, the value of x is $\dfrac{27}{4}$.

c) $\dfrac{16}{x} = \dfrac{x}{9}$

$(x)(x) = (9)(16)$ .... Cross multiplication and equate them.

$x^2 = 144$

$x = \sqrt{144}$

$x = -12$ or 12

Thus, the value of x can be -12 or 12.

d) $\dfrac{24}{x} = \dfrac{6}{7}$

$(x)(6) = (7)(24)$ .... Cross multiplication and equate them.

$\dfrac{6x}{6} = \dfrac{(7)(24)}{6}$ .... Divided both sides by 6.

$x = \dfrac{(7)(4)}{1}$ .... $24 \div 6 = 4$

$x = 28$

Thus, the value of x is 28.

e) $\dfrac{2x - 4}{x - 3} = \dfrac{6}{5}$

$6(x - 3) = 5(2x - 4)$ .... Cross multiplication and equate them.

$6x - 18 = 10x - 20$ .... Distributive property of multiplication over subtraction.

$6x - 10x - 18 + 18 = 10x - 10x - 20 + 18$...Collect like terms.

$-4x = -2$

$\dfrac{-4x}{-4} = \dfrac{-2}{-4}$ .... Divide both sides by -4

$x = \dfrac{1}{2}$

Thus, the value of x is $\dfrac{1}{2}$.

f) $\dfrac{x - 2}{6} = 2$

$\dfrac{x - 2}{6} = \dfrac{2}{1}$

$(6)(2) = (1)(x - 2)$ .... Cross multiplication and equate them.

$12 = x - 2$

$12 + 2 = x - 2 + 2$ .... Add 2 on both sides.

$14 = x$

$x = 14$

Thus, the value of x is 14.

More Examples

1) The present ages of David and Joseph are in the ratio of 3:5 respectively. If David's present age is 27. What is Joseph's present age?

Solution

Let David's present age be D and Joseph's present age be J.

David's present age is given, D = 27

$$\frac{D}{J} = \frac{3}{5}$$

$$\frac{27}{J} = \frac{3}{5}$$

(J)(3) = (5)(27)...Cross multiplication.

3J = 135

$$\frac{3J}{3} = \frac{135}{3}$$ ....Divide both sides by 3

J = 45

Thus, the present age of Joseph is 45.

2) When 200 dollars is divided for Sara and Julia in the ratio of 2:3 respectively. Then
   a) How much is Sara's share?
   b) How much is Julia's share?

Solutions

The sum of the ratio 2:3 is 2 + 3 = 5

a) Sara's share $= \frac{2}{5}$ of $200 = \left(\frac{2}{5}\right)(200)$

$$= 2 \times 40$$

Sara's share = 80

Thus, Sara's share is 80 dollars

b) Julia's share $= \frac{3}{5}$ of $200 = \left(\frac{3}{5}\right)(200)$

$$= 3 \times 40$$

Julia's share = 120

Thus, Julia's share is 120 dollars.

Sara's share is 80 dollars and Julia's share is 120 dollars.

3) From a total of 120 students in a class 40 are girls. Find the ratio of:-
   a) Girls to boys
   b) Girls to the students in a class.
   c) Boys to the students in a class.
   d) Boys and girls to the total number of students in a class.

   Solutions

Given that total number = 120 students

Number of girls = 40 students

Number of boys = 120 – 40 = 80 students.

a) Ratio of girls to boys $= \dfrac{40}{80} = \dfrac{1}{2} = 1:2$

b) Ratio of girls to total number of students $= \dfrac{40}{120} = \dfrac{1}{3} = 1:3$

c) Ratio of boys to total number of students in a class $= \dfrac{80}{120} = \dfrac{2}{3} = 2:3$

d) Ratio of boys and girls to the total number of students $= \dfrac{80+40}{120} = \dfrac{120}{120} = \dfrac{1}{1} = 1:1$

Thus, ratio of boys and girls to the total number of students = 1 = 1:1

4) The ratio of two numbers is 11:13 and their difference is 16. What are the two numbers?

Solution

Let x be the larger number and y be the smaller number.

The two expressions make simultaneous equation.

$\dfrac{y}{x} = \dfrac{11}{13}$

$y = \dfrac{11x}{13}$ ....(Value of y in terms of x)

Substitute the value of y in the second equation. (x – y) = 16

$x - \left(\dfrac{11x}{13}\right) = 16$ ....(Substituting $y = \dfrac{11x}{13}$)

$\dfrac{13x - 11x}{13} = 16$ ....(Taking LCM of 1 and 13)

$\dfrac{2x}{13} = 16$

(2x) (1) = (13)(16) ....Cross multiplication

2x = 208

$\dfrac{2x}{2} = \dfrac{208}{2}$ ....Divide both sides by 2.

x = 104

Thus, the value of x is 104.

To find the value of y substitute x = 104 in x – y = 16

x – y = 16

104 – y = 16

104 – 104 – y = 16 – 104 ....(Subtract 104 from both sides)

$-y = -88$

$$\frac{-y}{-1} = \frac{-88}{-1}$$ ....(Divide both sides by −1 to eliminate the negative sign)

$y = 88$

Thus, the value of y is 88.

Therefore, the two numbers are 104 and 88.

5) An electrician cuts electric wire of length 240m into 4 pieces in the ratio of 2:3:5:6. Find the lengths of the parts.

Solution

Let x be the length of a unit piece of electric wire, then 2x, 3x, 5x and 6x will be the four pieces of wires. Thus,

$2x + 3x + 5x + 6x = 240$..................................(Total length of wire)

$16x = 240$

$$\frac{16x}{16} = \frac{240}{16}$$ ....Divide both sides by 16.

$x = 15$

Thus, $2x = 2(15) = 30m$

$3x = 3(15) = 45m$

$5x = 5(15) = 75m$

$6x = 6(15) = 90m$

Therefore, the length of each pieces are 30m, 45m, 75m and 90m respectively.

6) Find the ratio of 20km to 36000m

Solution

Before we proceed to operate, we need to convert km to m or m to km i.e, both the numerator and the denominator must have the same unit of measurement.

Note: - The above statement holds true for operating any ratio.

In this case, let us convert km to meter.

1 km = 1000m

20 km =?

$$= \frac{(20\,km)(1000\,m)}{1\,km}$$ ....Cross multiplication

$= 20,000m$

Therefore, 20km = 20,000m

Thus, the ratio of 20km to 36,000m is

$$\frac{20,000\,m}{36,000\,m}$$

$$\frac{5}{9}$$

The ratio 20km to 36,000m will be $\frac{5}{9}$ or 5:9

Remark: The ratio of any numbers has no unit. (It is a pure number).

7) A line segment 40m long divided into two parts whose lengths have the ratio 3:5, find the lengths of the parts.

Solution: Let x be a unit length. Then, 3x and 5x will be the required parts. Thus,

$3x + 5x = 40$ ....(total length)

$8x = 40$

$\dfrac{8x}{8} = \dfrac{40}{8}$ ....(Divide both sides by 8)

$x = 5$

Therefore, $3x = 3(5) = 15$ and $5x = 5(5) = 25$

The length of the two parts is 15m and 25m.

8) If a, b and c are numbers such that a:b:c = 3:5:6 and b = 30, then what is the sum of the three numbers.

Solution: $\dfrac{a}{b} = \dfrac{3}{5}, \dfrac{b}{c} = \dfrac{5}{6}$, b = 30 (given)

Thus, $\dfrac{a}{b} = \dfrac{3}{5}$

$a = \dfrac{3}{5}b$ ....(Cross multiplication)

$a = \dfrac{3}{5}(30)$ ....(Substitute b = 30)

$a = \dfrac{90}{5}$

$a = 18$

$\dfrac{b}{c} = \dfrac{5}{6}$

$\dfrac{30}{c} = \dfrac{5}{6}$ ....(Substitute b = 30).

$5c = (30)(6)$..(Cross multiplication).

$5c = 180$

$\dfrac{5c}{5} = \dfrac{180}{5}$

$c = 36$

Thus, the three numbers are a = 18, b = 30 and c = 36.

The sum, a + b + c = 18 + 30 + 36 = 84

The same result can be found as follows.

Let the sum of the three numbers be y.

The given ratio of the numbers is a:b:c = 3:5:6

This implies that the ratio of b to a:b:c $= \dfrac{5}{3+5+6} = \dfrac{5}{14}$ which in turn equals to the ratio of b to y.

$\dfrac{5}{14} = \dfrac{b}{y}$ ....Cross multiplication

5y = 14b...cross multiplication

5y=14x30...b=30, ... given

$\dfrac{5y}{5} = \dfrac{30 \times 14}{5}$ ....dividing each term by 5

$y = \dfrac{420}{5} = 84$

Thus, the sum of the three numbers is 84.

The three numbers can be found as follows:

$\dfrac{a}{84} = \dfrac{3}{14}$ .... $\dfrac{a}{y} = \dfrac{a}{a+b+c}$

14 x a = 84 x 3..cross multiplication

$\dfrac{14a}{14} = 84x\dfrac{3}{14}$ .... dividing each term by 14

a = 18

b = 30 ....Given

$\dfrac{c}{84} = \dfrac{6}{14}$

14 x c = 6 x 84..cross multiplication

$\dfrac{14C}{14} = \dfrac{6 \times 84}{14}$ ....dividing each term by 14

c = 36

Thus, the three numbers are 18, 30 and 36.

9) Find two numbers such that their ratio is 5:7 and their sum is
   a) 60
   b) 72

Solutions: The sum of a parts, 5 + 7 = 12
   a) When their sum is 60

    (i) First part is $\dfrac{5}{5+7}$ of 60

        $= \dfrac{5}{12}(60)$

        $= 25$

    (ii) Second part is $\dfrac{7}{5+7}$ of 60

        $= \dfrac{7}{12}(60)$

= 35

Thus, the two parts are 25 and 35.

The same result can be found as follows:

Let x and y be the two numbers.

x / y = 5 / 7 .... given

5y = 7x .... cross multiplication

5y − 7x = 0 (a)

x + y = 60 (b) .... given

Solving the two equations a and b simultaneously gives x = 25 whereas y = 35.

b) When their sum is 72.

  i) First part is $\dfrac{5}{5+7}$ of 72

$$= \dfrac{5}{12}(72)$$

$$= 30$$

  ii) Second part is $\dfrac{7}{5+7}$ of 72

$$= \dfrac{7}{12}(72)$$

$$= \dfrac{(7)(72)}{12}$$

$$= 42$$

Thus, the two parts are 30 and 42.

The same result can be found as follows:

Let x and y be the two numbers.

x / y = 5 / 7 .... given

5y = 7x .... cross multiplication

5y − 7x = 0 (a)

x + y = 72 (b) .... given

Solving the two equations a and b simultaneously gives x = 30 whereas y = 42.

10) If $x : y = \dfrac{2}{5}$ and $y : z = \dfrac{3}{7}$, then find x:z.

Solution: $\dfrac{x}{y} = \dfrac{2}{5}$ and $\dfrac{y}{z} = \dfrac{3}{7}$

Thus, $\dfrac{x}{y} \cdot \dfrac{y}{z} = \dfrac{x}{z}$

$$\implies \dfrac{x}{z} = \dfrac{x}{y} \cdot \dfrac{y}{z} = \left(\dfrac{2}{5}\right)\left(\dfrac{3}{7}\right) = \dfrac{6}{35}$$

Therefore, $\dfrac{x}{z} = \dfrac{6}{35}$

12.2. Proportion

Proportion is defined as a part, a share, or a number considered in comparative relation to the whole.

Example:

A rope's length and weight below are examples of proportion.
If 25m of rope weighs 1kg, then
- 50m of the same rope weighs 2kg.
- 150m of the same rope weighs 6kg.
- 200m of the same rope weighs 8kg.
- 1000m of the same rope weighs 40kg.

Here, the ratio of proportionality is 25 to 1.

12.2.1. Direct and Inverse Proportionality.

Direct and inverse proportion are used to show how the quantities and amounts are related to each other. The symbol used to denote the proportionality is "α". Direct proportion occurs when an increase in one variable results in an increase in value of the other variable or a decrease in one variable results in a decrease in the other variable or the vice versa.

Inverse proportion occurs when an increase in value of one variable results with a decrease in value of the other variable and vice versa. Thus,

If x is directly proportional to y or if x varies directly with y, we write this as: -

$$x \propto y, \text{ or } \frac{x}{y} = k \Rightarrow x = ky$$

Where, k is called constant of proportionality. Meaning as x increases y also increases and as x decreases y also decreases.

12.2.2. Properties of Proportion

- The numbers a to b and c to d are proportional, if the ratio of the first two quantities is equal to the ratio of the last two quantities.

$$\text{i.e., a:b = c:d } \left(\frac{a}{b} = \frac{c}{d} \right).$$

- For the numbers a, b, c and d, If $\frac{a}{b} = \frac{c}{d}$, then: -

  - Each quantity in a proportion is called its term or its proportion.
  - The end terms a and d are called the extremes of proportion.
  - The middle terms b and c are called the means of proportion.

- The terms a, b, c and d are called the first, the second, the third and the fourth terms of proportion respectively.
- The product of the extremes equal to the product of the means.

  i.e., For a:b = c:d, $\dfrac{a}{b} = \dfrac{c}{d} \Rightarrow ad = bc$

- Invertendo property: For four numbers a, b, c and d, if a: b = c: d, then b: a = d:c, that is, if the two ratios are equal, then their inverse ratios are also equal.
- When shapes are "in proportion," their relative sizes are the same.
- If x is inversely proportional to y or if x is indirectly proportional to y, then it is written as:

  $x \propto \dfrac{1}{y}$, then xy = k, where k is the constant of proportionality.

When y increases x decreases and when y decreases x increases.

Examples

1) The area of a rectangle is given by the formula, A = LW, where L is the length and W is its width. Here, as the length of a rectangle increases, the area also increases and as the width of the rectangle increases, the area also increases and vice versa. In other words, the length and the width of a rectangle altogether and its area are examples of direct proportional terms.

2) The fuel consumption of a car (if the car is under normal operation) is directly proportional to the distance covered.

3) If $x \propto 2y$, then what is the constant of proportionality k?

   Solution

   $x \propto 2y$

   $\dfrac{x}{2y} = k$

   Thus, the constant of proportionality k is $\dfrac{x}{2y}$.

4) If 40 men can finish a job in 10 days, how long it takes for 16 men to finish the same job?

   Solution

   Let number of men = M and number of days = D.

   As the number of men increases, the number of the days to finish the same job decreases. In contrast to this, as the number of men decreases, the number of the days to finish the same job increases. Therefore, M and D are inversely related to each other.

   $M = \dfrac{1}{D}$

   Number of men and number of days are inversely proportional.

   MD = k

   From this, we can drive a general formula that: $M_1 D_1 = M_2 D_2$

Thus, given that, $M_1 = 40$, $D_1 = 10$, $M_2 = 16$ and $D_2 =$?

$M_1 D_1 = M_2 D_2$

$(40)(10) = (16)(D_2)$ ... Substituting the given value.

$400 = 16D_2$

$16D_2 = 400$

$\dfrac{16D_2}{16} = \dfrac{400}{16}$

$D_2 = 25$

Therefore 16 men can finish the same job in 25 days

5) 36 men can finish a piece of work in 12 days. How many men will be needed to finish it in 9 days?

Solution

$M_1 D_1 = M_2 D_2$

Given: $M_1 = 36$, $D_1 = 12$, $M_2 =$? and $D_2 = 9$

$M_1 D_1 = M_2 D_2$

$36(12) = 9(D_2)$ ....Substitution

$M_2 = \dfrac{(36)(12)}{9}$

$M_2 = 48$

Thus, 48 men will be needed.

6) If 63 students can complete their job in 8 hours, how long will it take for 21 students to complete the same job?

Solution

Let S be the number of students and T be the time taken to do their job.

Thus, $S_1 T_1 = S_2 T_2$

Given that, $S_1 = 63$, $T_1 = 8$, $S_2 = 21$ and $T_2 =$?

$S_1 T_1 = S_2 T_2$

$(63)(8) = (21)(T_2)$ .... Substitution

$T_2 = \dfrac{(63)(8)}{21}$

$T_2 = 24$ hours

Therefore, 21 students take 24 hours to complete the same job.

12.3. Rates

A rate is a ratio that compares two different related quantities in different units. A unit rate is a rate with a divisor one.

1) For examples, mile per hour, km per second, cents per pound etc.

12.3.1.Calculating the Unit Rate

In order to calculate the unit rate of two related quantities in different units, we divide the numerator to the denominator in such a way that the denominator becomes 1.

Examples:

1) 300 oranges were eaten by 30 people. This is a rate of
    • 300 oranges per 30 people.
    • 10 oranges per person.

2) A Toyota car manufacturer can produce 720 cars in 5 days. This is a rate of
    • 720 cars per 5 days.
    • 144 cars per day
    • $\dfrac{144 cars}{24 hours}$ = 6 cars per hour.

3) Elsa makes 48 pancakes every 16 minutes. This is a rate of
    • 48 pancakes per 16 minutes.
    • 3 pancakes per minutes.
    • 3 x 60 = 180 pancakes per hour.

Note: - When we compare two related quantities in a single unit quantity, it is called a unit rate.

Examples

•   If you ran 30 km in 2 hours, you ran on average of 15 km in one hour. Both of the ratios, 30 km in 2 hours and 15 km in one hour, are rates. However, the 15 km in 1 hour is a unit rate.
•   If a Toyota car manufacturer produces 1220 cars in ten days. It produces an average of 122 cars in one day. Both of the ratios, 1220 cars in 10 days and 122 cars in 1 day are rates. But 122 car in 1 day is a unit rate.

Examples

Answer each of the following questions

1) A Toyota car manufacturer can produce 210 Toyota car per day. How many days will it take to produce 3360 Toyota cars?
    Solution:
    The manufacturer can produce 210 cars in one day, 3360 cars can be produced by 3360 ÷ 210 = 16 days.
    Thus, 16 days will be needed to produce 3360 cars.

2) Elsa can type 160 characters per minute. How long will take for Elsa to write an essay with 32,160 characters?

Solution.

Elsa can write an essay in $\dfrac{32,160}{160}$ minutes which is 201 minutes or approximately 3.35 hrs.

3) Alex can deliver 1,640 newspapers in 10 hours. How many newspapers can Alex delivers in 24 hours?

Solution.

Let x be the required number of newspapers.

1,640 newspapers = 10 hrs.

$$x = 24 \text{ hrs.}$$

$$x = \dfrac{(1640)(24)}{10} \text{ newspapers.} \quad \dots\dots\dots\dots\text{Cross multiplication}$$

$$x = 3,936 \text{ newspapers.}$$

Thus, Alex can deliver 3,936 newspapers in 24 hrs.

4) Six gallons of gasoline cost $18.60. What is the price per gallon?

Solution:

$$\text{price per gallon} = \dfrac{\text{cost price}}{\text{gallon}}$$

$$\text{price per gallon} = \dfrac{18.60 \text{ dollar}}{6 \text{ gallon}}$$

$$= 3.10 \dfrac{\text{dollar}}{\text{gallon}}$$

Thus, the price per gallon is $3.10.

You can get the same answer by cross multiplication

6 gallons = $ 18.60

1 gallon = y...Cross multiplication

6 × y = 1 × 18.60

6y = 18.60

6y/6 = 18.60/6 ....Multiplying each side by 1/6.

y =$ 3.10

Thus, the price per gallon is $3.10.

5) Which of the following is cheaper to buy? 5 books for $33.10 or 16 books for $ 84.48

Solution

Price per book is

- $\dfrac{\$33.10}{5 \text{ books}} = \$ 6.62$

$$\bullet \quad \frac{\$84.48}{16\,\text{books}} = \$\,5.28$$

Thus, it is cheaper to buy 16 books for $84.48.

6) Steve walks 28 miles in 4 hours. How many miles does he walk per hour?
   Solution

   Steve can walk $\dfrac{28\,\text{miles}}{4\,\text{hrs}} = 7\,\text{miles/hr}$

   Thus, Steve walks 7 miles per hour.

12.4. Percentage

The term "percent" means "out of hundred." "Out of" mathematically means "divide by". The percentage is a fraction whose divisor (denominator) is 100.

For example, 25% (read as 'twenty five percent') meaning 25 divided by 100. or you can read as 25 out of 100.

We can represent percent as a ratio or fraction.

Examples:

1) 28% can be written as a fraction.

   $\dfrac{28}{100} = \dfrac{7}{25}$ or

   $= \dfrac{28}{100} = 0.28$ as a decimal.

2) 8% can be written as a fraction $\dfrac{8}{100} = \dfrac{2}{25}$ or
   $= 0.08$ as a decimal.

   Note: A percentage has no dimension (It is pure number). That is, it has no unit of measurement.

12.4.1. Notation of Percent

The notation n% is read as "n percent" meaning "n per one hundred." Notation for n% can be expressed either as a

(i) ratio	(ii) fraction
$n\% = \dfrac{n}{100}$	$n\% = n \times \dfrac{1}{100}$
Example:	Example
20%	25%
$= \dfrac{20}{100}$	$= \dfrac{25}{100}$

$$= \frac{1}{5} \qquad\qquad = \frac{1}{4}$$

(iii)   Decimal notation

$$n\% = n \times 0.01$$

Example

$$8\% = 8 \times 0.01$$
$$= 0.08$$

More examples

1)   What is $\frac{2}{5}$ as a percent?

Solution

Multiply $\frac{2}{5}$ by 100%

Thus, $\left(\frac{2}{5}\right)(100)\%$

$$\frac{200}{5}\%$$

$$= 40\%$$

2)   What is $\frac{3}{4}$ as a percent?

Solution.

Multiply $\frac{3}{4}$ by 100%

Thus, $\left(\frac{3}{4}\right)(100)\%$

$$\frac{300}{4}\%$$

$$= 75\%$$

3)   Calculate 25% of 80.

Solution

To find 25% of 80, multiply $\frac{25}{100}$ by 80.

$$\frac{25}{100}\ 80 = 20$$

Thus, 25% of 80 is 20.

4) Calculate 5% of 60.

Solution:

To find 5% of 60, multiply $\dfrac{5}{100}$ by 60.

$\left(\dfrac{5}{100}\right)(60) = 3.$

Thus, 5% of 60 is 3.

5) What is 16% of 320.

Solution

To find 16% of 320, multiply $\dfrac{16}{100}$ by 320.

16% of 320 = $\dfrac{16}{100}$ 320 = 51.2

6) What percent of 50 is 15?

Solution

Let x be the required number.

$\left(\dfrac{50}{100}\right) x = 15$ ... multiply $\dfrac{50}{100}$ by the required number (x)

$\dfrac{1}{2} x = 15$ ... $\left(\dfrac{50}{100} = \dfrac{1}{2}\right)$

x = 30

Thus, 30% of 50 is 15.

As to me: Let x be the required percent of the number.

50 (x) = 15 ... multiply 50 by the required number (x)

50x / 50= 15/ 50 ....Dividing both sides by 50.

x = 0.30 = 30/ 100 =30%

Thus, 30% of 50 is 15.

7) Isabella scored 24 out of 40 in chemistry test. What is her percentage score?

Solution

She scored 24 out of 40. To find her percentage score multiply $\dfrac{24}{40}$ by 100%

$\left(\dfrac{24}{40}\right)(100\%)$

$\left(\dfrac{24 \times 100}{40}\right)\%$

60%

Thus, Isabella's percentage score is 60%.

Or let y be the scored mark out of 100%.

Out of 40 = 24 mark scored

Out of 100 % = y

40 (y) = 24(100) %...................................Cross multiplication

40 (y)/40 = 24 (100%)/40 ....Dividing each side by 40

Y = 60% Thus, Isabella's percentage score is 60%.

8) If 45% of the price of used car is 1,800 dollars. What was its original full price?

Solution

Let x be the original full price of the car, then $\left(\dfrac{45}{100}\right) x = 1800$

45 x = (1800) (100) ... cross multiplication

$45 \ x \ / \ 45 = \dfrac{(1800)(100)}{45}$ ... Divide both sides by 45.

x = 4,000

Thus, the full price of the car is 4,000 dollars.

9) The regular price of a book is $45 in a book store. At a sale, the price of the book reduces with 20%. How much will the book cost after the discount?

Solution

To find the discount multiply the regular price by reduced percentage.

$(45) \left(\dfrac{20}{100}\right)$

= 45 x 20 /100 =900 / 100

= 9

Thus, the discount is $9. Therefore, the price of the book after discount is $45 – $9 = $36. i.e., The price of the book after discount is $36.

Or let y be the cost of a book after a discount.

100% of the price = $ 45

80% of the price = y

Cross multiplication yields:

80% x $45 = 100% x $y

80% x $45/100% = (100%) (y/100%)

y = 80 x $ 45 /100 = $ 360 / 100 = $ 36

The price of the book after discount is $36.

10) A dealer bought a used car 7,500 dollars and sold it for 6,000 dollars. What is his/her percent loss?

Solution:

Actual change in price = 7500 – 6000

= 1,500

Percent loss $= \dfrac{\text{Actual change in price}}{\text{original price}}$ x 100%

$$= \frac{1500}{7500} \times 100\%$$

$$= \frac{15 \times 100}{75}\%$$

$$= 20\%$$

Thus, the percent loss of a dealer is 20%.

11) What is 250% of 20.

Solution

Let x be the required amount.

Thus, $\dfrac{250}{100} = \dfrac{x}{20}$

$x = \left(\dfrac{250}{100}\right)(20)$ ... cross multiplication

$= (25)(2)$

$x = 50$

Hence 250% of 20 is 50.

Note: - You can multiply $\dfrac{250}{100}$ by 20

$\dfrac{250}{100} \times 20 = 50$

12) The shaded part of the circle shown below shows the total number of students who failed in physics. What is the percentage of students who failed in physics?

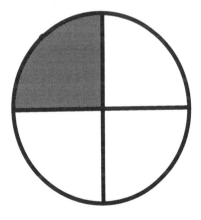

Fig. 12.1

Solution:

As you see from the picture, the circle is divided into four equal parts and ¼th of the circle is shaded, which is the number of students who failed in physics. Thus, the percentage of students who failed in physics is ¼th (100%) = 25%.

13) The amount in a savings account of Helen increased from \$644 to \$750. What is the percent increase of Helen's account?

Solution:

To find the increase

New amount – original amount = increase

\$750 – \$644 = \$106

$$\text{Percent increase} = \frac{\text{Increase}}{\text{original amount}} \times 100\%$$

$$= \frac{106}{644} \times 100\%$$

Percent increase ≈ 16.46%.

Thus, the percent increase of Helen's account is approximately 16.46%.

14) A book is on sale for a discount 10% of its original price. If the total cost, including the discount is \$45, what was the price of the book before discount?

Solution

Let x be the original price of the book.

$$\text{Discount} = (10\%)(x) = \frac{10}{100}x = 0.1x$$

Original price – Discount = Selling price

x – Discount = Selling price

$$x - 0.1x = 45$$

$$0.9x = 45$$

$$\frac{9x}{10} = 45$$

$$x = 45\left(\frac{10}{9}\right)$$

$$x = \frac{450}{9}$$

$$x = 50$$

Thus, the price of the book before discount was \$50.

15) Olivia and her three sisters received the same amount of money from their father to go to a cinema. Each girl spent 60 dollars. Afterward, the girls had 40 dollars altogether. What is the amount of money each girl received?

Solution:

Let x be the amount of money each girl received, then

There are 4 sisters, each girl spent 60 dollars and afterward 40 dollars altogether.

Thus, we can make an equation below; -

4(x-60) =40.........write an equation

4x-240=40 ....use distributive property

4x-240+240 =40+240 ……….. Add 240 on both sides

4x=280

$$\frac{4x}{4} = \frac{280}{4}$$

x = 70

Thus, each girl received 70 dollars.

Exercise 12.1 - 12.1.4

Answer each of the following questions.

1) Find the lowest term of each of the following fractions.

   a) $\frac{124}{360}$

   b) $\frac{142}{355}$

   c) $\frac{21}{84}$

   d) $\frac{105}{945}$

   e) $\frac{81}{64}$

   f) $\frac{72}{144}$

2) Find the value of the variable in each of the following ratios.

   a) x:6 = 8:5

   b) 16:x = 144:24

   c) k:81 = 27:33

   d) $\frac{16}{x} = \frac{x}{81}$

   e) 8:3 = x:10

   f) $\frac{3x-4}{x+2} = \frac{1}{2}$

3) The present age of Jack and Alexander is in the ratio of 2:3 respectively. If Jack's present age is 36, what is Alexander's present age?

4) When 450 dollars is divided for Daniel and Samuel in the ratio of 4:5 respectively. Then
   a) How much is Daniel's share?
   b) How much is Samuel's share?

5) From the total of 135 students in a class 45 are boys and the rest are girls. Find the ratio of
   a) Boys to girls
   b) Girls to boys
   c) Girls to the students in a class.
   d) Boys to the students in a class.
   e) Boys and girls to the total number of students in a class.

6) The ratio of two numbers is 17:21 and their difference is 28. What are the two numbers?

7) An electrician cuts electric wire of length 360 cm into 3 pieces in the ratio of 3:4:5. Find the length of each piece.

8) If $x \infty y^2$, then what is the constant of proportionality k?

9) If 24 men can finish their job in 8 days. How long it takes for 16 men to finish the same job?

10) If 54 students can complete their job in 6 hours. How long will take for 36 students to complete the same job?

11) A Toyota car manufacturer can produce 133 cars per day. How many days will it take to produce 27,797 Toyota cars?

12) Maria can type 45 words per minute. How long will it take for Maria to write an essay with 4,050 words?

13) Thomas can deliver 2040 newspapers in 10 hours. How long will it take for Thomas to deliver 6120 newspapers?

14) Eight gallons of gasoline cost $32.80, what is the price per gallon?

15) Which one is the best to buy? 6 books for $36.60, 8 books for $48.08 or 9 books for $54.009

16) Emily walks 21 miles in 3 hours. How many miles does she walks per hour?

17) Write each of the following percentage in the lowest term.
   a) 20%
   b) 24%
   c) 84%
   d) 88%
   e) 45%
   f) 17%
   g) 36%

18) Write each of the following in their decimal form.
   a) 17%

b) $\dfrac{33}{100}$

c) $\dfrac{27}{100}$

d) $\dfrac{234}{100}$

e) $\dfrac{27.4}{100}$

f) $\dfrac{11.43}{100}$

g) $\dfrac{1}{100}$

19) Write each of the following as a percentage:

a) $\dfrac{3}{5}$

b) $\dfrac{6}{25}$

c) $\dfrac{5}{4}$

d) $\dfrac{1}{5}$

e) $\dfrac{17}{25}$

f) $\dfrac{30}{60}$

g) 0.32

h) 1.5

i) $\dfrac{6}{32}$

20) What is 13% of 147?

21) Calculate 6% of 2400

22) What percent of 75 is 15?

23) What percent of 240 is 16?

24) Sandra scored 36 out of 60 in Biology test. What is her percentage score?

25) If 28% of the price of used car is 1260. What is the full price?

26) The regular price of a book is $36, in a book store. At a sale the price of the book reduces with 18%. How much will the book cost after the discount?

27) The shaded part of the circle below shows the total number of students in a class who scored below 50% in mathematics. What percent of the students in a class scored above 50%?

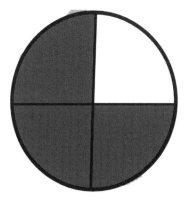

28) What percent of the square is shaded in the figure below?

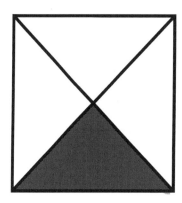

29) 26% of 204 is _____

30) A book is on sale for a discount of 25% of its original price. If the total cost including the discount is $150. What was the price of the book before discount?

31) A book is on sale for a discount of 24% of its original price. If the total cost excluding the discount is $760. What was the price of the book before discount?

32) David and his five brothers received the same amount of money from their mother to go to a restaurant. Each boy spent 80 dollars. Afterward, the boys had 60 dollars altogether. What is the amount of money each boy received?

CHAPTER 13

Graph of Square Root Function

13.1. Square Root Function

As defined earlier the square root function is defined by the equation $f(x) = \sqrt{x}$, where the domain is the set of all non-negative numbers and the range is all real numbers greater than or equal to zero.

The parent function of the square root function is given by: -

$f(x) = \sqrt{x}$

Domain of f(x) = {x: x ≥ 0}

Range of f(x) = {y: y ≥ 0}

Note: f(x) stands for 'y'

13.2. Graphing Square Root Function

The domain and range of the square root function written in interval notation as: -

Domain: [0, ∞)

Range: [0, ∞)

Examples

1) Graph the function: $y = \sqrt{x}$

   Solution

   First, complete table of values by taking some values of x and substituting them in the given equation so as to get their corresponding y values. Then plot these points on the x – y plane and connect them with a smooth curve.

$y = \sqrt{x}$	x	0	1	2	3	4	5	6
	y	0	1	1.4	1.7	2	2.3	2.5

   Remark:- Decimals are taken approximately for y values.

351

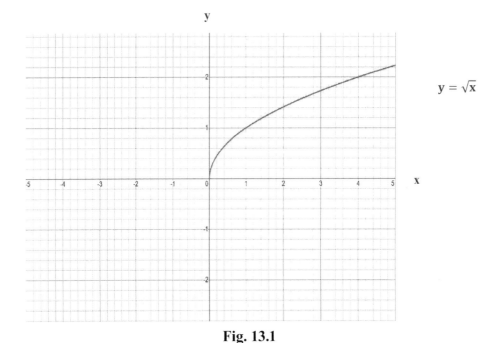

Fig. 13.1

Note: • The graph of the square root function is above the x-axis and is an increasing graph.
• Both the domain and the range of the function y = $\sqrt{x}$, is non-negative numbers.
• The square root function is neither even nor add.
The square root function has different characteristic, one of these characteristics is transformation. Let us see this characteristic.

13.3. Transformation of the Square Root Function.

In mathematics, transformation is the method of transforming the shape or size of an object using different types of rules. The square root function can be transformed in the same way as other functions. These are: - shifts, stretching (shrinking)and reflections. Let us see each of them using examples: -

13.4. Shift

13.4.1. Horizontal Shift

For the parent function f(x) = $\sqrt{x}$, if k is positive real number, y = f (x + k) shifts the graph of f(x) to the left k units and y = f (x – k) shifts the graph of f(x) to the right k units. This is known as the horizontal shifts of the graph of f(x).
Thus,
Let f be a function and k be a positive real number, then

• The graph of y = f (x + k) is the graph of y = f(x) shifted to the left k units.
• The graph of y = (x – k) is the graph of y = f(x) shifted to the right k units.

352

13.4.2. Horizontal Shift to the Left

Examples

1) Use the parent graph $f(x) = \sqrt{x}$ to obtain the graph of $g(x) = \sqrt{x+4}$.

Solution: Compare the equation $f(x) = \sqrt{x}$ and $g(x) = \sqrt{x+4}$. The equation of $g(x)$ is obtained by adding 4 to each value of x before taking the square root.

The graph of $g(x) = \sqrt{x+4}$ has the same shape as the graph of $f(x) = \sqrt{x}$, nevertheless, it is shifted to the left 4 units horizontally.

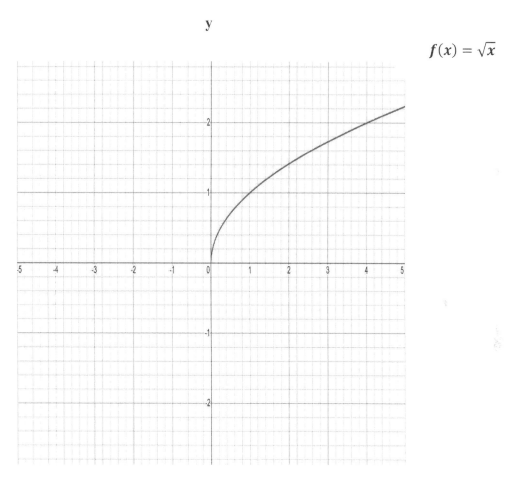

Fig. 13.2

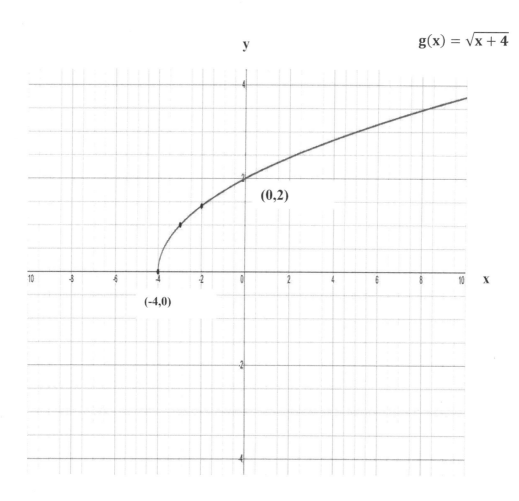

$$g(x) = \sqrt{x + 4}$$

Fig. 13.3

Graph of $g(x) = \sqrt{(x+4)}$ shifts the graph of $f(x) = \sqrt{x}$, 4 units to the left. i.e., subtract 4 from each x-value.

2) Use the parent graph $f(x) = \sqrt{x}$ to obtain the graph of $g(x) = \sqrt{x+3}$

Solution: - compare the equation $f(x) = \sqrt{x}$ with $g(x) = \sqrt{x+3}$. The equation of g(x) is obtained by adding 3 to each value of x before taking the square root.

The graph of g(x) shifts the graph of $f(x) = \sqrt{x}$, 3 units to the left.

The graph of $g(x) = \sqrt{x+3}$ has the same shape as the graph of $f(x) = \sqrt{x}$, but; it is shifted horizontally 3 units to the left. Thus,

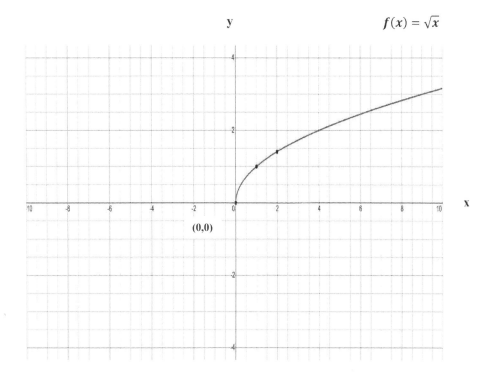

Fig. 13.4

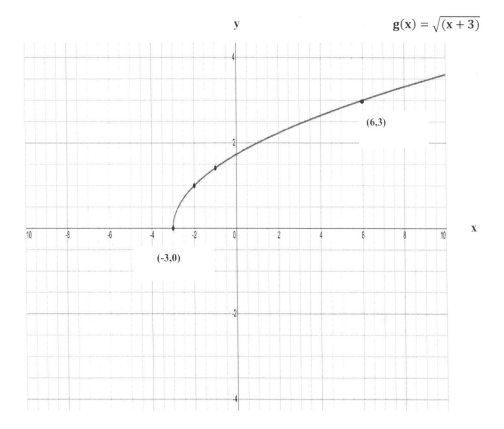

Fig. 13.5

355

Graph of $g(x) = \sqrt{x+3}$ shifts the graph of $f(x) = \sqrt{x}$, 3 units to the left.

13.4.3. Horizontal Shift to the Right

Examples 1) Use the parent graph $f(x) = \sqrt{x}$ to obtain the graph of $g(x) = \sqrt{x-5}$.

Solution: Compare the equation $f(x) = \sqrt{x}$ with $g(x) = \sqrt{x-5}$. The equation of $g(x)$ is obtained by subtracting 5 from each value of x before taking the square root.

The graph of $g(x) = \sqrt{x-5}$ has the same shape as the graph of $f(x) = \sqrt{x}$, however; it is shifted 5 units to the right horizontally. Thus,

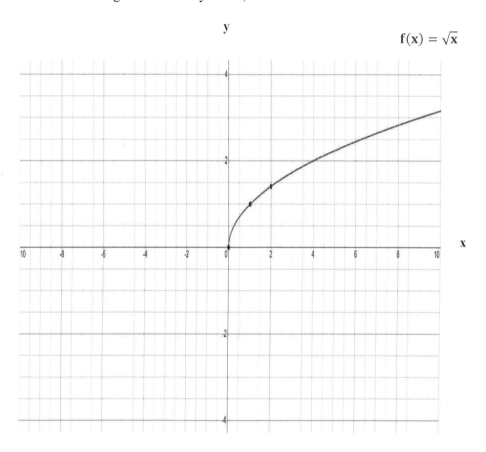

$$f(x) = \sqrt{x}$$

Fig. 13.6

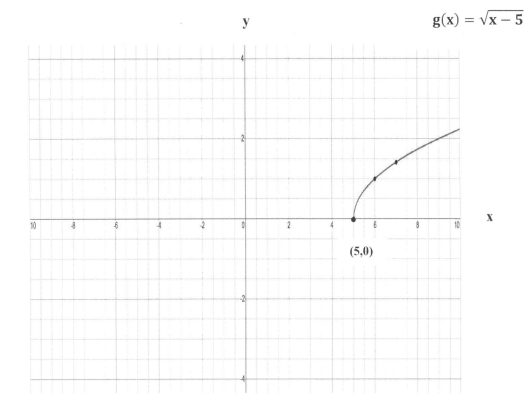

$$g(x) = \sqrt{x-5}$$

Fig. 13.7

Graph of $g(x) = \sqrt{x-5}$ shifts the graph of $f(x) = \sqrt{x}$ five units to the right.

2) Use the parent graph $f(x) = \sqrt{x}$ to obtain the graph of $g(x) = \sqrt{x-6}$.

Solution

Compare the equation $f(x) = \sqrt{x}$ with $g(x) = \sqrt{x-6}$. The equation of g(x) is obtained by subtracting 6 from each value of x before taking the square root.

The graph of $g(x) = \sqrt{x-6}$ has the same shape with the graph of $f(x) = \sqrt{x}$. However; it is shifted 6 units to the right horizontally. Thus,

$$f(x) = \sqrt{x}$$

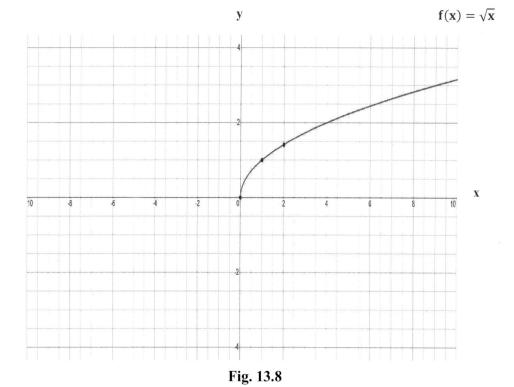

Fig. 13.8

$$g(x) = \sqrt{(x-6)}$$

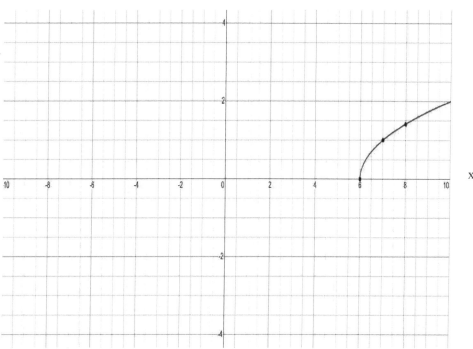

Fig. 13.9

358

Graph of $g(x) = \sqrt{x-6}$ shifts the graph of $f(x) = \sqrt{x}$ 6 units to the right.

Note: - The domain and the range of the square root function is the set of all non-negative numbers; i.e., $[0, \infty)$.

13.5. Stretching and Shrinking

As it is explained before, an object having one shape is transformed into a different shape vertically or horizontally without changing the basic shape of the graph. During stretching and shrinking, graphs remain rigid and proportionally the same.

- If we consider the graphs of $f(x) = \sqrt{x}$ and $g(x) = 2\sqrt{x}$ or $g(x) = 2f(x)$, thus for each value of x, the y-coordinate of g(x) is 2 times as large as the corresponding y-coordinate on the graphs of f(x). When we compare the graphs of f(x) and g(x), the graph of g(x) is a narrower graph because the values of y are rising faster. We say the graph of g(x) is obtained by vertically stretching the graph of f(x).

On the other hand, consider the equation of the graph of $f(x) = \sqrt{x}$ and $h(x) = \frac{1}{2}\sqrt{x}$ or $h(x) = \frac{1}{2}f(x)$. Here, for each value of x, the y-coordinate of h(x) is half of their corresponding y-coordinate on the graph of f(x). The result of the graph of h(x) is wider since the values of the y are rising slowly. In this case, we say that the graph of h(x) is obtained by vertically shrinking the graph of f(x). These types of transformation of the graphs can be generalized as follows.

Note: - For vertical stretching, y-values are multiplied by a factor of '**k**' and points get farther away from the x-axis whereas for vertical shrinking y-values are multiplied by a factor of **0<k<1** and points get closer to the x-axis.

13.6. Vertical Stretch and Shrink Graphs

Let f be a function and k a positive number.
- If k > 1, the graph of y = kf(x) is the graph of y = f(x) vertically stretched by multiplying each of its y-coordinates by k.

Stretching: k > 1

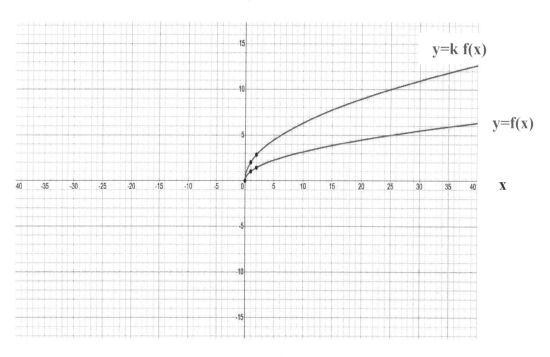

Fig. 13.10

Vertically Stretching

Shrinking: 0<k<1

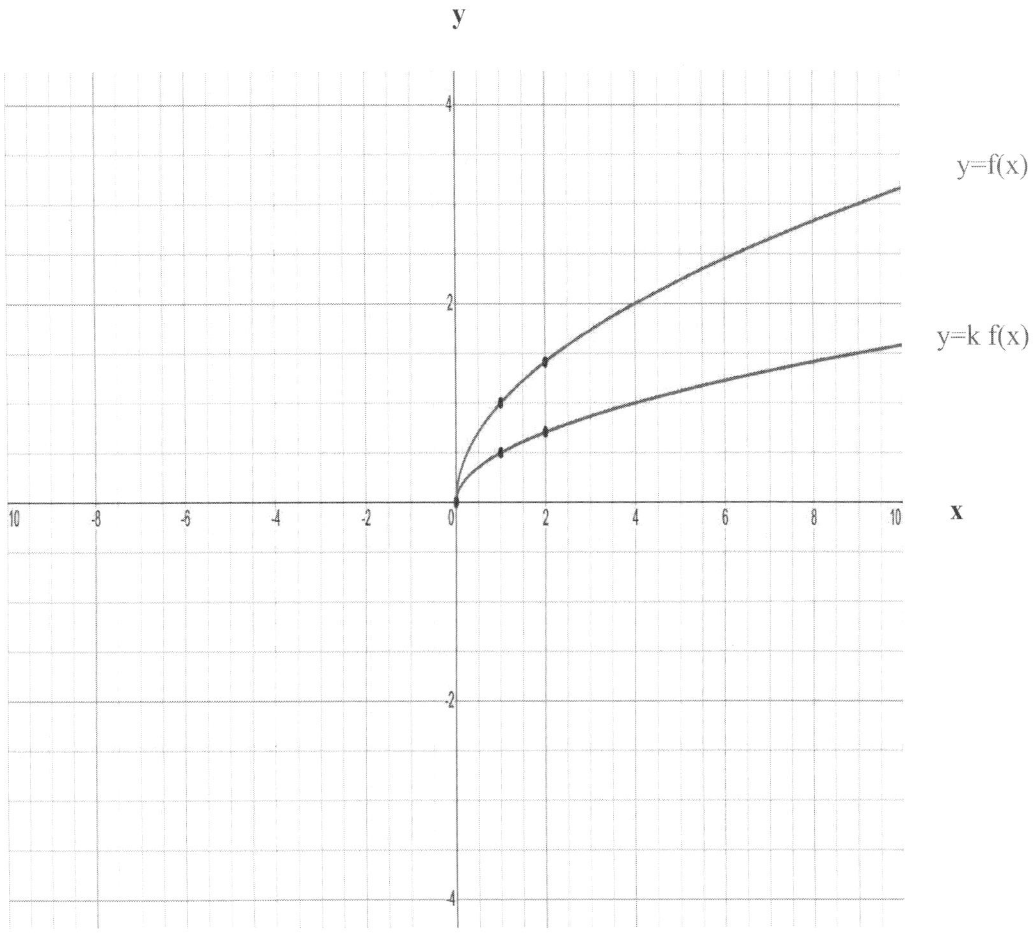

Fig. 13.11

Vertically Shrinking graph

Example: **Vertically stretching graph**

Use the graph of $f(x) = \sqrt{x}$ to obtain the graph of $g(x) = 2\sqrt{x}$.

Solution: The graph of $g(x) = 2\sqrt{x}$ is obtained by vertically stretching the graph of $f(x) = \sqrt{x}$.

- Graph of $g(x) = 2\sqrt{x}$ vertically stretch the graph of f, by multiplying each y value by 2.

Let us take some selected x values and make tables for $f(x) = \sqrt{x}$ and $g(x) = 2\sqrt{x}$

$f(x) = \sqrt{x}$

$y = \sqrt{x}$	x	0	1	4	9
	y	0	1	2	3

$g(x) = 2\sqrt{x}$					
$y = 2\sqrt{x}$	x	0	1	4	9
	y	0	2	4	6

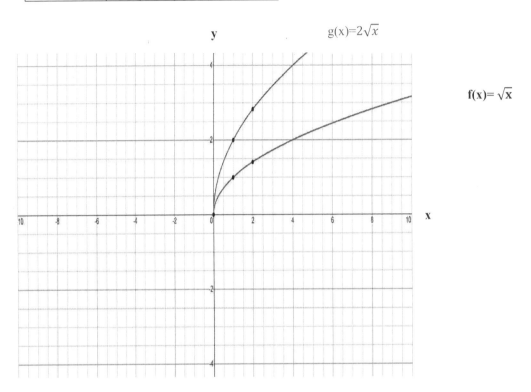

Fig. 13.12

The graph of $y = 2\sqrt{x}$ vertically stretching the graph of $y = \sqrt{x}$ by multiplying each y-coordinates by 2 and they move farther from the x-axis. This tends to make the graph steeper.

Example: - Vertically Shrinking Graph

Use the graph of $f(x) = \sqrt{x}$ to obtain the graph of $g(x) = \frac{1}{2}\sqrt{x}$.

Solution: - The graph of $g(x) = \frac{1}{2}\sqrt{x}$ is obtained by vertically shrinking the graph of $f(x) = \sqrt{x}$.

- Graph of $g(x) = \frac{1}{2}\sqrt{x}$ vertically shrinking the graph of $f(x)$ by multiplying each y-value by $\frac{1}{2}$.

Let us take some selected x values and make tables for $f(x) = \sqrt{x}$ and $g(x) = \frac{1}{2}\sqrt{x}$.

$y = \sqrt{x}$	x	0	1	4	9	
	y	0	1	2	3	

$y = \dfrac{1}{2}\sqrt{x}$	x	0	1	4	9
	y	0	0.5	1	1.5

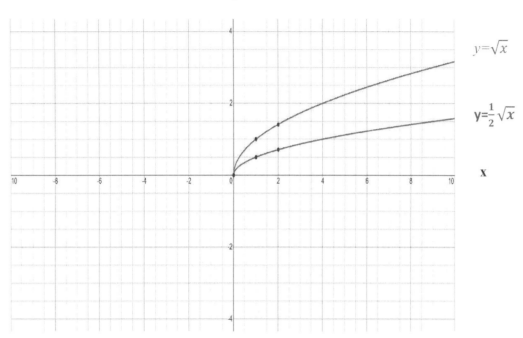

Fig. 13.13

Note:- For vertical stretching, y-values are multiplied by a factor of '**k**' and points get farther away from the x-axis whereas for vertical shrinking y-values are multiplied by a factor of **0<k<1** and points get closer to the x-axis. i.e., Vertically stretching is the stretching of the graph away from the x-axis and Vertical shrinking is the squeezing of the graph towards the x-axis.

In the above graph, observe that the graph of $y = \dfrac{1}{2}\sqrt{x}$ squeezes the graph of $y = \sqrt{x}$ towards the x-axis.

i.e., The graph of $y = \dfrac{1}{2}\sqrt{x}$ vertically shrinking the graph of $y = \sqrt{x}$ by multiplying each y coordinates by $\dfrac{1}{2}$ so that they move closer to the x-axis. This leads to make the graph flatter.

13.7. Horizontal Stretching and Shrinking Graph of Square Root Functions.

- A horizontal stretch is the stretching of the graph away from the y-axis.
- A horizontal shrinking is the squeezing of the graph towards the y-axis.

Note: - A horizontal stretch or shrink is the result of multiplying the x-value of function f(x) by a constant.

Let f be a function and k be a positive real number.

- If k > 1, the graph of y = f(kx) is the graph of y = f(x) horizontally shrinks by dividing each of its x-coordinates by k which moves the points closer to the y-axis. This is called a Horizontal Shrink.

  i.e., The point (a, b) on the graph of y = f(x) moves to a point $(\frac{a}{k}, b)$ on the graph of y = f(kx).

- If 0 < k < 1, the graph of y = f(kx) is the graph of y = f(x) horizontally stretched by dividing each of its x-coordinates by k, which moves the points farther away from the y-axis. This is called a Horizontal Stretch.

  i.e, The point (a, b) on the graph of y = f(x) moves to a point (ka, b) on the graph of y = $f\left(\frac{x}{k}\right)$.

Example: Horizontally shrinking a graph

Use the graph of y = $\sqrt{x}$ to obtain the graph of y = $\sqrt{2x}$

Solution: Let us take some values of x and make tables for y = $\sqrt{x}$ and y = $\sqrt{2x}$.

y = $\sqrt{x}$	x	0	1	4	9
	y	0	1	2	3

y = $\sqrt{2x}$	x	0	$\frac{1}{2}$	2	$\frac{9}{2}$
	y	0	1	2	3

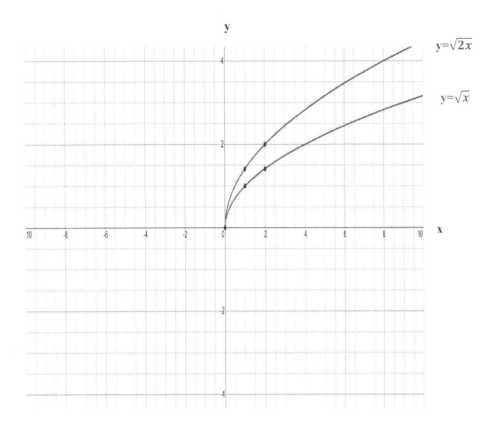

Fig. 13.14

Observe that the graph of $y = \sqrt{2x}$ is shrinking and close to the y-axis. Each of the values of $y = \sqrt{2x}$ is multiplied by a factor of $\dfrac{1}{2}$.

Example: Horizontal stretching of a graph

Use the graph of $y = \sqrt{x}$ to obtain the graph of $y = \sqrt{\dfrac{1}{2}x}$.

$y = \sqrt{x}$	x	0	1	4	9
	y	0	1	2	3

$y = \sqrt{\dfrac{1}{2}x}$	x	0	2	8	18
	y	0	1	2	3

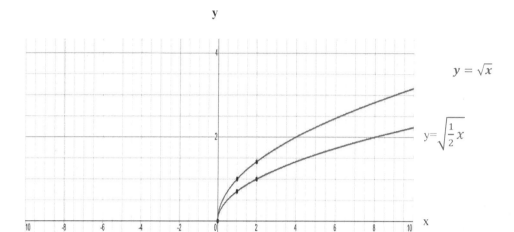

Fig. 13.15

Observe that the graph of $y = \sqrt{\dfrac{1}{2}x}$ is stretching and farther away from the graph of $y = \sqrt{x}$ or the y-axis. Each of the values of x for $y = \sqrt{\dfrac{1}{2}x}$ is multiplied by a factor of 2.

13.8. Reflection

A reflection is a transformation in which an object is reflected across a line, a point or a plane. Thus, the resulting image is flipped. The original object is called pre-image and the new figure created by a reflection is called Image. When reflecting an object in a line or in a point, the image is congruent to the pre-image. That means, the size and shape of the object does not change before and after the transformation. Note that the line of reflection is the perpendicular bisector between the pre-image and the image. It is possible for graphs to have mirror images with respect to the x-axis or y-axis.

The figures below are examples of reflection.

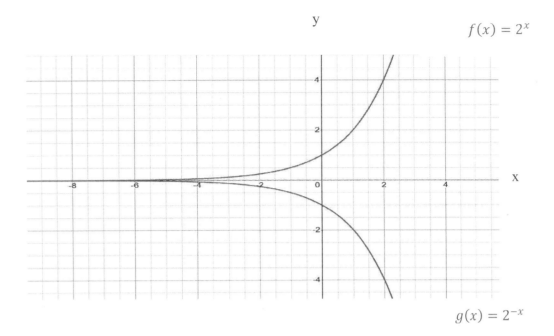

Fig. 13.16

Reflection about the x-axis

Mirror

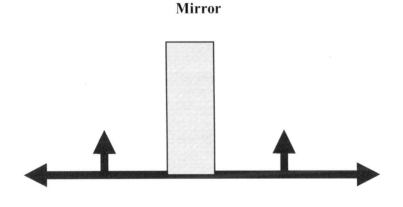

Fig. 13.17

Reflection about the y-axis

13.8.1. Reflection About the x-Axis

The graph of y = –f(x) is the graph of y = f(x) after reflected about the x-axis.
i.e., When reflecting across the x-axis, keep x the same and make y-negative.

- When the point (x, y) is reflected about the x-axis, the resulting point is (x, –y).

Examples 1) When the point (4, 6) is reflected about the x-axis, the resulting point is (4, –6).

367

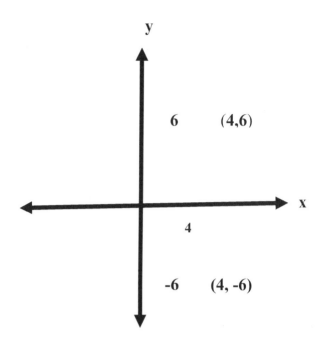

Fig. 13.18

Reflection across the x-axis

2) What is the reflection of the point (3, 5) about the x-axis?
Solution: - When the point (x, y) is reflected about the x-axis, keep x and make the value of y negative.
i.e., (x, y) → (x, –y) when reflected about the x-axis.
Thus, when the point (3, 5) is reflected about the x-axis, the resulting image will be (3, –5).

13.8.2. Reflection About the y-axis.

The graph of y = f(–x) is the graph of y = f(x) after reflected about the y-axis. i.e., when reflecting across the y-axis, keep the value of y the same and that of x-negative.

Examples 1) When the point (4, 2) is reflected across the y-axis, the resulting image is (–4, 2).

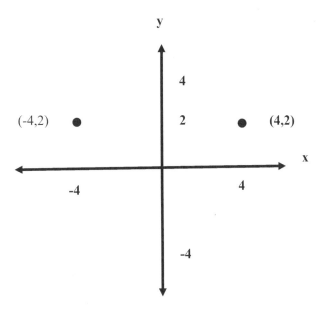

Fig. 13.19

Reflection across the y-axis

2)   What is the reflection of the point (3, 1) across the y-axis?
Solution: - When the point (3, 1) is reflected across the y-axis, keep the value of y and change that of x into negative.
i.e,   (3, 1) → (–3, 1)
Reflected across the y-axis.
Thus,

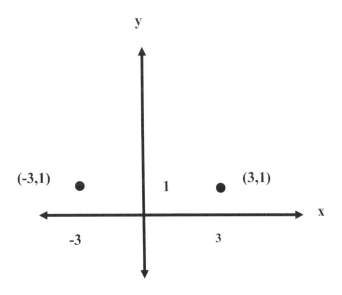

Fig. 13.20

Reflection across the y-axis

Examples: -

1) Draw the reflection of the graph of f(x) = $\sqrt{x}$ through the x-axis.

Solution: - When the coordinate of a point (a, b) is reflected through the x-axis, the resulting image is the point (a, –b). That means, the x value remains the same and the y value becomes negative. Therefore, g(x) = $-\sqrt{x}$ is the image of f(x) = $\sqrt{x}$ after reflected through the x-axis.

Let us make tables of value for f(x) = $\sqrt{x}$ and g(x) = $-\sqrt{x}$

f(x) = $\sqrt{x}$	x	0	1	4	9
	y	0	1	2	3

g(x) = $-\sqrt{x}$	x	0	1	4	9
	y	0	–1	–2	–3

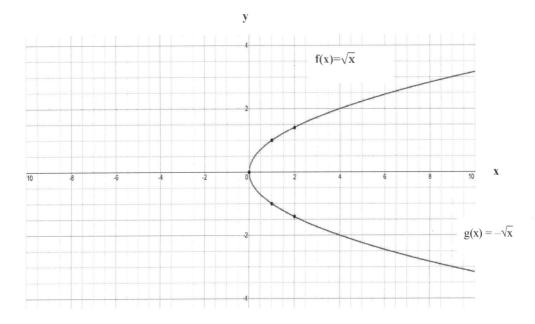

Fig. 13.21

Graph of f(x) and g(x)

2) Draw the reflection of the graph of f(x) = $\sqrt{x}$ through the y-axis.

Solution: When the point (a, b) is reflected across the y-axis, the resulting image is (–a, b). That is, the y value remains the same and the x-value becomes negative. Therefore, the reflection of f(x) = $\sqrt{x}$ through the y-axis is h(x) = $\sqrt{-x}$.

Let us make tables of value for f(x) and h(x).

$f(x) = \sqrt{x}$	x	0	1	4	9
	y	0	1	2	3

$h(x) = \sqrt{-x}$	x	-9	-4	-1	0
	y	3	2	1	0

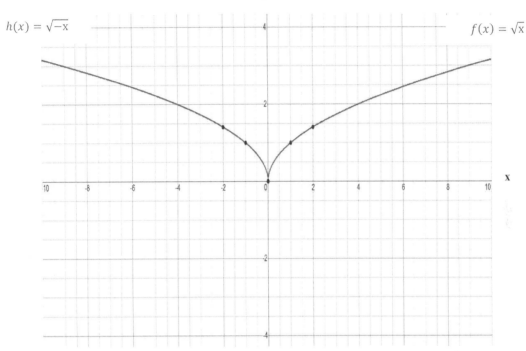

Fig. 13.22

Graph of f(x) and h(x)

$h(x) = \sqrt{-x}$ is the reflection of the graph of $f(x) = \sqrt{x}$ through the y-axis.

Note: - The rule of transformation of function also holds true for linear function, Quadratic function, cubic function etc.

Exercise 13.1-13.8.2

Answer each of the following questions.

1. Draw the reflection of the graph of $f(x) = 2\sqrt{x}$ across the x axis.
2. What is the reflection of the coordinate of the point (6, 2) across the y-axis?
3. What is the reflection of the coordinate of the point (4, 1) across the x-axis?
4. What is the reflection of the equation of $f(x) = \sqrt{3x}$ across the x-axis?

5. What is the reflection of the equation of $f(x) = 4\sqrt{x}$ across the y axis?
6. Draw the reflection of the graph of $f(x) = 2\sqrt{x}$ across the y-axis.
7. Draw the reflection of the graph of $f(x) = \sqrt{2x}$ across the x-axis.
8. Draw the reflection of the graph of $f(x) = \sqrt{2x}$ across the y-axis.
9. What is the reflection of the coordinate of the point (0, 0) or the origin across the x-axis?
10. What is the reflection of the coordinate of the point (0, 0) or the origin across the y-axis?

13.9. Domain and Range of Square Root Function

13.9.1. Domain of Square Root Function

The domain of a function f(x) is the set of all values (inputs) for which the function is defined and the range of a function is the set of all possible outputs (function values). Thus, if D is the domain of a function f, then the range of f is a collection of f(x) values for all x in the domain D.

- The domain of the square root function is the set of all possible input values (values of x) that result in a radicand (values of y) which is equal to or greater than zero.

13.9.2. Range of Square Root Function.

- The range of the square root function y = f(x) is the set of all possible values of y (output) taken for all values of x within the domain of x (input). Thus, the set of values of y are always non-negative numbers. That is, greater than or equal to zero.
Note: - Since the value of y (outputs) of a function depend on the value of x (inputs), we need to keep in focus the domain in order to determine the range of the function.

Examples:

1) Find the domain and range of $f(x) = \sqrt{x}$.
   Solution
   The radicand must be non-negative for any value of x in the domain. Thus, the value of x that is greater than or equal to zero makes $f(x) = \sqrt{x}$ true. Therefore, the solution set is: {x: $x \geq 0$}.
   That is, the domain D is {x: $x \geq 0$}
   In interval notation D is [0, ∞).
   - The range (output) of the function f(x) is the value of y when we substitute any value of x in the domain. Thus, the range (the value of y) is greater than or equal to zero. Therefore, the range is {y: $y \geq 0$}.
   In interval form the range is [0, ∞).

2) Find the domain and range of $y = \sqrt{x-3}$.
   Solution: The value of the radicand must be non-negative numbers for any value of x in the domain.

372

That is, $x - 3 \geq 0$ or $x \geq 3$

- Therefore, the domain is $x \geq 3$. In interval form, the domain is $[3, \infty)$.
- The range is the value of y: $\{y: y \geq 0\}$

In interval form: $[0, \infty)$.

3) Find the domain and range of $f(x) = \sqrt{8 - x}$

Solution: The value of the radicand must be non-negative number for any value of x in the domain. That is,

$8 - x \geq 0$

$8 - 8 - x \geq 0 - 8$ ... (subtract 8 from both sides)

$0 - x \geq -8$

$-x \geq -8$

$(-1)(-x) \geq -8(-1)$ ... (Multiply both sides by –1)

$x \leq 8$ ... ($\geq$ is flipped to $\leq$ when multiplied by negative number)

Thus, the domain of f(x) is $x \leq 8$.

The domain is $\{x: x \leq 8\}$

In interval form: $(-\infty, 8]$.

- The range is the value of y.

$\{y: y \geq 0\}$

In interval form: $[0, \infty)$.

4) Find the domain and range of $f(x) = \sqrt{x + 5}$.

Solution: The radicand and f(x) must be non-negative numbers. That is, $x + 5 \geq 0$.

$x + 5 \geq 0$

$x + 5 - 5 \geq 0 - 5$ ... (subtract 5 from both sides)

$x + 0 \geq -5$

$x \geq -5$

Thus, the domain of f(x) or the value of x is $x \geq -5$, i.e, domain: $\{x: x \geq -5\}$

In interval notation: $[-5, \infty)$.

- The range is the value of y.

Range: $\{y: y \geq 0\}$

In interval form: $[0, \infty)$.

Exercise 13.9-13.9.2

Answer each of the following questions

1) Find the domain and range of each of the following equation.

a) $f(x) = \sqrt{2x + 1}$

b) $f(x) = \sqrt{x - 9}$

c) $f(x) = \sqrt{x+4}$

d) $f(x) = \sqrt{6-x}$

e) $f(x) = \sqrt{5-10x}$

2) Write each of the following in interval notation.
   a) $x < 5$
   b) $6 - x \geq 24$
   c) $y \leq 4$
   d) $2 < x < 9$
   e) $2 < x$
   f) $-2 < x \leq 0$
   g) $3 < 6x < 12$
   h) $-1 < x < 4$

3) Choose the correct answer for each of the following questions.
   (i)   Which of the following is the domain of $f(x) = \sqrt{1 - 2x}$
         a) $(-\infty, \frac{1}{2})$ b) $(\frac{1}{2}, \infty)$ c) $(-\infty, \frac{1}{2}]$ d) $[\frac{-1}{2}, \frac{1}{2}]$
   (ii) Which of the following is the interval notation of $-3 < x \leq 4$?
         a) $[-3,4)$ b) $(-3,4)$ c) $[-3,4]$ d) $(-3,4]$
   (iii) Which of the following is the domain of the expression, $\sqrt{-x - 5} > 0$?
         a) All real numbers   b) $x \geq 0$ c) $y \geq 0$ d) $x < -5$
   (iv) The domain of a function is the value of 'x'.
         a) True   b) False
   (v)   The range of a function is the value of 'y'.
         a) True   b) False
   (vi) What is the domain of $\sqrt{(x - 4)} \geq 6$
         a) $x \leq 40$ b) $x \geq 40$ c) $y \leq 40$ d) $y \geq 40$
   (vii) $-2x > 6$ can be written as
         a) $x > 3$ b) $x < -3$ c) $x > -3$ d) $x < 3$
   (viii) The range of $f(x) = \sqrt{(x - 4)}$ is
         a) $x \geq 0$ b) All real numbers   c) $y \geq 0$ d) Doesn't have range
   (ix) The domain of $f(x) = 2\sqrt{(x - 4)} < 4$ is
         a) $x > 8$ b) $x < 8$ c) $x < -8$ d) $x > -8$
   (x)   What is the range of $f(x) = \sqrt{(x - 4)}$?
         a) All real numbers   b) $y \geq 0$ c) $y > 4$ d) Doesn't exist

CHAPTER 14

Statistics and Probability

14.1. Statistics

Statistics is a branch of applied mathematics dealing with data collection, organization, analysis, interpretation and presentation.

A statistician is someone who works with theoretical or applied statistics. Such people determine the type and size of the sample to be surveyed, after before the data are collected.

14.1.1. Data: - are the facts and figures that are collected, analyzed, and summarized for presentation and interpretation.

14.1.2. Raw Data: - Raw data, also known as a primary data, are data collected from the source.

Types of Data

14.1.3. Qualitative Data: It is a data that cannot be measured or expressed in numbers.

Examples: - smells, tastes, religion, intelligence.

14.1.4. Quantitative Data: - is a data that deals with numbers and things that can be measured objectively.

Quantitative data answers the questions, such as: "How many," How much," "How long" etc. Examples: - Weight, height, width, pressure, temperature, area, etc.

14.1.5. Variable: - A variable is any characteristics, number, or quantity that can be measured or counted. A variable may also be called a data item.

Examples: class grade, age, height, weight. etc.

14.2. Frequency Distribution Table

Frequency distribution table is a chart that summarizes values and their frequency. A frequency distribution has two columns. The first column lists all the values outcomes that happen in the data, and the second column lists the frequency of each outcome:

Example: - If we have the data 8, 2, 4, 3, 2, 5, 4, 8, 4, 9, 9, 9, 9.

The frequency distribution for the above data is shown below.

Value	2	3	4	5	8	9
Frequency	2	1	3	1	2	4

- In the above data, the value 2 occurs twice, thus the frequency is 2.
- The value 3 has a frequency 1.
- The value 4 has a frequency 3.
- The value 5 has a frequency 1.
- The value 8 has a frequency 2.
- The value 9 has a frequency 4

14.3. Measures of Location and Dispersion for Grouped Data.

When we make comparisons between groups of numbers in a given data, it is good to have a single value that is considered to be a good representative of each group. One such value is the average of the group. Colloquially, averages are also called measures of location or measures of central tendency. The most common measures of central tendency are, mean, (or arithmetic mean), median and mode.

14.4. Summation Notation

In mathematics, a summation notation or sigma notation is a method used to add a sequence of any kind of numbers in a concise way.

The summation is denoted by using $\sum$ notation, where $\sum$ is an enlarged capital Greek letter sigma.

Summation notation is written as below.

$$\sum_{i=1}^{n} x_i$$

n = upper limit .... (Where to stop)

Argument

    i = 1 starting point .... (Lower limit of summation)

                        .... (Index of summation)

$\sum$ = Greek letter "sigma"

Thus, sum of values of x from x_1 through x_n is given by: -

$$\sum_{i=1}^{n} x_i = x_1 + x_2 + x_3 + ... + x_n.$$

Example1) If we want to compute the sum of the first 12 natural numbers, we can write this obviously as:

$$1 + 2 + 3 + 4 + 5 + 6 + 7 + 8 + 9 + 10 + 11 + 12 = 78$$

In summation notation we can represent this as below: -

$$\sum_{i=1}^{12} x_i = 78$$

We read as, "The sum of all i." Where i runs from 1 to 12. The symbol i is called the "summation index."

Example 2: If we want to add the sequence $a_n = 2n + 1$, for $a \geq 1$,

we can write the sum $a_2 + a_3 + a_4 + a_5 + a_6$ as: -

$$\sum_{n=2}^{6} (2n+1) = (2(2) + 1) + (2(3) + 1) + (2(4) + 1) + (2(5) + 1) + (2(6) + 1)$$
$$= 5 + 7 + 9 + 11 + 13$$
$$= 45$$

The index variable 'i' is considered a 'dummy variable' meaning that it may be changed to any letter without affecting the value of the summation.

14.5. Mean

The mean $\bar{x}$ of a set of data is equal to the sum of the data items divided by the number of items contained in the data set.

$$\bar{x} = \frac{x_1 + x_2 + x_3 + x_4 + ... + x_n}{n} = \frac{\sum_{i=1}^{n} x_i}{n}$$

Suppose $x_1, x_2, x_3, ..., x_n$ is a set of data items, with frequencies $f_1, f_2, f_3, ..., f_n$ respectively, then

$$\bar{x} = \frac{f_1 x_1 + f_2 x_2 + f_3 x_3 + f_4 x_4 + ..., + f_n x_n}{f_1 + f_2 + f_3 + f_4 + ... + f_n} = \frac{\sum_{i=1}^{n} fnx_i}{\sum_{i=1}^{n} f}$$

Examples

1) Find the mean of the following data: 2, 6, 8, 14, 15
   Solution:
   First add the data
   $2 + 6 + 8 + 14 + 15 = 45$
   The number of data (frequency) is 5.
   Mean $= \dfrac{45}{5}$
   Mean $= 9$

2) The age of twelve students are: 24, 26, 30, 32, 30, 24, 24, 34, 38, 40, 34 and 42. Find the arithmetic mean of the data.

Solution

There are 12 data and the frequency of the data is 12.

$$\overline{x} = \frac{\sum x_i}{n} = \frac{24+26+30+32+30+24+24+34+38+40+34+42}{12}$$

$$= \frac{378}{12}$$

$$\overline{x} = 31.5$$

Thus, the mean of the given data is 31.5

3) Calculate the mean of 8, 12, 13, 14, 18, 22

Solution:

$$\overline{x} = \frac{8+12+13+14+18+22}{6}$$

$$\overline{x} = \frac{87}{6}$$

$$\overline{x} = 14.5$$

4) Below are the given data for Biology score of sixteen students out of hundreds.

76, 84, 76, 84, 66, 94, 84, 66, 94, 96, 84, 94, 92, 94, 94, 76

a) Prepare a frequency distribution table.

b) Find the mean of the data from the frequency distribution.

Solution: a) From the above raw data, we found the following frequency distribution which shows the biology score for sixteen students.

First, we have to list the data either in increasing or decreasing order and then list each value with the corresponding frequency on the distribution table.

Here, let us list the values in increasing order:-

66,66,76,76,76,84,84,84,84,92,92,94,94,94,94,94,96

V	66	76	84	92	94	96
f	2	3	4	1	5	1

b) The mean of the data from frequency distribution table is given by; -

$$\overline{x} = \frac{f_1 x_1 + f_2 x_2 + f_3 x_3 + f_4 x_4 + \dots, + f_n x_n}{f_1 + f_2 + f_3 + f_4 + \dots + f_n} = \frac{\sum_{i=1}^{n} f n x_i}{\sum_{i=1}^{n} f}$$

$$\frac{2 \times 66 + 3 \times 76 + 4 \times 84 + 1 \times 92 + 5 \times 94 + 1 \times 96}{2+3+4+1+5+1}$$

$$\overline{x} = \frac{1{,}354}{16}$$

$$\overline{x} = 84.625$$

Thus, the mean of the frequency distribution is given by: $\overline{x} = 84.625$

5) The following table is the frequency distribution of a certain population. Find the mean.

V	6	8	11	12	15
f	4	3	5	8	2

Solution:

To find the mean, we use the formula

$$M = \frac{\Sigma f_i x_i}{\Sigma f_i} = \frac{4 \times 6 + 3 \times 8 + 5 \times 11 + 8 \times 12 + 2 \times 15}{4+3+5+8+2}$$

$$M = \frac{229}{22}$$

$$M \approx 10.41$$

6) In a certain high school, the student took seven examinations of different subjects and scored: 74, 66, 55, 90, 86, and 92, in six of the subjects respectively. If his average score was 78, what was his score in the seventh subject?

Solution: -

Number of subjects were seven.

Average score = 78

Total sum of six subjects = 463

Let x be the score of the seventh subject, then,

$$\frac{74+66+55+90+86+92+x}{7} = 78$$

$$\frac{463+x}{7} = 78$$

463 + x = (78)(7) ....Multiply both sides by 7

463 + x = 546

463-463+x+= 546-463 ....Subtract 463 from both sides.

x = 83

Thus, the student's score of the seventh subject is 83

14.5.1. Properties of Mean

1. The set of numerical data has one and only one mean.
2. Mean is the most reliable measure of central tendency because it takes into account every item in the set of data.
3. If we add a constant number(N) to each number of a population having mean $\overline{x}$, the new mean will be $(\overline{x} + N)$.
4. If we subtract a constant number (N) from each number of populations having mean $\overline{x}$, the new mean will be $(\overline{x} - N)$.
5. If each member of a population function with mean is multiplied by N the new mean will be $\overline{x} N$.
6. If each member of a population function with mean $\overline{x}$ is divided by N, the new mean will be $\dfrac{\overline{x}}{N}$ or $\dfrac{1}{N}.(\overline{x})$.
7. The sum of the deviations of all values of the distribution from their arithmetic mean is zero. $\sum(x_i - \overline{x}) = 0$

Let us see some examples for properties (3-7).

Consider the number of books in the bag of seven students:

6,11,14,15,20,22,24

Here, the mean, $\quad \overline{x} = \dfrac{6+11+14+15+20+22+24}{7}$

$$\overline{x} = \dfrac{112}{7}$$

$$\overline{x} = 16$$

If we add two books for each student in a bag, the new mean will be: -

$$\dfrac{(6+2)+(11+2)+(14+2)+(15+2)+(20+2)+(22+2)+(24+2)}{7}$$

$$= \dfrac{126}{7}$$

$$= 18$$

Thus, the new mean is $\overline{x} + 2$, which is $16 + 2 = 18$, that is, the new mean is simply the previous mean plus two. $(16 + 2) = 18$.

- If we take away two books from each student, the new mean will be: -

$$\dfrac{(6-2)+(11-2)+(14-2)+(15-2)+(20-2)+(22-2)+(24-2)}{7}$$

$$= \dfrac{98}{7}$$

$= 14$

Thus, the new mean is the previous mean minus two. $(16 - 2) = 14$

- If we multiply the number of books in the bag of each student by two, the new mean will be: -

$$\frac{(6 \times 2) + (11 \times 2) + (14 \times 2) + (15 \times 2) + (20 \times 2) + (22 \times 2) + (24 \times 2)}{7}$$

$$= \frac{224}{7}$$

$$= 32$$

Thus, the new mean is simply the previous mean multiplied by 2. Thus, $16 \times 2 = 32$.

- If the number of books in the bag of each student is divided by 2, the new mean will be:

$$\frac{\frac{6}{2} + \frac{11}{2} + \frac{14}{2} + \frac{15}{2} + \frac{20}{2} + \frac{22}{2} + \frac{24}{2}}{7}$$

$$= \frac{56}{7}$$

$$= 8$$

Thus, the new mean will be the previous mean divided by two $= (16 \div 2) = 8$

- The sum of the deviation of all values of the distribution from their arithmetic mean is zero.

$$\frac{(x_1 - m) + (x_2 - m) + (x_3 - m) + (x_4 - m) + (x_5 - m) + (x_6 - m) + (x_7 - m)}{7}$$

$$\frac{(6 - 16) + (11 - 16) + (14 - 16) + (15 - 16) + (20 - 16) + (22 - 16) + (24 - 16)}{7}$$

$$= \frac{-10 + (-5) + (-2) + (-1) + 4 + 6 + 8}{7}$$

$$= \frac{-18 + 18}{7}$$

$$= \frac{0}{7}$$

$$= 0$$

14.6. The Median (Md)

Median is measure of central tendency. It is the middle number when the data is arranged in either ascending or descending order of magnitude. It is a half way point in a data set. When the data is arranged in order (called a data array), the median will be the middle value or will fall half way between the two values.

Examples:

1) Consider the following values which show the number of books sold by a bookstore for 13 days. What is the median of this distribution?

23, 35, 80, 35, 60, 79, 23, 67, 84, 35, 64, 79, 35

Solution

First arrange the data in an increasing or a decreasing order whichever you prefer.

23, 23, 35, 35, 35, 35, 60, 64, 67, 79, 79, 80, 84

Since the number of observations is an odd number.

$$\text{Median} = \left(\frac{n+1}{2}\right)^{th} \text{item} = \left(\frac{13+1}{2}\right)^{th} \text{item}$$

$$= \frac{14}{2}^{th} \text{item}$$

$$= 7^{th} \text{item.}$$

This shows us the median, the middle value, is the 7th item, counting the ordered data either from the left or from the right direction gives 60. Therefore, 60 is the median of the data.

2) Find the median of each of the following data: -

64, 44, 59, 47, 80, 36

Solution

First, let us write the data in an increasing order: 36, 44, 47, 59, 68, 80

Since the number of data is 6, n = 6, which is even. Thus, we are supposed to use the second formula, thus,

$$\text{Median} = \frac{\left(\frac{n}{2}\right)^{th} \text{item} + \left(\frac{n}{2}+1\right)^{th} \text{item}}{2}$$

$$\text{Median} = \frac{\left(\frac{6}{2}\right)^{th} \text{item} + \left(\frac{6}{2}+1\right)^{th} \text{item}}{2}$$

$$\text{Median} = \frac{3^{rd} \text{item} + (3+1)^{th} \text{item}}{2}$$

$$= \frac{3^{rd} \text{item} + 4^{th} \text{item}}{2}$$

$$= \frac{47+59}{2}$$

$$= \frac{106}{2}$$

$$Md = 53$$

Thus, the median, the middle value is 53.

Note: - We can list the data in a decreasing order and solve the given question in the same manner as above.

3) Find the median of each of the following data.

30, 27, 41, 27, 30, 23, 29, 54, 63

Solution:

First, list the data in increasing order.

23, 27, 27, 29, 30, 30, 41, 54, 63

Since the number of data is 9, which is an odd number, in this case n = 9, thus

$$\text{Median} = \left(\frac{n+1}{2}\right)^{th} \text{item}$$

$$= \left(\frac{9+1}{2}\right)^{th} \text{item}$$

$$= \left(\frac{10}{2}\right)^{th} \text{item}$$

$$Md = 5^{th} \text{item}$$

In the above data, which is written in an increasing order, the 5th item is 30. This is the Median or the middle value of the given data.

4) Consider the following values which show the number of cars sold by car dealers for 14 days. What is the median of the distribution?

31, 28, 10, 42, 19, 27, 42, 21, 6, 16, 26, 60, 33, 21

Solution

First, arrange the given data in ascending order.

6, 10, 16, 19, 21, 21, 26, 27, 28, 31, 33, 42, 42, 60

Since the number of observation (data) is 14, and which is an even number, n = 14.

$$\text{Median} = \frac{\left(\frac{n}{2}\right)^{th} \text{item} + \left(\frac{n}{2}+1\right)^{th} \text{item}}{2}$$

$$\text{Md} = \frac{\left(\dfrac{14}{2}\right)^{\text{th}} \text{item} + \left(\dfrac{14}{2} + 1\right)^{\text{th}} \text{item}}{2}$$

$$= \frac{7^{\text{th}} \text{item} + (7+1)^{\text{th}} \text{item}}{2}$$

$$= \frac{7^{\text{th}} \text{item} + 8^{\text{th}} \text{item}}{2}$$

$$= \frac{26 + 27}{2}$$

$$= \frac{53}{2}$$

$$\text{Md} = 26.5$$

Thus, the median or the middle value is 26.5

5) The following table is frequency - distribution of a certain population. Find the median.

Value	8	10	12	17	19
Frequency	4	3	6	2	1

Solution

The total number of observations is: (4 + 3 + 6 + 2 + 1) = 16. It is the total number of frequency. As we know, 16 is an even number. Since $\dfrac{16}{2} = 8$, then the required median is the 8th and the 9th values of the frequency distribution. When the data is arranged in an increasing or decreasing order, the first 4 members have a value 8, the 5th, 6th, and 7th members have value 10, from 8th up to 13th the values are 12, the 14th and 15th members have values 17. And the 16th value is 19. Thus;

$$\text{Median} = \frac{8^{\text{th}} \text{value} + 9^{\text{th}} \text{value}}{2}$$

$$= \frac{12 + 12}{2}$$

$$= \frac{24}{2}$$

$$\text{Md} = 12$$

Thus, the median or the middle value is 12.

6) Find the median of the following list of values: 15, 8, 7, 6, 4, 3.
Solution

384

First, arrange the values in increasing or decreasing order. In this case let's arrange in a decreasing order.

15, 8, 7, 6, 4, 3

The total number of observations is 6. It is obviously an even number. Thus, n = 6. Therefore, the median is given by: -

$$\text{Median} = \left[\frac{\left(\frac{n}{2}\right)^{th} \text{item} + \left(\frac{n}{2}+1\right)^{th} \text{item}}{2}\right]$$

$$\text{Median} = \frac{\left(\frac{6}{2}\right)^{th} \text{item} + \left(\frac{6}{2}+1\right)^{th} \text{item}}{2}$$

$$\text{Median} = \frac{(3)^{rd} \text{item} + (3+1)^{th} \text{item}}{2}$$

$$= \frac{3^{rd} \text{item} + 4^{rth} \text{item}}{2}$$

$$Md = \frac{7+6}{2}$$

$$Md = \frac{13}{2}$$

$$Md = 6.5$$

Note: - The 3rd item in the above list of decreasing order is 7 and the fourth item is 6. Thus, the median or the middle value for the given data is 6.5.

7) Find the median of the following list of values.

5, 3, 2, 4, 7, 9, 11, 24, 21

Solution

First arrange the given data in an increasing order.

2, 3, 4, 5, 7, 9, 11, 21, 24

The total number of observations is 9. It is an odd number. Thus, the median is given by:

$$\text{Median} = \left(\frac{n+1}{2}\right)^{th} \text{item}$$

$$= \left(\frac{9+1}{2}\right)^{th} \text{item}$$

$$= \left(\frac{10}{2}\right)^{th} \text{item}$$

$$Md = 5^{th} \text{item}$$

The fifth item in the data above written in an increasing order is 7. Thus, the median or the middle value is 7.

14.7. The Mode

Mode: The mode of a set of data is one of the measures of central tendency, and it is the value in the data which appears most frequently in the set of values. It is possible to have no mode, one mode, or more than one mode in a given variable.

Note: -

- A set of values which has only one mode is called Unimodal.
- A set of values which has two modes is called Bimodal.
- A set of values which has more than two modes is called Multimodal.
- To find the mode from a frequency distribution table, we can take the value which corresponds to the highest frequency.

Examples:

1) Determine the mode of the following data set.
   3, 4, 3, 2, 7, 8, 3, 2, 5, 3, 11, 3
   Solution
   As you see from the data, 3 occurs five times. It is the most frequent value among others. Therefore, the mode of the data is 3. That is, Mode = 3

2) Find the mode of the following data set.
   2, 5, 7, 9, 12, 2, 5, 4
   Solution
   As you see from the data, 2 and 5 occur twice.. Thus, the mode of the given data set are 2 and 5. That means, the data is bimodal. Because it has 2 and 5 as modes.

3) Find the mode of the following data set.
   3, 5, 7, 8, 9, 4
   Solution: -
   As you see from the data, each value occurs once. That means, the frequency of each value is the same. Thus, the data does not have mode. (No mode)

4) Determine the mode of the following data set.
   6, 8, 4, 6, 7, 8, 6, 7, 4, 8, 7, 9, 11
   Solution
   As you see from the data, 6 occurs three times, 7 occurs three times and 8 occurs three times, they have the highest and same frequency which is 3. Thus, the modes of the given data are 6, 7 and 8.
   This means, the given data has 3 modes (6, 7 and 8). It is trimodal data.

Such type of mode is also called Multimodal.

5) Consider the following table which shows the frequency distribution of a population function v.

v	4	8	9	10	12
f	6	19	7	2	4

What is the mode (modal value) of v?

Solution

As you see from the table, the highest frequency is 19 whose corresponding value is 8. Therefore, 8 is the mode of the given distribution of a population function.

6) Consider the following table which shows the frequency distribution of a population function v.

v	−5	0	4	6	8	9
f	10	10	10	10	10	10

What is the mode (modal value) of v?

Solution

As you see from the table of data, the frequency of each value is the same; which is 10. Thus, the distribution of a population function v doesn't have mode. (No mode).

14.8. Measure of Dispersion

Measure of dispersion is the extent to which a distribution is stretched or squeezed. It indicates how dispersed a set of observation is. In other words, it tells how far the measures of central tendency are reliable.

The common measures of dispersion or measures of variation are: - range, variance, standard deviation, mean absolute deviation (MAD) and median absolute deviation.

14.9. Range

Range: - is defined as the difference between the maximum and the minimum values in a data set.

$Range = x_{max} - x_{min}$

Where 'x' stands for the value in a data set.

Examples:-

1) Find the range of the following data.

9, 2, 5, 4, 7, 26, 13

Solution

The maximum value of the data is 26. The minimum value of the data is 2

Thus, $x_{max} = 26$ and $x_{min} = 2$

$Range = x_{max} - x_{min}$

$= 26 - 2$

Range $= 24$

The range of the given data is 24.

2) Find the range of the following data.

10, 6, 28, 7, 44, 33

Solution

The maximum value of the data is 44.

The minimum value of the data is 6.

Thus, $x_{max} = 44$ and $x_{min} = 6$

Range $= x_{max} - x_{min}$

$\qquad = 44 - 6$

Range $= 38$

Thus, the range of the given data is 38.

Note: - If the values of the given data are all equal, then the range of that data is zero.

3) What is the range of the following data?

6, 6, 6, 6, 6, 6, 6

Solution: - As you see from the given data, all members are equal (the same). The range (the distance from the highest to the lowest) is zero.

Thus; Range $= x_{max} - x_{min}$

$\qquad = x - x$

$\qquad = 6 - 6$

Range $= 0$

4) Find the range of the following frequency distribution.

v	3	7	11	14	16	24
f	10	6	9	11	8	23

Solution

From the frequency distribution table, we see that the maximum value is 24 and the minimum value is 3.

Range $= x_{max} - x_{min}$

$\qquad = 24 - 3$

Range $= 21$

Thus, the range of the given data is 21.

14.10. Standard Deviation

Standard Deviation is a measure of the amount of variation or dispersion of a set of values. It is the most commonly used measure of dispersion. The value of the standard deviation tells how closely the values of a data set are clustered around the mean. In general, a low standard deviation indicates that the values tend to be close to the mean of the set, while a large value of standard deviation for a data set indicates that the values are spread out over a wider range around the mean.

14.11. Variance

Variance is a measure of dispersion of a set of values from the center of their distribution. The more spread the data, the larger the variance is in relation to the mean. Mathematically, variance is defined as the sum of the squared differences between each data point and the mean divided by the number of data values. Here, below is the mathematical formula of variance.

If x_1, x_2, x_3, ..., x_n are n observed values, then the variance for the sample data is given by:-

$$\text{Variance } (s^2) = \frac{(x_1 - \bar{x})^2 + (x_2 - \bar{x})^2 + ... + (x_n - \bar{x})^2}{n}$$

$$= \frac{\sum_{n=i}^{n}(x_i - \bar{x})^2}{n}$$

Where: $\bar{x}$ = The mean value of the observation.
s^2 = Variance
x_i = The value of ith element.
n = The number of elements in the population.
Note:- The quantities $x - \bar{x}$ in the above formula are the deviations of x from the mean.
- Variance is always non-negative, as squares of numbers are always positive (or zero).
- A variance value of zero indicates that all values with in a set of values are identical.
- If x_1, x_2, x_3, ..., x_n are values with corresponding frequencies f_1, f_2, f_3, ..., f_n, then the variance is given by:

$$S^2 = \frac{f_1(x_1 - \bar{x})^2 + f_2(x_2 - \bar{x})^2 + ... + f_n(x_n - \bar{x})^2}{\sum fi}$$

$$S^2 = \frac{\sum_{i=1}^{n} fi(x_i - \bar{x})^2}{\sum_{i=1}^{n} fi}$$

- Standard deviation is the positive square root of variance.
Standard deviation (Sd) $=\sqrt{\text{Variance}}$

$$SD = \sqrt{\frac{\sum_{i=1}^{n}(x_i - \bar{x})^2}{n}}$$

14.11.1.

Steps to find the variance and standard deviation for a given data.

1. Calculate the mean of the data.
2. Find the deviation of each value from the mean and square it.
3. Add the squared deviation.
4. Divide the sum obtained in step 3 by the number of frequency (n) to get a variance.
5. The square root of the result (i.e., variance) found in step 4 gives a standard deviation.

Examples:

1) Find the variance and standard deviation of the following data.
   11, 3, 5, 2, 9, 12, 14, 16
   Solution
   Step-1
   Calculate the mean ($\overline{x}$)

   $$\overline{x} = \frac{11+3+5+2+9+12+14+16}{8}$$

   $$\overline{x} = \frac{72}{8}$$

   $$\overline{x} = 9$$

   The mean is 9.

   Step-2

   Calculate the deviation of each value from the mean and square it.
   $\overline{x} = 9$

x	$(x - \overline{x})$	$(x - \overline{x})^2$
11	$(11 - 9) = 2$	$2^2 = 4$
3	$(3 - 9) = -6$	$(-6)^2 = 36$
5	$(5 - 9) = -4$	$(-4)^2 = 16$
2	$(2 - 9) = -7$	$(-7)^2 = 49$
9	$(9 - 9) = 0$	$0^2 = 0$
12	$(12 - 9) = 3$	$3^2 = 9$
14	$(14 - 9) = 5$	$5^2 = 25$
16	$(16 - 9) = 7$	$7^2 = 49$

   Step-3
   Add the squared deviation.
   $\Sigma(x - \overline{x})^2 = 4 + 36 + 16 + 49 + 0 + 9 + 25 + 49$
   $\Sigma(x - \overline{x})^2 = 188$
   Step-4
   Divide the sum obtained in step-3 by the number of frequency (n).

$$V = \frac{\sum (x - \bar{x})^2}{n}$$

$$\text{Variance} = \frac{188}{8}$$

$$V = 23.5$$

Thus, the variance is 23.5

Standard deviation is the positive square root of the variance.

Hence, **Standard Deviation** $= \sqrt{\text{Variance}}$

$$SD = \sqrt{23.5}$$
$$SD \approx 4.85$$

The standard deviation of the data is 4.85.

2) Calculate the variance and standard deviation of the following data.

2, 4, 9, 3, 7, 11

Solution

First find the mean ($\bar{x}$) of the data.

$$\bar{x} = \frac{2+4+9+3+7+11}{6}$$

$$= \frac{36}{6}$$

$$\bar{x} = 6$$

$$\text{Variance} = \frac{(2-6)^2 + (4-6)^2 + (9-6)^2 + (3-6)^2 + (7-6)^2 + (11-6)^2}{6}$$

$$V = \frac{(-4)^2 + (-2)^2 + (3)^2 + (-3)^2 + 1^2 + 5^2}{6}$$

$$V = \frac{16+4+9+9+1+25}{6}$$

$$V = \frac{64}{6}$$

$$V \approx 10.7$$

Thus, the variance is approximately equal to 10.7

Standard deviation is a positive square root of variance, hence

$$\textbf{SD} = \sqrt{\text{Variance}}$$

$$\textbf{SD} \approx \sqrt{10.7}$$

$$\textbf{SD } 3.27$$

Thus, the standard deviation is approximately equal to 3.27.

3) Calculate the variance and standard deviation of the following data.
1, 2, 1, 7, 2, 1, 8, 5, 1, 2, 7, 1, 4, 2, 7, 4, 1, 8, 12

Solution

Let us make a table for the given distribution.

v	1	2	4	5	7	8	12
f	6	4	2	1	3	2	1

Mean $(\overline{x}) = \dfrac{(1\times6)+(2\times4)+(4\times2)+(5\times1)+(7\times3)+(8\times2)+(12\times1)}{19}$

$\overline{x} = \dfrac{76}{19}$

$\overline{x} = 4$

Variance $= \dfrac{6(1-4)^2 + 4(2-4)^2 + 2(4-4)^2 + (5-4)^2 + 3(7-4)^2 + 2(8-4)^2 + (12-4)^2}{19}$

$V = 194/19$

$V \approx 10.21$

Thus, the variance of the given data is equal to 10.21
The standard deviation is the positive square root of variance.

SD $= \sqrt{\text{Variance}}$

SD $= \sqrt{10.21}$

SD $\approx 3.195 \approx 3.2$

Thus, the standard deviation is approximately equal to 3.2

4) Find the variance and standard deviation of the following distribution.

v	5	2	4	6	8
f	2	4	6	3	5

Solution
Calculate the mean

Mean $(\overline{x}) = \dfrac{(5\times2)+(2\times4)+(4\times6)+(6\times3)+(8\times5)}{20}$

$= \dfrac{10+8+24+18+40}{20}$

$\overline{x} = \dfrac{100}{20}$

$$\overline{x} \quad 5$$

Variance $= \dfrac{2(5-5)^2 + 4(2-5)^2 + 6(4-5)^2 + 3(6-5)^2 + 5(8-5)^2}{2+4+6+3+5}$

$$V = \dfrac{2(0) + 4(-3)^2 + 6(-1)^2 + 3(1)^2 + 5(3)^2}{20}$$

$$V = \dfrac{0 + 36 + 6 + 3 + 45}{20}$$

$$V = \dfrac{90}{20}$$

$$V = \dfrac{9}{2}$$

$$V = 4.5$$

Thus, the variance of the given data is 4.5

Standard deviation is the positive square root of variance.

SD = √Variance

$= \sqrt{4.5}$

SD= 2.12

Thus; the standard deviation is approximately equal to 2.12.

5) Given below is a frequency distribution of a population function v.

v	−3	−2	0	5	3	10
f	6	2	3	8	4	2

a) What is the mean?
b) What is the mode?
c) What is the median?
d) What is the range?
e) What is the variance?
f) What is the standard deviation?
g) What percent of the population is negative?
h) What percent of the population is non-negative value?

Solutions

a) Mean $(\overline{x}) = \dfrac{(-3 \times 6) + (-2 \times 2) + (0 \times 3) + (5 \times 8) + (3 \times 4) + (10 \times 2)}{6 + 2 + 3 + 8 + 4 + 2}$

$$\bar{x} = \frac{-18 + (-4) + 0 + 40 + 12 + 20}{25}$$

$$\bar{x} = \frac{-22 + 72}{25}$$

$$\bar{x} = \frac{50}{25}$$

$$\bar{x} = 2$$

The mean of the given data is 2.

b) The mode is the highest frequent value. From the given data above, the highest frequent value is 5. Thus, the mode is 5.
[The frequency of the value 5 is 8 and which is the highest frequency of the given data, the mode, which is the highest frequency of the given data is 5.]

c) In the given frequency distribution of the table, the values are listed neither in an increasing nor in decreasing order. Before we determine the median, we have to list the values as an increasing order. Thus,

v	−3	−2	0	3	5	10
f	6	2	3	4	8	2

As you see in the above table the values are listed as increasing order so that it is easy to determine the median, the middle value of the given data.
The total number of values in the frequency distribution is 25, which is an odd number. Thus, the median for n = 25 is given as follows: -

$$\text{Median} = \left(\frac{n+1}{2}\right)^{th} \text{item, where n} = 25$$

$$= \left(\frac{25+1}{2}\right)^{th} \text{item}$$

$$= \left(\frac{26}{2}\right)^{th} \text{item}$$

$$\text{Median} = 13^{th} \text{item}$$

The 13th item in the data is the value 3. So, the median, the middle value of the given data is 3.

d) The range is the difference between the maximum and the minimum values.
$$\text{Range} = x_{max} - x_{min}$$
$$= 10 - (-3)$$
$$= 10 + 3$$

Range = 13

Thus; the range of the given value is 13.

e) Variance $= \dfrac{\sum\limits_{i=1}^{n} fi(x_i - \bar{x})^2}{\sum\limits_{i=1}^{n} fi}$

The mean, $\bar{x}$ of the data is 2.

$\bar{x} = 2$

Variance $= \dfrac{6(-3-2)^2 + 2(-2-2)^2 + 3(0-2)^2 + 8(5-2)^2 + 4(3-2)^2 + 2(10-2)^2}{25}$

$V = \dfrac{6(25) + 2(16) + 3(4) + 8(9) + 4(1) + 2(64)}{25}$

$V = \dfrac{150 + 32 + 12 + 72 + 4 + 128}{25}$

$V = \dfrac{398}{25}$

$V = 15.92$

Thus; the variance of the given data is 15.92

f) The standard deviation is the positive square root of variance.

Standard Deviation $= \sqrt{\text{Variance}}$

$\textbf{SD} = \sqrt{15.92}$

$\textbf{SD} \approx \textbf{3.99}$

Thus, the standard deviation is approximately equal to 3.99

g) Total number of frequencies is the sum of all frequency $(\sum f)$.

$\sum f = 6 + 2 + 3 + 8 + 4 + 2 = 25$

There are $(6 + 2) = 8$ negative valued data.

Thus; $\dfrac{8}{25} \times 100\% = 32\%$

Thus, 32% of the frequency distribution is negative value.

h) There are $(3 + 8 + 4 + 2) = 17$ non-negative valued data, thus,

$\dfrac{17}{25} \, 100\% = 68\%$

Thus, 68% of the frequency distribution is non-negative valued.

Exercise 14.1-14.11.1

1) The weight of 7 students in kilograms is as follows.
   84, 90, 72, 68, 48, 24, 62
   a) What is the mean of the data?
   b) What is the median of the given data?
   c) What is the mode of the given data?
   d) What is the range of the given data?
   e) What is the variance of the given data?
   f) What is the standard deviation of the given data?
   g) What will happen to the mean, if each student loses 7 kilograms?
   h) What will happen to the mean, if each students gain 5 kilograms?

2) Below is the frequency distribution of the banana eaten by the students.
   26, 2, 2, 6, 5, 2, 16, 2, 5, 4, 7
   a) What is the mean of the data set?
   b) What is the median of the data set?
   c) What is the mode of the data set?
   d) What is the range of the data set?
   e) What is the variance of the data set?
   f) What is the standard deviation of the data set?
   g) What percent of the students eat banana less than 5?
   h) What percent of the students eat banana equal to 5?
   i) What percent of the students eat banana greater than 5?

3) Given below is the frequency distribution of the population to the nearest year of the ages of certain 9th grade class students.

v	14	16	18	19	20	21	26	28
f	4	3	7	6	4	8	1	1

   a) What is the mean of the given frequency distribution?
   b) What is the median of the given frequency distribution?
   c) What is the mode of the given frequency distribution?
   d) What is the range of the given frequency distribution?
   e) What is the variance of the given frequency distribution?
   f) What is the standard deviation of the given frequency distribution?
   g) What percent of the students in the class is younger than the modal age?
   h) What percent of the students in the class is older than the mean?
   i) What percent of the students in the class equal to the modal age?

4) In a final examination, a student took six examinations and scored 64, 84, 66, 58, and 90 in five subjects. If his average score was 70. What was his score in the sixth subject?

5) Calculate the mean, mode, median, range, variance and standard deviation of the following distribution.

v	3	5	8	9	10	11
f	4	6	3	4	5	8

6) The mean grade of 12 students was 84. If three of the students with grade of 72, 64 and 82 are cancelled what should be the mean of the remaining data?

7) Given below is the frequency distribution of a certain data: -

v	−3	−1	0	4	6	8
f	5	3	6	2	1	3

Find
a) The mean
b) The mode
c) The median
d) The range
e) The variance
f) The standard deviation
g) What percent of the frequency distribution valued less than zero?
h) What percent of the frequency distribution valued greater than zero?
i) What percent of the frequency distribution valued equal to zero?

8) Consider the following values which show the age of the students in a class.
0, 20, 18, 16, 20, 15, 16, 13, 19, 16, 18, 20, 16, 19, 15, 15
a) Prepare a frequency distribution table.
b) Find the mean age of the students from the frequency distribution table.
c) Find model age of the students from the frequency distribution table.
d) Find median age of the students from the frequency distribution table.
e) Find the variance of the data from the frequency distribution table.
f) Find the standard deviation of the data from the distribution table.

9) Find the mode of each of the following data.
a) 7, 4, 3, 2, 1, 2, 3, 2.
b) 4, 8, 6, 3, 2, 1

c)

v	−2	4	6	8	9
f	6	8	13	2	7

10) The mean of five numbers is 66.2, if four of the numbers are 60, 54, 50, and 75. What is the value of the fifth number?

14.12. Probability

Probability is the branch of mathematics that deals with a type of ratio where we compare events how many times an outcome can occur compared to all possible outcomes.

Before we proceed to the next section, we need to know the meaning of some terms.

- An experiment is any activity having an observable result (outcome).
- An outcome is any result obtained in an experiment.
- A sample space (s) is a set that contains all possible outcomes of an experiment.
- An event is any set of a sample space.
- Each outcome in a sample space is called an element or a member of the sample space.
- An experiment is said to be random experiment, if it has more than one possible outcome.
- If an experiment has only one possible outcome, then it is known as deterministic experiment.
- An experiment is composed of one or more trials, a trial with two mutually exclusive outcomes is known as Bernoulli trial.
- A population is the set representing all measurements of interest to the sample collector.
- If P denotes a probability and A, B and C denote specific events. The P(A) denotes the probability of Event A occurring.
- Possibility set of an experiment is a set which contain all the possible outcomes of an experiment.
- Probability of an Event A can be expressed by a real number P where $0 \leq P(A) \leq 1$, for any event A.
- The probability of an impossible event is zero.
- The probability of an event that is certain to occur is 1.
- The complement of an Event A, denoted by $\bar{A}$, consists of all outcomes in which Event A does not occur.

Example
- A "fair" coin is tossed twice at random.
a) What are the possible outcomes?
b) What is the probability of getting at least one head?
c) What is the probability of getting the same face?
d) What is the probability of an event getting at least one tail?

Solutions
a) The possible outcomes are: HH, HT, TH, TT. Thus, the possible outcomes equal to 4.
b) The number of outcomes having at least one head are: HH, HT, TH. Which is equal to 3.

Therefore, the probability of getting at least one head is equal to $\frac{3}{4}$.

c) The possible outcomes are: HH, HT, TH, TT. The total number of favorable outcomes for this event is 4.

Thus, HH, TT have the same face and the probability of getting the same face is $\frac{2}{4} = \frac{1}{2}$

d) The number of outcomes having at least one tail are: HT, TH, TT. Thus, the possible outcome having at least one tail is 3. Therefore, the probability of getting at least one tail is $\frac{3}{4}$.

Note: -

In tossing a coin, if the coin is fair, the two possible outcomes have an equal chance of occurring. In this case, we say that the outcomes are EQUALLY LIKELY.

Note: -

In tossing a coin, if the coin is fair, the two possible outcomes have an equal chance of occurring in this case, we say that the outcomes are EQUALLY LIKELY.

- If we consider an experiment of tossing three coins and observing the sequence of head and tails. In this experiment the possibility set is: -

  {(H, H, H), (H, H, T), (H, T, H), (H, T, T), (T, H, T), (T, H, H), (T, T, H), (T, T, T)}, thus, there are 8 different possible outcomes in this experiment.

14.13. Probability of an Event (E).

If S is possibility set of an experiment and if each element of S is equally likely to occur, then the probability of the event E is given by P(E), where: -

$$P(E) = \frac{\text{Number of Elements in E}}{\text{Number of Elements in S}}$$

$$P(E) = \frac{n(E)}{n(S)}$$

Examples

1) A box contains 3 white and 4 black balls. If one ball is drawn at random, what is the probability of getting a
   a) white ball?
   b) black ball?
   Solution
   Let W be a white ball appears and B be a black ball appears.
   The total number of events is $(3 + 4) = 7$

   Thus, a) $P(W) = \frac{n(W)}{n(S)} = \frac{3}{7}$

   Where S is total number of events.

   Thus, probability of getting the white ball is $\frac{3}{7}$,

b) $P(W) = \dfrac{n(B)}{n(S)} = \dfrac{4}{7}$

Thus, probability of getting black ball is $\dfrac{4}{7}$.

2) If a number is to be selected at random from 1 through the 12, what is the probability that the number is: -
   a) Even
   b) Odd
   c) Divisible by 2.
   d) Divisible by 5.

Solution:
S= {1, 2, 3, 4, 5, 6, 7, 8, 9, 10, 11, 12}
a) Even is the Event E = {2, 4, 6, 8, 10, 12}

$$P(\text{Even}) = \dfrac{\text{number of even}}{\text{Total number}}$$

$$= \dfrac{6}{12}$$

$$P(E) = \dfrac{1}{2}$$

Thus, the probability of getting even number = $-$.

b) Odd is the Event E = {1, 3, 5, 7, 9, 11}

$$P(\text{Odd}) = \dfrac{\text{number of odd}}{\text{total number}}$$

$$= \dfrac{6}{12}$$

$$= \dfrac{1}{2}$$

Thus, the probability of getting odd number = $\dfrac{1}{2}$.

c) Divisible by 2 is the Event E = {2, 4, 6, 8, 10, 12}.

$$P(\text{Divisible by 2}) = \dfrac{\text{Number of divisible by 2}}{\text{total number}}$$

$$= \dfrac{6}{12}$$

$$= \dfrac{1}{2}$$

Thus, the probability of the number divisible by 2 is $\dfrac{1}{2}$.

d) Divisible by 5 is the Event E = {5, 10}.

$$P(\text{Divisible by 5}) = \frac{\text{Number of divisible by 5}}{\text{Total number}}$$

$$= \frac{2}{12}$$

$$= \frac{1}{6}$$

Thus, the probability that the number is divisible by 5 is $\frac{1}{6}$.

3) If you toss a fair dice, what is the probability to see:
   a) An even number.
   b) The number 5.
   c) The number that is greater than 4.
   d) The number that is less than 1.
   e) The number that is a multiple of 2.

Solution

Our possibility set is S = {1, 2, 3, 4, 5, 6}.

a) Even is the Event E = {2, 4, 6}

$$P(\text{Even}) = \frac{P(E)}{P(S)} = \frac{3}{6} = \frac{1}{2}$$

Thus, the probability to see even number is $\frac{1}{2}$

b) The possibility set is {5}.

Probability to see the number 5 $= \frac{n(E)}{n(S)}$

$$= \frac{1}{6}$$

Note: - A dice is a small cube which has one up to six spots or numbers on its six sides. Each side has only one such number.

c) The possibility set that the numbers greater than 4 is {5, 6}.

The probability to get the number that is greater than 4 $= \frac{n(E)}{n(S)} = \frac{2}{6} = \frac{1}{3}$

d) In our possibility set S = {1, 2, 3, 4, 5, 6}, there is no number less than 1. Thus, the possibility set is 0. Therefore, probability that is less than 1 $= \frac{n(E)}{n(S)} = \frac{0}{6} = 0$

e) From our total possibility set S = {1, 2, 3, 4, 5, 6}, the numbers that are multiple of 2 are 2, 4 and 6.

Thus, our possibility set is {2, 4, 6}.

Therefore, the probability to get a number which is a multiple of $2 = \dfrac{n(E)}{n(S)} = \dfrac{3}{6} = \dfrac{1}{2}$

4) Two coins are tossed. What is the probability that two tails are obtained?
   Solution
   • Let S be our sample space and it is given by S = {(H, T), (H, H), (T, H), (T, T)}.
   Where H represents for head and T represents for tail.
   • Let E be the event "Two tails are obtained."
   E = {(T, T)}, Thus,

   $P(E) = \dfrac{n(E)}{n(S)} = \dfrac{1}{4}$. (Probability of obtaining two tails)

   (Meaning the probability of getting two tails at a time when simultaneously tossed is (T,

   T) $= \dfrac{1}{4}$

5) A box contains 8 black balls. One ball is drawn at a random. Find the probability of getting:
   a) a black ball
   b) a red ball

   Solution
   a) The box contains all black balls; so, we are sure that black will occurring any draw.
      Thus, the probability of getting a black ball is one.

      That is, $P(B) = \dfrac{n(B)}{n(S)} = \dfrac{8}{8} = 1$

   b) Obviously, the box contains no red balls. The chance of getting red ball is impossible and
      the probability is zero. Thus,

      $P(R) = \dfrac{n(R)}{n(S)} = \dfrac{0}{8} = 0$

      i.e., The probability of getting red ball is zero.

6) If three coins are tossed, then what is the probability to see tail in succession?
   Solution: - The three events are independent events.

   Probability to see tail in three succession is equal to: $\dfrac{1}{2} \times \dfrac{1}{2} \times \dfrac{1}{2} = \dfrac{1}{8}$

14.14. Types of Events

14.14.1. Simple Event (or Single Event): - is an event with a single outcome.

Example: In a toss of one coin, getting "tail" is a Simple Event.

14.14.2. Compound Event: - is an event consisting two or more simple events (with two or more outcomes).

Example: - Rolling a die and getting an odd number less than 6 is {1, 3, 5}.

14.15. Occurrence or Non-occurrence of an Event.

Occurrence or non-occurrence of the event are two possibilities during a certain experiment. Example: - If a die is thrown, then the sample space: S = {1, 2, 3, 4, 5, 6}. Let E be the event of getting even number. Thus E = {2, 4, 6}. If we throw the die and getting 4, then $4 \in E$; meaning E has occurred. If we throw the die again and getting 2, then $2 \in E$; meaning E has occurred. If in the other trial, the outcome is 5, then as $5 \notin E$, we say that E has not occurred.

14.16. Complement of an Event

The complement of an Event A is the set of all outcomes that is not A. The complement of an Event A is usually denoted as A′ or $\bar{A}$. (A′, read as: A - complement) or ($\bar{A}$, read as: A bar).

Example: If a coin is tossed, it can either land showing "head" or "tail. This is because the coin cannot show both head and tail at the same time. As there are no other possible outcomes which is not represented between these two, they are complement of each other.

14.16.1. Complement Rule

The probability of an event and its complement always has the sum equal to 1. (An event either occurs or doesn't occur.) Thus,
$$P(A) + P(A') = 1$$
$$P(A') = 1 - P(A)$$

Examples:

1) A fair die is thrown. What is the probability of getting an odd number?
   Solution: Sample space S = {1, 2, 3, 4, 5, 6}
   The favorable outcomes are: 1, 3 and 5.
   $$P(O') + P(O) = 1$$
   $$P(O') = 1 - P(O)$$
   Where 'O' stands for an odd number.
   $$P(O') = 1 - \frac{1}{2}$$
   $$P(O') = \frac{1}{2}$$

2) When tossing a fair die, the probability of not getting a number 5 is given by:
   Solution
   $$P(\bar{5}) + P(5) = 1$$

$$P\left(\overline{5}\right) = 1 - P\left(5\right), \text{ where } P\left(5\right) = \frac{1}{6}$$

$$= 1 - \frac{1}{6}$$

$$= \frac{6-1}{6}$$

$$P(\overline{5}) = \frac{5}{6}$$

Thus, probability of not getting 5 is equal to $\frac{5}{6}$.

14.17. Algebra of Events

Algebra of events is a set of events related to one another by the operations like and, or, and not.

14.17.1. Exhaustive Events: are events where at least one of them must necessarily occur every time when the experiment is performed.

Example: When we throw a die, the sample space is; S = {1, 2, 3, 4, 5, 6}. From this, events: {1}, {2}, {3}, {4}, {5}, {6} are exhaustive events. Moreover; the events {1, 2}, {3, 4}, {4, 5, 6} are also exhaustive events where this experiment is performed. Thus;
{1} ∪ {2} ∪ {3} ∪ {4} ∪ {5} ∪ {6} ∪ {4, 5, 6} = {1, 2, 3, 4, 5, 6}
All events shown above are exhaustive because they form a complete sample space by themselves. More generally, Events E_1, E_2, E_3, ..., E_n form a set of exhaustive events of a sample space S where E_1, E_2, E_3, ..., E_n are subsets of S and $E_1 \cup E_2 \cup E_3 \cup ... \cup E_n = S$

14.17.2. Mutually Exclusive Events

Mutually Exclusive Events are two or more events that cannot happen at the same time. The Venn-diagram below shows two mutually exclusive events A and B.

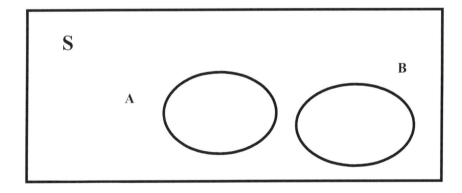

Fig. 14.1

The Venn-diagram above displays the sample space 'S' and two mutually exclusive events A and B, which are disjoint; meaning they don't have any common events. So that it is not possible for the events A and B to occur at the same time.

Examples

1) When a coin is tossed once, either we get head or tail but we cannot get both at the same time. Hence, {H} and {T} are mutually exclusive events.

2) Suppose a die is rolled,
   E_1 = Getting an even number.
   E_2 = Getting a multiple of 3.
   E_1 and E_2 are not mutually exclusive because 6 is an even and a multiple of 3 at the same time.

14.18. Exhaustive and Mutually Exclusive Events.

14.18.1. Exhaustive Event is an event when one of the events must occur from the list of events.

14.18.2. Mutually Exclusive Events: are two or more events that cannot occur at the same time.

  i.e., Two or more events are known to be mutually exclusive if they have no common outcomes.
  Examples: Consider the following two events 'A' and 'B' below:
      A= {1,5, 9,11}
      B = {0,2,4,6,8}
  Events A and B are called mutually exclusive since they have no common element in them.
      i.e., A∩B= {   }
  * Two or more events are called non - mutual exclusive events, if they share common outcomes
  Example: Let event: P = {1, 2, 3} and
      Q = {1,4, 5}
  Events P and Q are non-mutual exclusive events since they share common element.
      i.e., P∩Q= {1}

14.19. Independent Events: - are the two events whose happening or not happening of one event does not affect the happening or not happening of the other.

  Example: - When we throw two coins simultaneously, the event of getting a head on the first coin and the event of getting a head on the second coin are independent.

14.20. Dependent Events: - Two events are said to be dependent, if the happening or not happening of one event affects the happening or not happening of the other event.

Example: - If a card is chosen at random from a standard deck of 52 cards, without replacing it, then the result of drawing a second card is dependent on the first draw.

Exercise 14.12-14.20

Answer each of the following question

1) A bag contains 6 red, 8 black, and 5 white balls. One ball is picked up randomly. What is the probability that it is neither red nor white?

2) If two dice are thrown, then what is the probability to see number 5 on the upper faces of both dice?

3) If you toss a fair dice, what is the probability to see: -
   a) The number 2.
   b) An odd number.
   c) An even number.
   d) A number less than 6.
   e) A number greater than 4.
   f) A number that is a multiple of 2.

4) If a "fair" coin is tossed twice at random:
   a) What are the possible outcomes?
   b) Give the sample space.
   c) Give the event of H appearing on the second throw.
   d) Give the event at least one T appearing.

5) A fair die is thrown, what is the probability of getting a multiple of 2?

6) A fair die is thrown, what is the probability of not getting a number 6?

CHAPTER 15
SOLVED WORD PROBLEMS

15.1. Solved Algebra Word Problems

Algebraic word problems are questions that require translating sentences into their equivalent mathematical equations. Thus, if a problem is stated in English sentence, then these given sentences can be translated into algebraic equations. Such equations may have one or more variable(s) that are going to be solved. Usually, these variables represent an unknown quantity in a real-life background.

Examples:

15.1.1. Addition

- The sum of a number and 8, can expressed in algebraic expression as: $x + 8$
- Nine is added to a number can expressed in algebraic expression as: $x + 9$
- A number increased by ten is expressed in algebraic expression as: $x + 10$
- Five more than a number is expressed as an algebraic expression as: $x + 5$
- Twice a number is increased by 5 is expressed as an algebraic expression as: $2x + 5$
- Y plus nine: $y + 9$
- Five more than twice a number is twenty is expressed as an algebraic expression as: $5 + 2x = 20$

15.1.2. Subtraction

- A number decreased by ten: $x - 10$
- A number decreased by ten equals eight: $x - 10 = 8$
- 3 less than a number: $x - 3$
- X minus 4: $x - 4$
- Six is subtracted from a number: $x - 6$
- The difference between x and 4: $x - 4$
- A number minus 5 is equal to six: $x - 5 = 6$

15.1.3. Multiplication

- Five times a number: $5x$
- The product of a number and six: $6x$

- Three times a number: 3x
- The product of a number and 5 equals 7: 5x = 7

15.1.4. Division

- A number is divided by eight: $\dfrac{x}{8}$

- Ten divided by a number: $\dfrac{10}{x}$, where $x \neq 0$

- Twenty divided by twice of a number: $\dfrac{20}{2x}$, where $x \neq 0$.

- The quotient of a number to six is 5: $\dfrac{x}{6} = 5$

- A number is divided by seven equals eight: $\dfrac{x}{7} = 8$

Examples of solved word problems.

1) The sum of two numbers is 64. If the larger is divided by the smaller, the quotient is 4 and the remainder is 14. What are the numbers?

Solution:

Let x be the smaller number and 64 – x be larger number.

Thus, $\dfrac{\text{Larger number}}{\text{smaller number}} = 4 + \dfrac{14}{\text{smaller number}}$

$$\dfrac{64 - x}{x} = 4 + \dfrac{14}{x}$$

$$\dfrac{64 - x}{x} = \dfrac{4x + 14}{x} \quad \text{...............(LCM = x)}$$

$$64 - x = 4x + 14 \text{...............(x is cancelled).}$$

$$-x - 4x = 14 - 64 \text{...............(collect like terms)}$$

$$-5x = -50$$

$$\dfrac{-5x}{-5} = \dfrac{-50}{-5} \quad \text{..................(Divide both sides by –5)}$$

$$x = 10$$

Thus, the smaller number is 10. To find the larger number, substitute x = 10 in 64 – x; This means 64 – 10 = 54. Hence the larger number is 54. Therefore, the two numbers are 54 and 10.

2) Six more than eight times a number is forty-six. Find the number

Solution

Let x be the required number.

Six more than eight times a number = 8x + 6, Thus, 8x + 6 = 46

$8x + 6 - 6 = 46 - 6$..............................Subtract 6 from both sides.

$\qquad 8x = 40$

$\qquad \dfrac{8x}{8} = \dfrac{40}{8}$ ....Divide both sides by 8.

$\qquad x = 5$

Thus, the required number is 5.

Note: Consecutive Numbers are numbers that follow one another continuously in the sequence form. Suppose, if each number is being one more than the previous number, this fact can be represented by x, (x + 1), (x + 2), (x + 3), ...

Examples

Consecutive positive numbers

$\qquad$ 1, 2, 3, 4, 5, ...

Consecutive negative numbers

$\qquad$ ..., –5, –4, –3, –2, –1

Consecutive odd numbers

$\qquad$ 1, 3, 5, 7, 9, 11, ...

Consecutive even numbers

$\qquad$ 0, 2, 4, 6, 8, ...

3) The sum of two consecutive integers is 95, what are the numbers?

Solution

Let x be the smaller number, then (x + 1) will be the larger number. Thus,

$\qquad x + (x + 1) = 95$

$\qquad 2x + 1 = 95$

$\qquad 2x = 95 - 1$

$\qquad 2x = 94$

$\qquad \dfrac{2x}{2} = \dfrac{94}{2}$ ....Divide both sides by 2.

$\qquad x = 47$

The smaller number is 47, to find the larger number substitute x = 47 in (x + 1), which is (47 + 1) = 48.

Thus, the two numbers are 47 and 48.

4) The product of two consecutive odd integers is 143. What are the two numbers?

Solution

Let x be the smaller number and (x + 2) the larger number.

Why the larger number is (x + 2)?

This is because each consecutive odd number has a difference of 2.

Example: 1, 3, 5, 7, 9, ...

$\qquad x(x + 2) = 143$

$\qquad x^2 + 2x = 143$

$x^2 + 2x - 143 = 0$

$x^2 + 13x - 11x - 143 = 0$ ... Factorization.

$x(x + 13) - 11(x + 13) = 0$

$(x - 11)(x + 13) = 0$

$x - 11 = 0$ or $x + 13 = 0$

$x = 11$ or $x = -13$

When $x = 11$, $(x + 2) = (11 + 2) = 13$

When $x = -13$, $(x + 2) = (-13 + 2) = -11$

Therefore, the two numbers are 11 and 13 or –13 and –11.

5) The sum of three consecutive integers is 276. What are the three integers?

Solution

Let x be the smallest number, $(x + 1)$ be the next number and $(x + 2)$ be the largest number. Thus,

$x + (x + 1) + (x + 2) = 276$

$3x + 3 = 276$

$3x = 276 - 3$

$3x = 273$

$\dfrac{3x}{3} = \dfrac{273}{3}$ ....Divide both sides by 3.

$x = 91$

The first number is 91. To find the second number, substitute $x = 91$ in $(x + 1)$ which is $91 + 1 = 92$; and to find the third number, substitute $x = 91$ in $(x + 2)$ which is $(91 + 2) = 93$.

Therefore,

The first number is 91.

The second number is 92 and the third number is 93.

i.e, 91, 92 and 93.

6) Find four consecutive positive integers whose sum is 342.

Solution

Let x be the first, $(x + 1)$ the second, $(x + 2)$ the third and $(x + 3)$ the fourth numbers.

i.e, The first number: x

The second number: x + 1

The third number: x + 2

The fourth number: x + 3

Thus, the sum of the number is:

$x + (x + 1) + (x + 2) + (x + 3) = 342$

$4x + 6 = 342$

$4x = 342 - 6$

$4x = 336$

$\dfrac{4x}{4} = \dfrac{336}{4}$

$x = 84$

The first number is 84, the second number (x + 1) is (84 + 1) = 85, the third number (x + 2) = 86 and the fourth number (x + 3) is (84 + 3) = 87.

i.e, 84, 85, 86 and 87

7) Twelve less than six times a number is 42. Find the number.

Solution

Let x be the required number, then translating the sentence into equation and solving for it gives:

$6x - 12 = 42$

$6x - 12 + 12 = 42 + 12$ .... Add 12 on both sides.

$6x = 54$

$\dfrac{6x}{6} = \dfrac{54}{6}$ .... Divide both sides by 6.

Thus, the required number is 9.

8) Two times the difference of a number and 3 is equal to the number increased by 10. What is the number?

Solution

Let x be the required number, then translating the sentence into equation gives:

$2(x - 3) = x + 10$

Solving for x results in:

$2x - 6 = x + 10$

$2x - x = 10 + 6$ .... Collecting like terms

$x = 16$

Thus, the required number is 16.

9) Ten times the sum of a number and 5 is equal to 20 times a number decreased by 8. What is the number?

Solution

Let x be the required number. Translating the sentence into equation yields:

$10(x + 5) = 20(x - 8)$

Solve for x

$10(x + 5) = 20(x - 8)$

$10x + 50 = 20x - 160$ .... Distribute property.

$10x - 20x = -160 - 50$ .... Collecting like terms

$-10x = -210$

$\dfrac{-10x}{-10} = \dfrac{-210}{-10}$ .... Divide both sides by –10

$x = 21$

Thus, the required number is 21.

10) Find two consecutive odd integers such that three times the smaller exceeds the larger by 8.

Solution

Let x be the smaller odd number, then (x + 2) is the larger odd number. Translating the sentence into equation results in:

3x - (x + 2) = 8

3x – x – 2 = 8

2x - 2 = 8

2x = 2 + 8...Collecting like terms

2x = 10

$x = \dfrac{10}{2}$

x = 5

Thus, the smaller odd number is 5. To find the larger odd number, substitute x = 5 in (x + 2) which is 5 + 2 = 7

Hence, the two consecutive odd numbers are 5 and 7.

11) Sophia sold three times as much oranges in the morning than in the afternoon. If she sold 480 kgs. of oranges that day, how many kilograms did she sell in the morning and how many kilograms in the afternoon?

Solution

Let x be the number of kilograms she sold in the afternoon. Then, she sold 3x kilograms in the morning. Thus, the total oranges sold will be: -

x + 3x = 480

4x = 480

$\dfrac{4x}{4} = \dfrac{480}{4}$

x = 120

Thus, Sophia sold 120 kgs oranges in the afternoon and 3 × 120 = 360 kgs oranges in the morning.

12) A number is 12 more than another. The sum of twice the smaller plus four times the larger is 66. What are the two numbers?

Solution

Let x be the smaller number. Then the larger number is 12 more than the smaller one. That means, the larger number is (x + 12).

When we translate these sentences into equation, we get:

2x + 4 (x + 12) = 66

2x + 4x + 48 = 66

6x + 48 = 66

6x = 66 – 48

6x = 18

$\dfrac{6x}{6} = \dfrac{18}{6}$ ....Divide both sides by 6.

x = 3

Thus, the smaller number is 3. To find the larger number, substitute the value of x = 3 into (x + 12). It means: (3 + 12) = 15.

i.e., The smaller number is 3 and the larger number is 15.

13) A farmer has dogs and hens on his farm with 56 heads and 174 legs. What is the number of hens and the number of dogs?

Solution

Let D be the number of dogs and H be the number of hens. It is obvious that each dog and each hen have one head. Thus,

(*) D + H = 56

Each dog has four legs and each hen has two legs. Thus,

(**) 4D + 2H = 174

Combine (*) and (**). Then, solve for D and H.

i.e., $\begin{cases} D + H = 56 \\ 4D + 2H = 174 \end{cases}$

Thus, after solving this simultaneous equation, D=31 and H=25.

Therefore, the number of dogs are 31 and the number of hens are 25.

14) The sum of a number and its square is 132. What is the number?

Solution: Let x be the required number, then

$x^2 + x = 132$

$x^2 + x - 132 = 0$

$x^2 + 12x - 11x - 132 = 0$ ... Factorizing

$x(x + 12) - 11(x + 12) = 0$

$(x - 11)(x + 12) = 0$

$x - 11 = 0$ or $x + 12 = 0$

$x = 11$ or $x = -12$

Thus, the required number is 11 or –12.

15) The sum of the squares of two numbers is 117 and their product is 54. Find the two numbers.

Solution

Let x and y be the two required numbers. Then,

(*) $x^2 + y^2 = 117$ ... (sum of squares of two numbers)

(**) $xy = 54$ ... (product of two numbers)

Combining (*) and (**) and then solving for x and y gives:

$\begin{cases} x^2 + y^2 = 117 \\ xy = 54 \end{cases}$

$xy = 54$

$x = \dfrac{54}{y}$

Solve for y by substituting $x = \dfrac{54}{y}$ in equation (*).

$x^2 + y^2 = 117$

$$\left(\frac{54}{y}\right)^2 + y^2 = 117$$

$$\frac{2916}{y^2} + y^2 = 117$$

$$y^2\left(\frac{2916}{y^2} + y^2\right) = 117 \,(y^2) \dots\dots\dots\dots\text{Multiply both sides by } y^2$$

$2916 + y^4 = 117y^2$

$y^4 - 117y^2 + 2916 = 0 \dots\dots\dots\dots\text{(Take } 117y^2 \text{ to the left side and change the sign)}$

$y^4 - 81y^2 - 36y^2 + 2916 = 0 \dots\dots\dots\text{Factorizing.}$

$y^2(y^2 - 81) - 36\,(y^2 - 81) = 0$

$(y^2 - 36)\,(y^2 - 81) = 0$

$y^2 - 36 = 0 \text{ or } y^2 - 81 = 0$

$y^2 = 36 \text{ or } y^2 = 81$

$y = \pm\sqrt{36}, \; y = \pm\sqrt{81}$

$y = \pm6 \text{ or } y = \pm9$

Thus, to find the value of x, substitute the value of y in $x = \dfrac{54}{y}$

$x = \dfrac{54}{y}$, when $y = 6$, $x = \dfrac{54}{6} = 9$, (9, 6)

when $y = -6$, $x = \dfrac{54}{-6} = -9$, (−9, −6)

$x = \dfrac{54}{y}$, when $y = 9$, $x = \dfrac{54}{9} = 6$, (6, 9)

$x = \dfrac{54}{y}$, when $y = -9$, $x = \dfrac{54}{-9} = -6$, (−6, −9)

Thus, the two numbers can be (x, y), (9, 6) or (−9, −6) and (6, 9) or (−6, −9).

The same results can be found as follows:

The original equation is $y^4 - 117y^2 + 2916 = 0$

$(y^2)^2 - 117y^2 + 2916 = 0$

Let $y^2 = c$.

$y^4 - 117y^2 + 2916 = 0 \dots\dots\dots\dots\text{can be rewritten as } c^2 - 117c + 2916 = 0$

$= c^2 - 36c - 81c + 2916 = 0$

$= c\,(c - 36) - 81(c - 36) = 0$

$= (c - 36)\,(c - 81) = 0$

$= c = 36 \text{ or } c = 81$

Since $c = y^2$, we can say $y^2 = 36$ or $y^2 = 81$

$(y^2)^{1/2} = (36)^{1/2}$ or $(y^2)^{1/2} = (81)^{1/2}$

$y = \pm 6$ or $y = \pm 9$

For the remaining steps, follow the same ways as above.

Thus, to find the value of x, substitute the value of y in $x = \dfrac{54}{y}$

$x = \dfrac{54}{y}$, when y = 6, $x = \dfrac{54}{6} = 9$, (9, 6)

when y = −6, $x = \dfrac{54}{-6} = -9$, (−9, −6)

$x = \dfrac{54}{y}$, when y = 9, $x = \dfrac{54}{9} = 6$, (6, 9)

$x = \dfrac{54}{y}$, when y = −9, $x = \dfrac{54}{-9} = -6$, (−6, −9)

Thus, the two numbers can be (x, y), (9, 6) or (−9, −6) and (6, 9) or (−6, −9).

16) Think of a number, triple the number. Subtract 8 from the result and divide the answer by 4. The quotient will be 22. What is the number?

Solution

Let x be the number, then tripled the number = 3x. Subtracting 8 from the result = 3x − 8.

Dividing the answer by 4 = $\dfrac{3x-8}{4}$. The quotient of overall operation = $\dfrac{3x-8}{4} = 22$.

Thus, solving for x yields:

$\dfrac{3x-8}{4} = 22$

$(4)\dfrac{(3x-8)}{4} = 22 \times 4$ ....Multiply both sides by 4.

$3x − 8 = 88$

$3x − 8 + 8 = 88 + 8$ ....Add 8 on both sides.

$3x = 96$

$\dfrac{3x}{3} = \dfrac{96}{3}$ ....Divide both sides by 3.

$x = 32$

Thus, the required number is 32.

17) The sum of two positive numbers is 8 and the sum of their squares is 34. What are the two numbers?

Solution: Let the two required numbers be a and b. Then, their sum: a + b = 8.

The sum of their product: $a^2 + b^2 = 34$

    (*) $a + b = 8$

    (**) $a^2 + b^2 = 34$

Combine (*) and (**) and solve for a and b.

    $a + b = 8$

    $a = (8 - b)$

Substitute $a = (8 - b)$ in the second equation (**) $a^2 + b^2 = 34$

    $(8 - b)^2 + b^2 = 34$; where $a = 8 - b$

    $64 - 16b + b^2 + b^2 = 34$

    $2b^2 - 16b + 64 = 34$

    $2b^2 - 16b + 64 - 34 = 0$

    $2b^2 - 16b + 30 = 0$

    $\dfrac{1}{2}(2b^2 - 16b + 30) = 0\left(\dfrac{1}{2}\right)$ .... Multiply both sides by $\dfrac{1}{2}$.

    $b^2 - 8b + 15 = 0$

    $b^2 - 3b - 5b + 15 = 0$ .... Factorizing

    $b(b - 3) - 5(b - 3) = 0$

    $(b - 5)(b - 3) = 0$

    $b - 5 = 0$ or $b - 3 = 0$

    $b = 5$ or $b = 3$

To find the value of a substitute the values of b in $a = 8 - b$,

    $a = 8 - b$

When $b = 3$, $a = 8 - 3$, $a = 5$, $(a, b) = (5, 3)$

When $b = 5$, $a = 8 - 5$, $a = 3$, $(a, b) = (3, 5)$

Thus, the two numbers are 5 and 3 or 3 and 5.

18) Alex chose a number, multiplied by 4, then subtracted 124 from the result and got 96. What was the number he chose?

    Solution: Let x be the number he chose, then multiplied by 4. 4x, subtracted 124 from the result 4x – 124 and the result is 96; 4x – 124 = 96. And solve for x.

    $4x - 124 = 96$

    $4x - 124 + 124 = 96 + 124$

    $4x = 220$

    $\dfrac{4x}{4} = \dfrac{220}{4}$

    $x = 55$

    Thus, the required number is 55.

19) Kimberly has 80 dollars, which is eight dollars more and three times what David has. How much dollars do David have?

    Solution

    Let x be the money David's has, then,

    $8 + 3x = 80$

$3x = 80 - 8$

$3x = 72$

$\dfrac{3x}{3} = \dfrac{72}{3}$

$x = 24$

Thus, David has 24 dollars.

20) The sum of the digits of a two-digit number is 14. When we interchange the digits, it is found that the resulting new number is greater than the original number by 36. What is the two-digit number?

Solution: Let x be the digit at tens place, then the digit at one place will be (14 – x). Thus, the original two-digit number is 10x + (14 – x). After interchanging the digits, the new number is equal to 10(14 – x) + x, based on the equation: -

$\quad$ 10x + (14 – x) + 36 = 10 (14 – x) + x

$\quad$ original digits $\qquad$ interchanged digits

Solve for x

$\quad$ 10x + (14 – x) + 36 = 10(14 – x) + x

$\quad$ 10x – x + 14 + 36 = 140 – 10x + x

$\quad$ 9x + 50 = 140 – 9x

$\quad$ 9x + 9x = 140 – 50......................................Collect like terms.

$\quad$ 18x = 90

$\quad$ $\dfrac{18x}{18} = \dfrac{90}{18}$ ....Divide both sides by 18.

$\quad$ x = 5

Hence, to find the original number, substitute x = 5 in the original digits

$\quad$ 10x + (14 – x)

$\quad$ 10(5) + (14 – 5)

$\quad$ 50 + 9

$\quad$ 59

Therefore, the two-digit number is 59.

21) The number of hours that were left in the day was one-fifth of the number of hours already passed. How many hours were left in the day?

Solution

Let h be the hours left and p be the hours passed.

Given that: Hours left = $\dfrac{1}{5}$ (hours passed)

$\quad$ Hours left = $\dfrac{1}{5}$ p

$\quad$ h = $\dfrac{1}{5}$ p

Hours already passed + Hours left = Total Hours in a day

$$h + p = \text{Total hours in a day}$$

$$h + p = 24$$

$$\frac{1}{5}p + p = 24$$

$$\frac{p + 5p}{5} = 24 \dots\dots\dots\dots\dots\dots\dots\dots\dots\dots\dots\text{LCM}$$

$$\frac{6p}{5} = 24$$

$$5\left(\frac{6p}{5}\right) = 24(5)\dots\dots\dots\dots\dots\dots\dots\dots\dots\text{Multiply both sides by 5.}$$

$$6p = 120$$

$$p = \frac{120}{6}$$

$$p = 20$$

Thus, the number of hours already passed is 20 hrs and the number of hours left is

$$\frac{1}{5}p = \frac{1}{5}(20) = 4 \text{ hrs.}$$

i.e., The number of hours left is 4 hrs.

22) Alex chose a number. Multiplying it by 3 and then subtracting 121 from the result got 5. What was the number he chose?

Solution

Let x be the number Alex chose, then

$$3x - 121 = 5$$

$$3x = 5 + 121$$

$$3x = 126$$

$$x = \frac{126}{3}$$

$$x = 42$$

23) A woman paid 50 dollars for each day she cleans an office and forfeits 5 dollars for each day she is late. If at the end of 30 days she earns 840 dollars. How many days was she late?

Solution: Let x be number of days late;(30-x) be number of days she cleaned; Thus,

Amount earned–Amount forfeited = Net earned

$$50(30 - x) - 5x = 840$$

$$1500 - 50x - 5x = 840 \dots\dots\dots\text{Use distributive property}$$

$$1500 - 55x = 840$$

$$1500 - 1500 - 55x = 840 - 1500 \dots\dots\dots\text{Subtract 1500 from both sides}$$

$$-55x = -660$$

$$\frac{-55x}{-55} = \frac{-660}{-55} \dots\dots\dots\dots\text{Dividing both sides by } -55$$

x= 12

Thus, a woman was late for 12 days.

15.2. Solved Age Word Problems

Age word problems are algebra word problems that deal with the ages of the people. Age word problems basically involves comparing two people's ages at different points in time. Thus, the time may be at present, in the past or in the future. This lesson involves age word problems that can be solved using one variable and age word problems that can be solved using two or more variables. If the age problem involves for a single person, then it can be solved using one variable and if the age problem involves the ages of two or more people it can be solved with two or more variables. Let's see the following examples.

1) Jhon is 20 years younger than Robert, in two years, Robert will be twice as old as Jhon. What are the present ages of Jhon and Robert?

Solution

• Let the present age of Robert be R, then we can solve Jhon's age in terms of R. Thus, the first sentence tells us the present ages (current ages). i.e., Jhon is 20 years younger than Robert.

If Robert's age is R, then Jhon's age (which is 20 years younger) will be (R – 20). And the second sentence tells us two things: -

1) The age difference for both Jhon and Robert increases by two years.

2) In two years, Robert will be twice the age of Jhon. in two years.

To understand it easily, we can make age problem chart as shown below.

Person	Present Age	Age difference (in two years)
Jhon	R – 20	(R – 20) + 2
Robert	R	R + 2

From the last statement, we can drive the equation to solve the age of Jhon and Robert.

$R + 2 = 2 (R - 20 + 2)$

$R + 2 = 2 (R - 18)$ .... Distributive Property

$R + 2 = 2R - 36$

$R + 2 - 2 = 2R - 36 - 2$ .... (subtract 2 from both sides)

$R = 2R - 38$

$R - 2R = 2R - 2R - 38$ .... (subtract 2R from both sides)

$-R = -38$

$R = 38$ .... (Multiplying both sides by –1)

Thus, the present age of Robert is 38 years.

Jhon's present age is R – 20

(See from the table)

Substitute R = 38 in R – 20 = 38 – 20 = 18

Jhon's present age is 18 years.

2) Alex is 14 years younger than David. If the sum of their age is 28. How old is David?

Solution:-

Let x be the age of Alex, since Alex is 14 years younger than David, we have to add 14 years to Alex's age to denote David's age.

That is, David's age is (x + 14), the sum of their age is 28. Thus

$\quad$ x + (x + 14) = 28

$\quad$ 2x + 14 = 28

$\quad$ 2x = 28 – 14

$\quad$ 2x = 14

$\quad$ x = 7

Alex's age is 7 years

David's age is (x + 14) = (7 + 14) = 21 years.

3) Sofia is 10 years older than James. Ten years ago, she was twice as old as he. What is the present age of James?

Solution: - Let x be the present age of James (smaller number).

Since Sofia is 10 years older than James, Sofia's present age is (x + 10). Ten years ago, Sofia was twice as old as James. Thus, from the above statements we can drive the equation.

$\quad$ (x + 10) – 10 = 2(x – 10)

$\quad$ x = 2x – 20

$\quad$ x – 2x = –20

$\quad$ –x = – 20

$\quad$ x = 20

James's present age is 20 years.

Note: - Sofia's present age is 30 years.

4) David is 41 years old. In seven years, he will be four times as old as his sister. How old is his sister now?

Solution

Let x be the age of David's sister. In seven years, the age of David is four times as old as the age of his sister. Thus,

In seven years, David's age is 41 + 7 and his sister age will be (x + 7).

$\quad$ 4 (x + 7) = 41 + 7

$\quad$ 4x + 28 = 48

$\quad$ 4x = 48 – 28

$\quad$ 4x = 20

$\quad$ x = $\dfrac{20}{4}$

$\quad$ x = 5

Thus, the present age of David's sister is 5 years.

5) Barbara's present age is 65 years and Charlotte's present age is 15 years. In how many years will Barbara's age be three times as old as Charlotte's age?

Solution: Let x be the required number of years. From the above word problem, we can create mathematical relation between their age. Thus,

Barbara's age after x years = 65 + x

Charlotte's age after x years = 15 + x

After x years Barbara's age is three times Charlotte's age. That is,

$$65 + x = 3(15 + x)$$
$$65 + x = 45 + 3x$$
$$65 - 45 = 3x - x$$
$$20 = 2x$$
$$x = 10$$

Therefore, in 10 years, Barbara will be three times as old as Charlotte.

6) Eight years ago, Robert was four times as old as John. Now Robert is only three times as old as John. Find their present ages.

Solution: Let x be the present age of John (smaller age), then Robert's present age will be 3x. Eight years ago John's age was x – 8.

Eight years ago, Robert's age was (3x – 8)

Eight years ago, since Robert's age was four times that of John's age, we have:

$$3x - 8 = 4 (x - 8)$$
$$3x - 8 = 4x - 32$$
$$3x - 4x = -32 + 8$$
$$-x = -24 \text{ (Multiply both sides by } -1)$$
$$x = 24$$

Thus, present age of John is 24 years and present age of Robert is 3 x 24 = 72 years.

7) Sofia is 5 times as old as Emma. After 7 years, Sofia will be four times as old as Emma. Find the present ages of Sofia and Emma.

Solution: - Let Emma's present age be x years. Then, Sofia's present age will be 5x.

After 7 years, Emma's age = x + 7. Since Sofia's age after seven years is four times as old as Emma's age it will be 4 (x + 7).

Thus, 5x + 7 = 4 (x + 7)

$$5x + 7 = 4x + 28$$
$$5x - 4x = 28 - 7$$
$$x = 21$$

Hence, Emma's present age is 21 years and Sofia's present age is 5 x 21 = 105 years.

Note: - When we do age related word problems, we use plus sign for the phrase "after x years" i.e., for the future and minus sign for the phrase "x years ago" i.e., for the past.

8) Susan is 6 years older than Maria. Next year Susan will be 3 times as old as Maria. Find their present ages.

Solution: Let x be the present age of Susan. Then, the present age of Maria is (x – 6).

Next year, Susan's age will be (x + 1) years and Maria's age will be (x – 6) + 1 = x – 5 years.

Again, for the following year, since Susan's age will be 3 times as old as that of Maria, we can write this relation as an equation: x + 1 = 3 (x – 5)

$$x + 1 = 3 (x - 5)$$
$$x + 1 = 3x - 15$$
$$x - 3x = -1 - 15$$
$$-2x = -16$$
$$x = \frac{-16}{-2}$$
$$x = 8$$

Thus, the present age of Susan is 8 years and the present age of Maria is $(x - 6) = (8 - 6) = 2$ years.

9) Linda's age now is six times it was 15 years ago. How old is Linda now?
   Solution: - Let x be the present age of Linda.
   Linda's age 15 years ago: $x - 15$
   Linda's age is now six times as it was 15 years ago; this can be written as an equation:

   $$x = 6 (x - 15)$$
   $$x = 6x - 90 \dots\dots\dots\dots\dots\dots\dots\dots\text{Distributive property.}$$
   $$x - 6x = 6x - 6x - 90 \dots\dots\dots\dots\text{Subtract 6x from both sides.}$$
   $$-5x = -90$$
   $$\frac{-5x}{-5} = \frac{-90}{-5} \dots\dots\dots\dots\dots\dots\text{Divide both sides by } -5$$
   $$x = 18$$

   Thus, Linda's present age is 18 years.

10) William is 29 years old and he is also 7 years older than his brother, James. How old is James?
    Solution
    Let W be the age of William and J be the age of James. From the given problem, we can write the equation: -

    $$W = 29 \dots\dots\dots\dots\dots\dots\dots\dots\text{(Present age of William)}$$
    $$W = J + 7 \dots\dots\dots\dots\dots\dots\dots\text{(William is 7 years older than James).}$$
    Substitute 29 into the second equation and solve for J.

    $$W = J + 7$$
    $$29 = J + 7$$
    $$29 - 7 = J + 7 - 7 \dots\dots\dots\dots\dots\text{Subtract 7 from both sides.}$$
    $$22 = J$$
    Thus, James is 22 years old.

11) Thirteen years from now, Emma will be five times as old as she was nineteen years ago. How old is Emma now?
    Solution

Let E represents Emma's present age. From the given information, we can write the following equation.

$$E + 13 = 5(E - 19)$$
$$E + 13 = 5E - 95 \dots\dots\dots\dots\dots\dots\dots\dots\dots \text{Distributive property}$$
$$E + 13 - 13 = 5E - 95 - 13 \dots\dots\dots\dots \text{Subtract 13 from both sides.}$$
$$E = 5E - 108$$
$$E - 5E = 5E - 5E - 108 \dots\dots\dots\dots\dots \text{Subtract 5E from both sides.}$$
$$-4E = -108$$
$$\frac{-4E}{-4} = \frac{-108}{-4} \dots\dots\dots\dots\dots\dots\dots\dots \text{Divide both sides by } -4.$$
$$E = 27$$

Thus, Emma is 27 years old now.

12) The sum of the ages of Ethan and Mason is 49. In three years, Ethan will be four times as old as Mason. How old are Ethan and Mason now?

Solution

Let E be the present age of Ethan and M be the present age of Mason. Thus, from the word problem, it is given that

$$E + M = 49 \dots\dots\dots\dots\dots\dots\dots\dots\dots \text{The first statement}$$
$$E + 3 = 4(M + 3) \dots\dots\dots\dots\dots\dots\dots \text{The last statement.}$$

Combining the two equations and solving for E and M results:

$$E = 49 - M$$

Substitute the value of E in the second equation.

$$E + 3 = 4M + 12$$
$$(49 - M) + 3 = 4M + 12$$
$$52 - M = 4M + 12$$
$$-M - 4M = 12 - 52 \dots\dots\dots\dots\dots\dots \text{Collect like terms.}$$
$$-5M = -40$$
$$\frac{-5M}{-5} = \frac{-40}{-5} \dots\dots\dots\dots\dots\dots\dots \text{Divide both sides by } -5.$$
$$M = 8$$

To find the value of E, substitute M = 8 in either of the equation.

$$E + M = 49$$
$$E + 8 = 49$$
$$E + 8 - 8 = 49 - 8 \dots\dots\dots\dots\dots\dots \text{Subtract 8 from both sides.}$$
$$E = 41$$

Thus, Ethan's present age is 41 and Mason's present age is 8.

13) Sophia is four years younger than Charlotte. Twenty-two years ago, Charlotte's age was 15 years more than half the age of Sophia. How old are they now?

Solution

Let x be Charlotte's present age. Since Sophia is four years younger than Charlotte, Sophia's present age is (x – 4).

- Twenty-two years ago, is past time and so that Charlotte's age was (x – 22) and Sophia's age was (x – 4 – 22) which is equal to (x – 26). Thus,

Twenty two years ago Charlotte's age $= \frac{1}{2}$(Twenty's two years ago Shophia's age) $+ 15$

Substitute the age that was given 22 years ago for both Charlotte and Sophia results:

$$(x - 22) = \frac{1}{2}(x - 26) + 15$$

$$x - 22 = \frac{1}{2}x - 13 + 15$$

$$x - 22 = \frac{1}{2}x + 2$$

$$x - \frac{1}{2}x = 2 + 22 \dots\dots\dots\dots\dots\dots\dots\dots\text{Collect like terms.}$$

$$\frac{1}{2}x = 24$$

$$x = 24 \times 2$$

$$x = 48$$

Thus, Charlotte's age is 48 years and Sophia's age is (x – 4) = (48 – 4) = 44

14) Isabella is 32 years old. Her son David is 12 years old. In how many years will Isabella be double her son's age?

Solution

Given that Isabella's present age is 32 and David's present age is 12.

Let x be the required number of years. Thus,

$$32 + x = 2 (12 + x)$$

$$32 + x = 24 + 2x \dots\dots\dots\dots\dots\dots\dots\dots\text{Distributive property.}$$

$$x - 2x = 24 - 32 \dots\dots\dots\dots\dots\dots\dots\dots\text{Collecting like terms.}$$

$$-x = -8$$

$$(-1)(-x) = -8(-1) \dots\dots\dots\dots\dots\dots\dots\text{Multiply both sides by } -1$$

$$x = 8$$

Hence, in eight years, Isabella be double her son's age.

Isabella's age	David's age
32 + x	12 + x
32 + 8	12 + 8
40 years	20 years

15) Ethan is nine years older than his sister Nora. If the sum of Ethan and Nora's ages is seventy-five. How old are Ethan and Nora?

Solution

Let E represents Ethan's present age and let N be Nora's present age. Thus,

$$E = N + 9$$

E + N = 75

Substitute the value of E in the second equation.

E + N = 75

(N + 9) + N = 75

2N + 9 = 75

2N = 75 – 9

2N = 66

$\dfrac{2N}{2} = \dfrac{66}{2}$ ....Divide both sides by 2.

N = 33

E = N + 9

E = 33 + 9

E = 42

Thus, Ethan's present age is 42 years and Nora's present age is 33 years.

16) The sum of the present ages of Mason and William is 142 years and the difference of their ages is 34 years. How old are Mason and William now?

Solution

Let M be Mason's present age and W be William's present age. Then from the word problem.

M + W = 142

M – W = 34

Combining these two equations and solve for M and W.

$$+\begin{cases} M + W = 142 \\ M - W = 34 \end{cases}$$ ....Adding vertically.

$$2M + 0 = 176$$

$$2M = 176$$

$$\dfrac{2M}{2} = \dfrac{176}{2}, M = 88$$

Substitute M = 88 in either of the above equation.

M + W = 142

88 + W = 142

88 – 88 + W = 142 – 88

W = 54 years

Thus, Mason's present age is 88 years and William's present age is 54 years.

17) The present ages of Charlotte and her brother are in the ratio of 3:2 and the ratio of their ages will be 7:5 after 10 years. Then, what are the present ages of Charlotte and her brother?

Solution

Let the present age of Charlotte be C and the present age of her brother be B. Then

$$\begin{cases} \dfrac{C}{B} = \dfrac{3}{2} \\ \dfrac{C+10}{B+10} = \dfrac{7}{5} \end{cases}$$

$$\frac{C}{B} = \frac{3}{2}$$

$$C = \frac{3}{2}B$$

Substitute $C = \dfrac{3}{2}B$ in the second equation

$$\frac{\frac{3}{2}B + 10}{B + 10} = \frac{7}{5}$$

$$\frac{3B + 20}{2B + 20} = \frac{7}{5}$$

15B + 100 = 14B + 140 ....Cross multiplication

15B − 14B = 140 − 100

B = 40

Substitute B = 40 in

$$C = \frac{3}{2}B$$

$$C = \frac{3}{2}(40)$$

$$C = 60$$

Thus, Charlotte's present age is 60 years and her brother's present age is 40 years

15.3. Solved Geometry Word Problems

Different types of geometry word problems along step-by-step solutions will help students to practice lots of measurement skills in the same field. In solving geometry word problems, most of the time, you will be asked to solve problems involving geometric relationships like perimeter, length, width or area. Such types of problems are very common in real world construction works and architecture. This lesson also helps to demonstrate how algebra and geometry are integrated. Here, we use algebraic equations to explain measures of geometric figures and solve those problems under discussion.

Examples: -

1) The length of a rectangle is 8 units longer than twice its width. If the area of a rectangle is 24 sq units, then what is the perimeter of a rectangle?

Solution: Let L be the length of a rectangle and W be its width.

To find the required perimeter P, first we have to find the length and the width of a rectangle.

Given that: - L = 2W + 8

$$A = LW$$
$$A = 24$$

Substitute the value of L in A = LW.

$$A = W (2W + 8)$$

W (2W + 8) = 24, where A = 24 is given.

$$2W^2 + 8W = 24$$
$$2W^2 + 8W - 24 = 0$$
$$\frac{1}{2}(2W^2 + 8W - 24) = 0\left(\frac{1}{2}\right) \ldots\ldots\ldots\ldots\ldots \text{Multiply both sides by } \frac{1}{2}.$$

$$W^2 + 4W - 12 = 0$$
$$W^2 - 2W + 6W - 12 = 0$$
$$W (W - 2) + 6 (W - 2) = 0$$
$$(W + 6) (W - 2) = 0$$

W + 6 = 0 or W - 2 = 0

W = -6 or W = 2

Since no negative width, we take W = 2 units.

To find the length L, substitute the value of W = 2 into L = 2W + 8.

$$L = 2(2) + 8$$
$$L = 4 + 8$$
$$L = 12 \text{ units}$$

The perimeter of a rectangle is given by the formula: P = 2L + 2W

$$P = 2L + 2W$$
$$P = 2(12) + 2(2)$$
$$P = 24 + 4$$
$$P = 28 \text{ units}$$

Thus, the perimeter of a rectangle is 28 units.

2) The length of a rectangle is 2 units more than three times its width. The perimeter of the same rectangle is 164 units. Find the length and width of the rectangle.

Solution: -

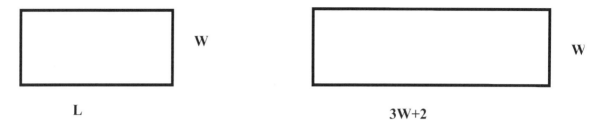

W	W
L	3W+2

Fig. 15.1   **Fig. 15.2**

The perimeter of a rectangle = 2L +2W

    P = 2L + 2W

Given that: L = 3W + 2

         P = 164

        P = 2L + 2W

Thus, substituting the given values give us,

    P = 2L + 2W

    164 = 2 (3W + 2) + 2W

    164 = 6W + 4 + 2W

    164 = 8W + 4

    164 – 4 = 8W

    8W = 160

$$\frac{8W}{8} = \frac{160}{8}$$

    W = 20 units

To find the length L, substitute W = 20 into: L = 3W + 2

    L = 3(20) + 2

    L = 60 + 2

    L = 62 units

Thus, the length of a rectangle is 62 units and its width is 20 units.

3) The legs of a right triangle have lengths of x units and (x + 3) units. The hypotenuse has length $\sqrt{89}$ units. What is the area of the triangle?

Solution

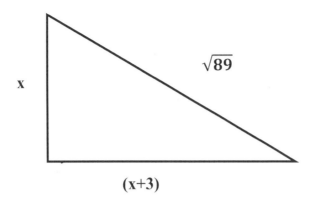

Fig. 15.3

Pythagoras theorem states:

$a^2 + b^2 = c^2$

$x^2 + (x + 3)^2 = \left(\sqrt{89}\right)^2$

$x^2 + x^2 + 6x + 9 = 89$

$2x^2 + 6x + 9 = 89$

$2x^2 + 6x + 9 - 89 = 0$

$2x^2 + 6x - 80 = 0$

$\dfrac{1}{2}(2x^2 + 6x - 80) = 0\left(\dfrac{1}{2}\right)$ ....Multiply both sides by $\dfrac{1}{2}$.

$x^2 + 3x - 40 = 0$

$x^2 - 5x + 8x - 40 = 0$

$x(x - 5) + 8(x - 5) = 0$

$(x + 8)(x - 5) = 0$

$x + 8 = 0$ or $x - 5 = 0$

$x = -8$ or $x = 5$

As negative value (i.e $x = -8$) cannot be the length of the triangle, $x = 5$ is the correct value. Thus,

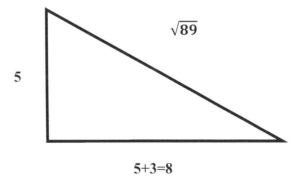

5+3=8

Fig. 15.4

Area of a right triangle $= \dfrac{1}{2}$ b h

$A = \dfrac{1}{2}$ (8) (5)

A = 20 sq units

4) The side of a square exceeds the side of another square by 5 cm and the sum of the area of the two squares is 37 sq cm
Find the dimensions of the squares.
Solution: Let x be the side of the smaller square. The side of a square which exceed by 5 is (x + 5).

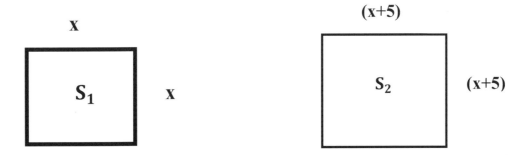

X

S_1

X

(x+5)

S_2

(x+5)

Fig. 15.5	Fig. 15.6

Area of $S_1 = x^2$
Area of $S_2 = (x + 5)^2$
$S_1 + S_2 = x^2 + (x + 5)^2$
$37 = x^2 + (x + 5)^2$
$x^2 + (x + 5)^2 = 37$
$x^2 + x^2 + 10x + 25 = 37$
$2x^2 + 10x + 25 - 37 = 0$
$2x^2 + 10x - 12 = 0$

$\dfrac{1}{2}(2x^2 + 10x - 12) = 0\left(\dfrac{1}{2}\right)$ ....Multiply both sides by $\dfrac{1}{2}$.

$x^2 + 5x - 6 = 0$
$x^2 - x + 6x - 6 = 0$
$x(x - 1) + 6(x - 1) = 0$
$(x + 6)(x - 1) = 0$
$x + 6 = 0$ or $x - 1 = 0$
$x = -6$ or $x = 1$
The correct value of x is 1.

Thus, the length of the smaller square is 1 cm. To find the length of the larger square, substitute x = 1 into x + 5 which is 1 + 5 = 6. Therefore, the length of the larger square is 6 cm.

i.e., Each dimension of the length or the width of the smaller square is 1 cm whereas that of the larger square is 6 cm.

5) The square whose side is (x − 3) and an isosceles triangle whose sides are x, x and 8 have the same perimeters. Find the value of x and the perimeter of a triangle or a square.

Solution

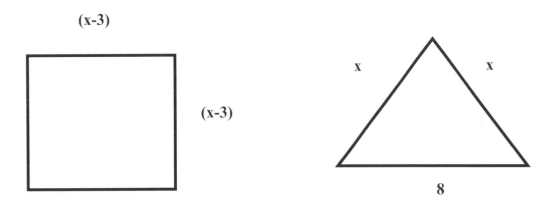

(x-3)

(x-3)

Fig. 15.7   **Fig. 15.8**

Perimeter of square = (x − 3) + (x − 3) + (x − 3) + (x − 3)
  P(S) = 4x − 12
Perimeter of triangle = x + x + 8
  P(T) = 2x + 8
Perimeter of square = Perimeter of triangle
  4x − 12 = 2x + 8
  4x − 2x = 8 + 12
  2x = 20
  $x = \dfrac{20}{2}$

  x = 10
Thus, the value of x is 10.
Perimeter of square = 4x − 12
  = 4(10) − 12
  = 40 − 12
  P(S) = 28
Since perimeter of square is equal to perimeter of triangle, the required perimeter is 28
i.e, x = 10 and perimeter = 28.

6) An isosceles triangle has a perimeter of 70 units. If two of its equal sides have lengths of 8 more than the third side. What is the length of each side?

Solution: Let x be the length of the third side, then (x + 8) will be the length of the side of one of the two equal sides, thus,

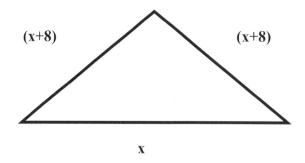

Fig. 15.9

Perimeter of a triangle is the sum of all sides.
Thus,
P = (x + 8) + (x + 8) + x
$\quad$ (x + 8) + (x + 8) + x = 70
$\quad$ 3x + 16 = 70
$\quad$ 3x = 70 – 16
$\quad$ 3x = 54
$\quad$ x = $\dfrac{54}{3}$
$\quad$ x = 18 units
The length of equal sides is (x + 8) = 18 + 8 = 26 units
Therefore, the lengths of the three sides are: 26 units, 26 units and 18 units.

7) The area of a rectangle is 24 sq cm and its perimeter is 20 cm. What is the length and width of a rectangle?
Solution: Let L be the length of a rectangle and W be the width of a rectangle
Area of a rectangle = LW
Perimeter of a rectangle = 2L + 2W
Given that: A = 24 sq cm
$\qquad$ P = 20 cm
$$\begin{cases} LW = 24 \\ 2L + 2W = 20 \end{cases}$$

$$\begin{cases} LW = 24 \\ \dfrac{2L}{2} + \dfrac{2W}{2} = \dfrac{20}{2} \end{cases}$$ ....Divide both sides by 2.

$$\begin{cases} LW = 24 \\ L + W = 10 \end{cases}$$

$$\overline{L + W = 10}$$

$$L = 10 - W$$

Substitute L = 10 – W into LW = 24

(10 – W) W = 24

$10W - W^2 = 24$

$-W^2 + 10W = 24$

$-W^2 + 10W - 24 = 0$

$(-1)(-W^2 + 10W - 24) = 0 \, (-1)$ ....Multiply both sides by –1

$W^2 - 10W + 24 = 0$

$W^2 - 6W - 4W + 24 = 0$ ....Factorize

W (W – 6) – 4 (W – 6) = 0

(W – 4) (W – 6) = 0 ....Taking common factor.

W – 4 = 0 or W – 6 = 0

W = 4 or W = 6

To find L, substitute values of W in L = 10 – W

When W = 4, L = 10 – 4 = 6, L = 6

When W = 6, L = 10 – 6 = 4, L = 4

Therefore; W = 4 cm and L = 6 cm or

W = 6 cm and L = 4 cm

8) A 24 m electric cable is cut into two pieces such that the first is three times larger than the second. Find the length of each piece.

Solution: Let x be the length of the shorter piece. Then, 3x will be the larger piece.

Thus, x + 3x = 24

$$4x = 24$$

$$x = \frac{24}{4}$$

$$x = 6$$

The length of the shorter piece is 6 cm and the length of the larger piece is 3x = 3(6) cm = 18 cm.

9) One leg of a right-angled triangle exceeds the other leg by 4 and the hypotenuse is 8 more than the smaller leg. Find the dimensions of the right-angled triangle:

Solution

Let x be the length of the smaller leg, so that (x + 4) is the other leg and the hypotenuse will be (x + 8).

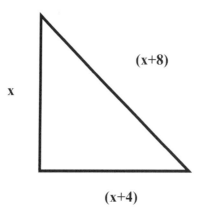

Fig. 15.10

Using Pythagoras theorem

$a^2 + b^2 = c^2$

$x^2 + (x + 4)^2 = (x + 8)^2$

$x^2 + x^2 + 8x + 16 = x^2 + 16x + 64$

$2x^2 + 8x + 16 = x^2 + 16x + 64$

$2x^2 - x^2 + 8x - 16x + 16 - 64 = 0$

$x^2 - 8x - 48 = 0$

$x^2 - 12x + 4x - 48 = 0$

$x (x - 12) + 4 (x - 12) = 0$

$(x + 4) (x - 12) = 0$

$x + 4 = 0$ or $x - 12 = 0$

$x + 4 = 0$ or $x - 12 = 0$

$x = -4$ or $x = 12$

$x = -4$ is invalid because the value of x is negative. The correct value of x is 12.

Therefore,

$x = 12$ units

The length of the smaller leg is 12 units.

The length of the other leg is $(x + 4) = 12 + 4 = 16$ units and

The length of the hypotenuse is $(x + 8) = (12 + 8) = 20$ units.

The dimensions of a triangle are 12 units, 16 units and 20 units.

10) If the height of a triangle is six centimeters less than the length of the base, if the area of the triangle is 36 sq cm. Find the length of the base and the height.

Solution: Let b be the base and h be the height of the triangle, then given that:-

$h = b - 6$

$A = 36$ sq cm.

Area of a triangle $= \dfrac{1}{2} b h$

Substituting $h = b - 6$ into $A = \dfrac{1}{2} b h$ yields:

$A = \frac{1}{2} b (b - 6) = \frac{1}{2} b (b - 6) = 36 ....(A = 36 \text{ given})$

$= 2\left[\frac{1}{2} b(b - 6)\right] = 36(2) ....\text{Multiply both sides by 2}$

$b(b - 6) = 72$
$b^2 - 6b = 72$
$b^2 - 6b - 72 = 0$
$b^2 + 6b - 12b - 72 = 0$
$b(b + 6) - 12 (b + 6) = 0$
$(b - 12) (b + 6) = 0$
$b - 12 = 0 \text{ or } b + 6 = 0$
$b = 12 \text{ or } b = -6$

The correct value of b is 12.
i.e., The base of a triangle is 12 cm.
To find the height, substitute b = 12 cm into h = b – 6
   i.e, h = 12 – 6 = 6 cm
Thus, the base of a triangle is 12 cm and the height is 6 cm.

15.4. Solved Motion Word Problems

1) Two metro trains leave the same station at the same time, one going east and the other west. The east bound train average speed is 60 km/hr., while the west bound train average speed is 50 km/hr. In how many hours will they be 550 km apart?

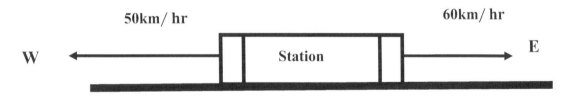

Total distance = 550 km

Let t be the time traveled by each train.
Distance = rate × time
   D = rt
   For East bound train:
Let D_1 = the distance traveled by East bound train.
   rate = 60 km/hr.
   time = t (Both trains depart at the same time)
   $D_1 = 60t$
For West bound train:
Let D_2 = the distance traveled by West bound train.
   rate = 50 km/hr. and time = t

$D_2 = 50t$

Though the distance traveled by both the east and the west bound trains are different due to their rate differences, at unknown time t, they apart 550 kms. That means, the sum of D_1 and D_2 is 550 Km.

$D_1 + D_2 = 550$

$60t + 50t = 550$ .... Substituting $D_1 = 60t$ and $D_2 = 50t$

$110t = 550$

$t = \dfrac{550}{110}$

$t = 5$ hrs.

In 5 hours, the two trains will apart 550 km. Recall that at this time, the distance traveled by East bound train = 60 (t) = 60(5) = 300Km while the distance traveled by West bound train = 50(t) = 50 (5) = 250Km.

2) How long will it take a car traveling 72 km/hr to go 216 kms?

Solution: Distance = rate x time

$D = rt$

Given: D = 216 km

$r = 72$ km/hr

$t = ?$

$D/r = rt/r$ .... Dividing both sides by r

$t = \text{---}$

$t = \dfrac{216\,km}{72\,km/hr}$

$t = 3$ hrs.

Thus, it will take 3 hrs. for the car to travel 216 km at 72 km/hr.

3) Two buses B_1 and B_2 traveling towards each other left from two different station that 1050 miles apart at 4pm. If the average speed of B_1 is 80 mph and that of B_2 is 70 mph.

a) At what time they meet?

b) Where will they meet?

Solution

a) Let t be the time in hours each bus travels before they meet and D_1 and D_2 be the distance traveled by B_1 and B_2 respectively.

Distance= speed × time

$D_1 = 80t$ and $D_2 = 70t$

$D_1 + D_2 =$ **Total distance**

$80t + 70t = 1050$

After t hours the total distance traveled by the two buses is given by:-

$D_1 + D_2 =$ Total distance

$80t + 70t = 1050$

$$150t = 1050$$
$$t = \frac{1050}{150}$$

t = 7 hrs.

a) They meet at 4:00+7:00= 11p.m.
b) They meet at a distance of: -
$D_1 = 80t$
= 80(7)
D_1 =560miles from initial position of B_1 **or**
D_2 =70t
= 70(7)
D_2 = 490 miles from initial position of B_2

4) Alex left his school at 9 am traveling along route at 25 mph. At 11 am, his sister Sofia left her school and started after him on the same road at 35 mph. At what time did Sofia catch up to Alex?

Solution: Let t be the time, r be the rate and d be the distance, thus,
d = rt, t = for Alex
t − 2 = for Sofia

Alex   9a.m. $\boldsymbol{d_1=25t}$

11a.m.   **?**

$\boldsymbol{d_2=35(t-2)}$

Sofia

Alex and Sofia traveled the same distance but different time. Sofia left the school 2 hrs. after her brother. Thus,
$d_1 = d_1$
25t = 35 (t − 2)
25t = 35t - 70
25t − 35t = −70
−10t = −70
$t = \frac{-70}{-10}$

t = 7 hrs.

Sofia catches up her brother in 7 − 2= 5hrs., i.e., 11 + 5= 16hrs.
We are asked at what time? So that
9 am + 7 hrs. = 16, that is
16 − 12 = 4 pm
Therefore, Sofia catches up her brother at 4 pm.

5) A turtle can walk 0.075 kilometer in an hour. The turtle is 0.005 km away from a pond. At this speed, how long will it take the turtle to reach the pond?

Solution:

Speed (v) of a turtle = 0.075 km/hr.

Distance (s) = 0.005 km

$s = vt$, where t is time

$$t = \frac{s}{v}$$

$$t = \frac{0.005 \text{ km}}{0.075 \text{ km} / \text{hr}}$$

$t = 0.0667$ hrs.

$t = 4.002$ minutes

The turtle takes 4.002 minutes to reach in the pond.

6) City X and City Y are located 500 miles from each other. At 6 pm a train T_1 leaves City X with a speed of 55 mph and travels towards City Y. At the same time, a train T_2 leaves City Y traveling towards City X with a speed of 45 mph. Find
   a) At what time ill they meet?
   b) Where will they meet?

Solutions

a) Distance = (rate) (time)

$D = rt$

$$t = \frac{D}{r}$$

$r_1 = 55$ mph

$r_2 = 45$ mph

D= 500miles

Where d_1 is the distance travelled by T_1 and d_2 is the distance travelled by T_2

$d_1 + d_2 = D$

Distance travelled by T_1 + Distance travelled by T_2 = 500miles

$r_1 t + r_2 t = D$

$55t + 45t = 500$

$100t = 500$

$100t = 500$

$t = \dfrac{500}{100}$

$t = 5$ hrs.

Thus, the two trains left at 6 p.m. at the same time and they meet at 11p.m.
(6 pm + 5 hrs.) = 11 pm

b) $d_1 = r_1 t$ or $d_2 = r_1 t$
    = 55 (5)   = 45 (5)
    = 275 miles   = 225 miles

Thus, the two trains meet at a distance 275 miles from the initial position of 'X' or at a distance 225miles from initial position of 'Y'.

7) Helen took a trip to visit her sister. Her sister lives 224 miles away. She drove in town at an average speed of 40 mph, then she drove on interstate highway at an average speed of 60 mph. The total trip took four hours. Find
    a) How long Helen traveled on the interstate?
    b) How long Helen traveled in town?
    c) How far did Helen drive on the interstate highway?
    d) How far did Helen drive in town?

Solutions
a) Given that
    Total distance = 224 miles
    Rate in town = 40 mph
    Rate on interstate = 60 mph
    d = rate × time
    d = rt
    The problem contains two-parts-trip that is in town and on interstate highway. That is

In town distance   Interstate distance

Let d be the distance traveled on interstate and 224 – d will be the distance traveled in town. Let t be time taken to travel interstate and 4 – t will be the time taken to travel in town.

224-d   **d**

Total distance= 224miles

Let us make a table

	Distance	rate	time
In town	224 – d	40	4 – t
Interstate	d	60	t

Thus,

In town distance = rt

$224 - d = 40 (4 - t)$

$224 - d = 160 - 40t$

$224 - 160 = -40t + d$

$64 = -40t + d$

$d = 64 + 40t$

Interstate distance = rt

$d = 60t$

We have two equations: d = 60t and d = 64 + 40t. So, we can equate the two equations:

$60t = 64 + 40t$

$60t - 40t = 64$

$20t = 64$

$t = \dfrac{64}{20}$

$t = 3.2$ hrs.

a) Helen traveled on interstate for 3.2 hours.

b) Helen traveled in town for 4 hrs. – 3.2 hrs. = 0.8 hrs.

      i.e., 48 minutes

c) Helen traveled interstate highway

$d = 60t$

$ = 60 (3.2)$

$d = 192$ miles

d) Helen traveled in town

$d = 40 (4 - t)$

$d = 40 (4 - 3.2)$

$d = 40 (0.8)$

$d = 32$ miles

8) Two buses leave Metro station at the same time and travel in opposite directions. One bus travels with an average speed of 45 mph and the other bus travels with an average speed of 55 mph. When will they be 624 miles apart?

Solution

Distance = rate × time

$d = rt$

1st bus: $d = 45t$

2nd bus: $d = 55t$

Distance = $45t + 55t$

$624 = 100t$

$t = \dfrac{624}{100}$

$t = 6.24$ hrs.

Thus, the two buses apart 624 miles in 6.24 hrs.

9) If a woman travels with a speed of $\dfrac{3}{5}$ times of her original speed and she reached her school 20 minutes late to the fixed time, then what is the time taken with her original speed?

Solution

Let the original speed v_1 be x, then v_2 which is $\dfrac{3}{5}$ times of her original speed will be $\dfrac{3}{5}$x.

Let the original time taken be t, then the time t_2 will be (t + 20)

Distance = rate × time

$d = rt$

$d_1 = v_1 t$

$d_1 = xt$

$d_2 = v_2 t$

$d_2 = \dfrac{3}{5}x\,(t + 20)$

$d_1 = d_2$

$xt = \dfrac{3}{5}x\,(t + 20)$

$t = \dfrac{3}{5}(t + 20)$ ....x is cancelled

$5t = 3(t + 20)$ ....Cross Multiplication

$5t = 3t + 60$

$5t - 3t = 60$

$2t = 60$

$t = \dfrac{60}{2}$

$t = 30$ minutes

Thus, the time taken with her original speed is 30 minutes.

10) A man saves 8 minutes by increasing his speed by 25%. What is the time taken to cover the distance at his usual speed?

Solution: Let a man's usual speed be $v_1 = v$

His new increased speed $v_2 = v + 0.25v$; $v_2 = 1.25v$

(i.e, $25\%v = 0.25v$)

Time for the first distance $= t$

Time for the second distance $= t - 8$

$d = vt$ $d_2 = 1.25v \, (t - 8)$

$d_1 = vt$

$d_1 = d_2$

$vt = 1.25v \, (t - 8)$

$t = 1.25 \, (t - 8)$ ... v is cancelled.

$t = 1.25t - 10$

$t - 1.25t = -10$

$-0.25t = -10$

$t = \dfrac{-10}{-0.25}$

$t = 40$ minutes

Thus, the time taken to cover the distance at his usual speed is 40 minutes.

11) A car travels along a straight road 120 km east then 75 km west. Find the distance and displacement of the car.

Solutions: To find the distance traveled, the direction of the path does not matter; thus,

Distance $= 120$ km $+ 75$ km $= 195$ km

∴ The distance traveled by the car is 195 km.

To find the displacement, the direction of the travel is important. (Since displacement is a vector quantity.)

Note:- In physics, vector is a quantity that has both magnitude and direction.

Examples, Displacement, Velocity, Acceleration etc.

Therefore, Displacement $= 120$ km E $+ 75$ km W

                     $= 120$ km E $- 75$ km E

      Displacement $= 45$ East

12) An athlete runs around a rectangular track with length 400 m and width 500 m. After he runs around the rectangular track three times, an athlete backs to the starting point. What is the distance and displacement of an athlete?

Solution

- Distance $=$ Perimeter $= 2L + 2W$

              $= 2(400) + 2(500)$

              $= 800 + 1000$

              $= 1800$ m

1800 m is a single trip. An athlete runs three times this trip. So, the total distance he runs equal to $(1800 \times 3) = 5400$ m.

- Since an athlete starting point and ending point are the same the displacement is zero.

Exercise 15.1-15.4

Answer each of the following questions

1) Find two consecutive numbers such that their sum is 43.
2) David sold three times as much oranges in the morning than in the afternoon. If he sold 320 kgs of oranges that day. How many kilograms did he sell in the morning and how many in the afternoon?
3) The sum of two numbers is 132, and their difference is 56, what are the two numbers?
4) One number is 8 more than another. The sum of three times the smaller plus twice the larger is 66. What are the two numbers?
5) The sum of two consecutive odd numbers is 184. What are the two numbers?
6) Six years from now, Alex will be 56 years old. In 12 years, the sum of the ages of Alex and David will be 132. What is the present age of David?
7) Sofia's father is four times as old as Sofia. In 6 years, Sofia will be one-third as old as her father. What is the present ages of Sofia and her father?
8) An isosceles triangle has a perimeter of 96. If two of its equal sides are 6 more than the third side. What are the lengths of each side?
9) Find the length and width of a rectangle such that its perimeter is 50 m and its area is 144 sqm.
10) The sides of a triangle are in the ratio of 2:4:5 and if the perimeter is 44. What is the length of each side of the triangle?
11) How long will it take a car traveling 84 km/hr to go 336 kms?
12) Two trains leave train stations at the same time. One traveling east at an average speed of 50 km/hr. while the other train is traveling west with an average speed of 60 km/hr. In how many hours will they be 990 kms apart?
13) Find the area of a square whose perimeter is 56 meters.
14) A car travels along a straight road 200 km east then 75 km west. Find the distance and displacement of the car.
15) An athlete runs around a square track field with side 400 meters. After running twice around a square track, he returned to his starting point. What is the distance and the displacement of an athlete?
16) Emily and Luna who live 40 miles apart start at 2 pm to walk towards each other at a rate of 4 mph and 6 mph respectively.
    a) After how many hours will they meet each other?
    a) At what time do they meet?
17) Find the length of each side of a square whose perimeter is 28 meters.
18) Find the length of each side of an equilateral triangle whose perimeter is 54 meters.

ANSWER TO EXERCISES

CHAPTER-1

Exercise 1.1-1.3.4

(i) a) Variable

   b) Equation

   c) Quantity

   d) Constant

   e) Numerical Expression.

   f) Algebraic Expression.

   g) Operations.

   h) Solution Set (true set)

   i) Open Sentence

   j) Closed Sentence.

(ii) a) Six more than a number.

   b) Four less than a number.

   c) - Eight times a number.

     - The product of eight and a number.

     - A number multiplied by eight.

   d) - Three times a number divided by two.

     - The quotient of three times a number n and 2.

   e) Twelve divided by a number.

(iii) a) $m + 6$

   b) $5x$

   c) $y - 10$

   d) $\dfrac{4m}{13}$

   e) $6y - 4$

   f) $7x + 3$

   g) $\dfrac{10}{8 - 5x}$

   h) $(m + 4) - 2$

   i) $6n - 5$

(iv) 1) C
   2) A
   3) E
   4) C
   5) E
   6) C
   7) A
   8) B
   9) E
   10 E

Exercise 1.4-1.4.2

1) 20
2) 31
3) 124
4) 6
5) 9
6) 8
7) 43
8) 2
9) 375
10) 12

Exercise 1.5-1.7

1) C
2) A
3) C
4) B
5) B
6) D
7) B
8) C
9) C
10) C

Exercise 1.8

1) B
2) D
3) C
4) A
5) C
6) C

7) D
8) A
9) C
10) A

Exercise 1.9-1.9.4.1

1) B
2) A
3) B
4) C
5) A
6) C
7) B
8) B
9) C
10) A

CHAPTER-2

Exercise 2.1-2.3

1) E
2) C
3) B
4) C
5) B

Exercise 2.4-2.5

1) C
2) B
3) B
4) A
5) C

Exercise 2.6

1) $x = \dfrac{1}{3}$

2) $x = \dfrac{21}{8}$

3) $m = 1$

4) $x = \dfrac{6}{13}$

5) $x = \dfrac{14}{17}$

6) $x = \dfrac{13}{14}$

7) $x = 2$

8) $x = \dfrac{37}{28}$

9) $x = -$

10) $x = \dfrac{3}{8}$

Exercise 2.7

1) $x = \dfrac{17}{9}$

2) $x = \dfrac{45}{11}$

3) $x = \dfrac{7}{12}$

4) $x = \dfrac{80}{21}$

5) $x = 22$

6) $x = \dfrac{21}{31}$

7) $x = \dfrac{4}{5}$

8) $x = -5$

9) $x = \dfrac{12}{7}$

10) $x = 10$

Exercise 2.8

1) a) Quadrant I
   b) Quadrant II
   c) Quadrant III
   d) Quadrant IV
   e) Lies on the y-axis.

f) Lies on the x-axis.

g) Lies at the origin.

2) a) (a, –4)

b) (8, 0)

c) (–5, 3)

d) (0, 6)

e) (0, –3)

f) (–3, 2)

g) (–8, –3)

h) (0, 0)

3) A)

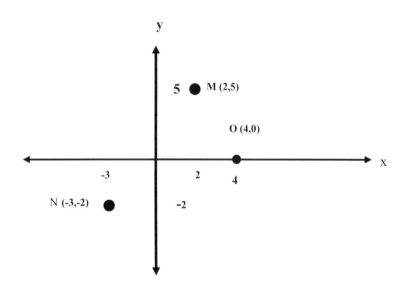

B)

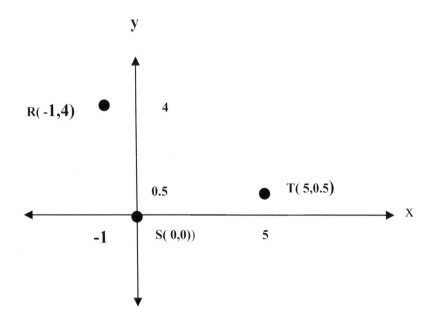

C)

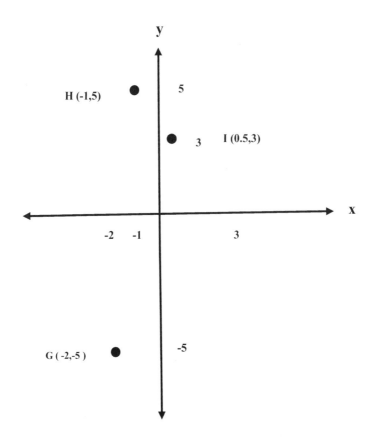

Exercise 2.9-2.10

1) a) 1

   b) 5

   c) $\dfrac{1}{4}$

   d) $\dfrac{-2}{3}$

   e) No slope (undefined)

   f) 0

2) a) $\dfrac{3-b}{4-a}$

   b) $\dfrac{7-2b}{3-a}$

   c) $\dfrac{2y-5}{x+5}$

450

Exercise 2.11

a) slope = –2
   y-intercept = (0, 3)
   x-intercept = $(\frac{3}{2}, 0)$

b) slope = $\frac{1}{3}$

   y-intercept = (0, –8)
   x-intercept = (24, 0)

c) slope = 2
   y-intercept = (0, 0)
   x-intercept = (0, 0)

d) slope = 3
   y-intercept = (0, 7)
   x-intercept = $(\frac{-7}{3}, 0)$

e) slope = $\frac{2}{3}$

   y-intercept = (0, 0)
   x-intercept = (0, 0)

f) slope = $\frac{-5}{7}$
   y-intercept = $(0, \frac{8}{7})$
   x-intercept = $(\frac{8}{5}, 0)$

g) slope = 6
   y-intercept = (0, 12)
   x-intercept = (–2, 0)

Exercise 2.12-2.12.2

1) a) $y = \frac{-1}{4}x + \frac{11}{4}$

   b) $y = \frac{-1}{7}x + \frac{19}{7}$

   c) $y = -2x + 2$

   d) $y = \frac{-7}{9}x + \frac{8}{3}$

e) $y = \dfrac{3}{2}x - 3$

f) $y = 4x - 3$

g) $y = x$

2) a) $y = 4x + 10$

b) $y = 3x + 3$

c) $y = x - 6$

d) $y = 2x - 16$

e) $y = -4x - 23$

Exercise 2.13

1) a) $y = -2x + 1$

x	−2	−1	0	1	2
y	5	3	1	−1	−3

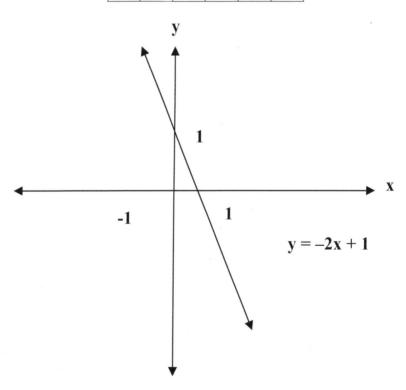

b) $y = \dfrac{1}{2}x$

x	–2	–1	0	1	2	3
y	–1	–0.5	0	0.5	1	1.5

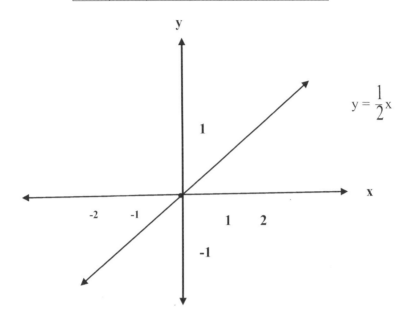

$y = \dfrac{1}{2}x$

c) $x + y = 1$
     $y = -x + 1$

x	–2	–1	0	1	2	3
y	3	2	1	0	–1	–2

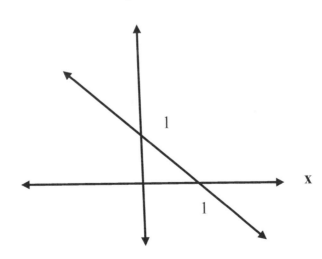

d)   y = 2x

x	−2	−1	0	1	2
y	−4	−2	0	2	4

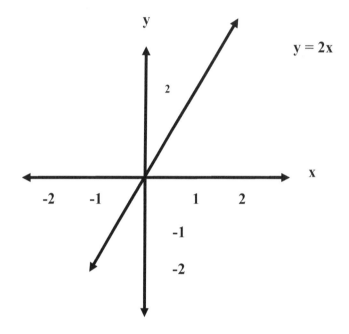

2)   x = 2
3)   y = 2

Exercise 2.14-2.15.2

a)   Perpendicular
b)   Parallel
c)   Perpendicular
d)   Parallel
e)   Neither parallel nor perpendicular.
f)   Perpendicular
g)   Neither parallel nor perpendicular.

CHAPTER -3

Exercise 3.1-3.4

1)   A
2)   A
3)   D
4)   B
5)   B

6) D
7) B
8) A
9) C
10) C

CHAPTER -4

Exercise 4-1-4.2

A) 1) $\{x: x \geq 2\}$
   2) $\{x: x < 4\}$
   3) $\{x: x \leq -6\}$
   4) $\left\{x : x \geq \dfrac{3}{4}\right\}$
   5) $\left\{x : x \leq -\dfrac{1}{2}\right\}$

B) 1) $(-\infty, 3]$
   2) $(-\infty, 2]$
   3) $\left(\dfrac{9}{4}, \infty\right)$
   4) $\left(\dfrac{-8}{3}, \infty\right)$
   5) $(-\infty, -9]$

C) 1)

2)

3)

4)

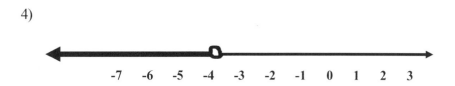

5)

6)

7)

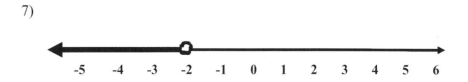

8)

Exercise 4.3-4.4

1)

S.S = {x: −1 < x < 4}

2)

S.S = {x: x ∈ R}

3)

S.S = {x: −1 ≤ x ≤ 3}

4)

S.S = {x: x ≤ 0 ∪ x ≥ 1}

5)

S.S = {x: x ≤ −4 ∪ x > −1}

6)

S.S = {x: 1 < x ≤ 2}

7)

S.S = {   }

8)

S.S = {x: −4 < x < 3}

9)

S.S = {x: x ∈ R}

10)

S.S = {x: −2 < x < 3}

Exercise 4.5

1) {1, 7}
2) {−4, 7}
3) { }
4) { }
5) $\left\{\dfrac{-2}{3}\right\}$

6) $\left\{\dfrac{-13}{6}, \dfrac{5}{2}\right\}$

7) $\left\{\dfrac{1}{4}, \dfrac{3}{4}\right\}$

8) $\left\{\dfrac{1}{2}\right\}$

9) {3, 17}
10) { }

Exercise 4.6

1)

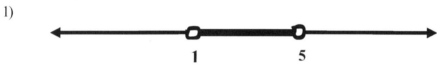

S.S = {x: 1 < x < 5}

2)

S.S $= \left(-\infty, \dfrac{1}{2}\right) \cup \left(\dfrac{7}{2}, \infty\right)$

3)

S.S = {x: x ≤ − 4 or x ≥ 4}

4)

S.S = {x: x ∈ R}

    = (−∞, ∞)

5)

S.S = {x: –3 ≤ x ≤ 5}

6)

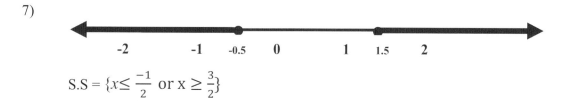

S.S = {x: –7 < x < –1}

7)

S.S = $\{x \le \frac{-1}{2}$ or $x \ge \frac{3}{2}\}$

8)

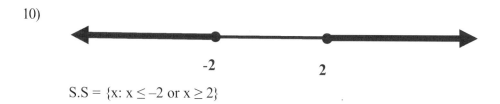

S.S = {x: –6 ≤ x ≤ 6}

9) Since |2x| is always positive and –3 is negative. |2x| is always greater than –3. So the inequality cannot be true. Thus, no solution.
    S.S = {   }

10)

S.S = {x: x ≤ –2 or x ≥ 2}

Exercise 4.7–4.8

1)

x	0	1	2	3	4
y	2	1	0	1	2

y

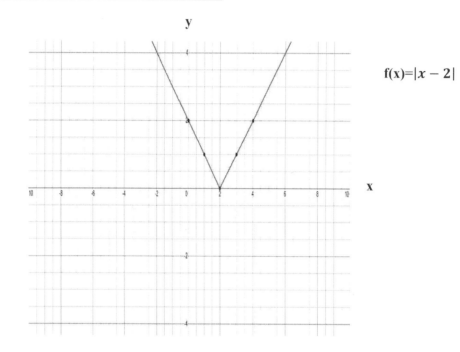

$f(x)=|x-2|$

2)

x	−2	−1	0	1	2
y	−4	−2	0	−2	−4

y

$f(x)=-|2x|$

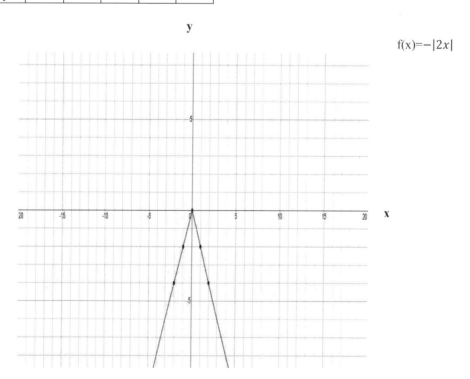

460

3)

x	–2	–1	0	1	2	3	4
y	4	3	2	1	2	3	4

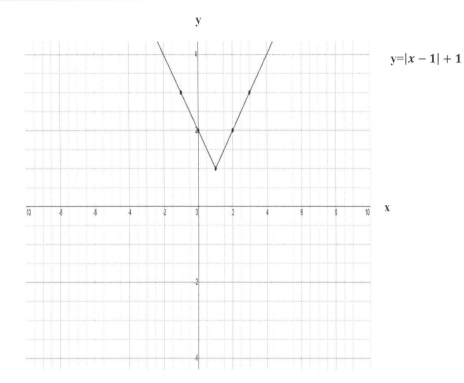

$y=|x-1|+1$

4)

x	–2	–1	0	1	$\dfrac{3}{2}$	2	3	4	5
y	7	5	3	1	0	1	3	5	7

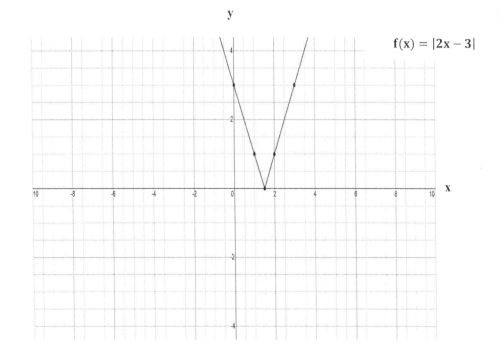

$f(x) = |2x - 3|$

5)

x	−4	−3	−2	−1	0	1	2	3
y	−2	−1	0	−1	−2	−3	−4	−5

y

y= − |x + 2|

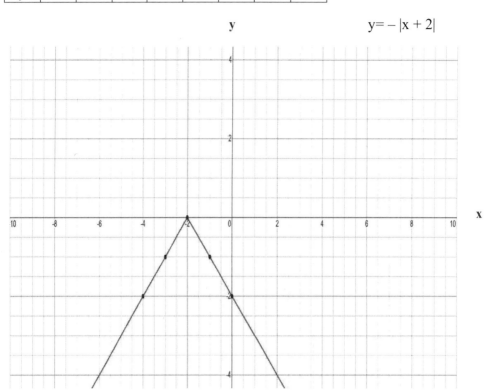

Exercise 4.9-4.9.2

A) 1)

y

$x + y < 2$

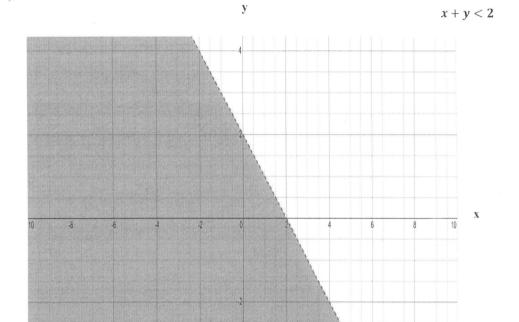

2)

y

$x + y > -3$

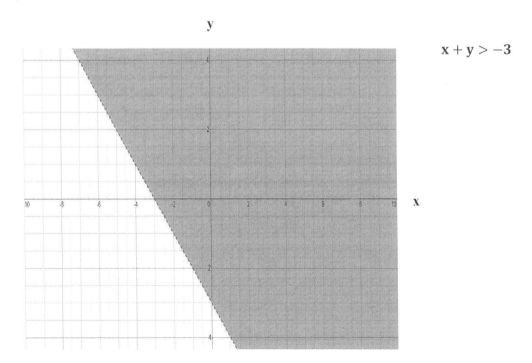

3)

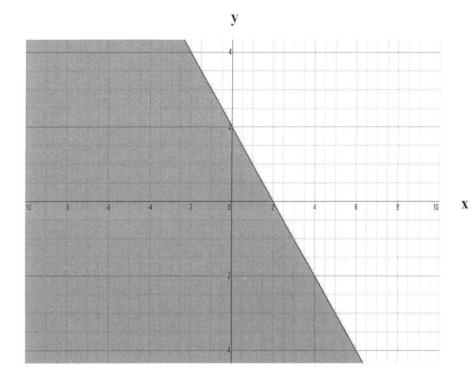

$$y \leq -x + 2$$

4)

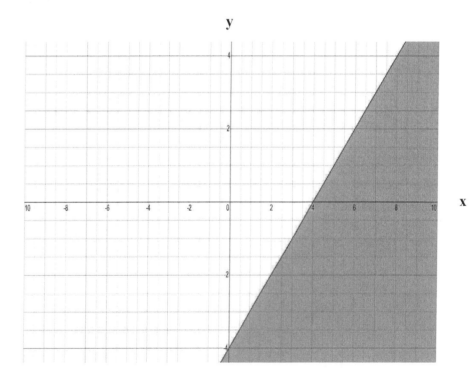

$$x - y \geq 4$$

5)

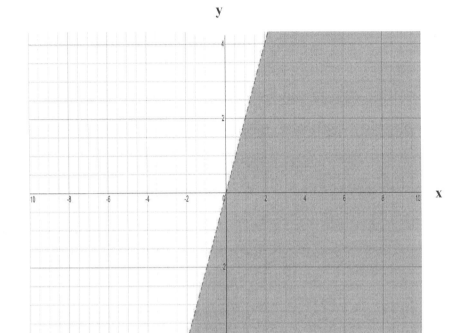

$$y < 2x$$

6)

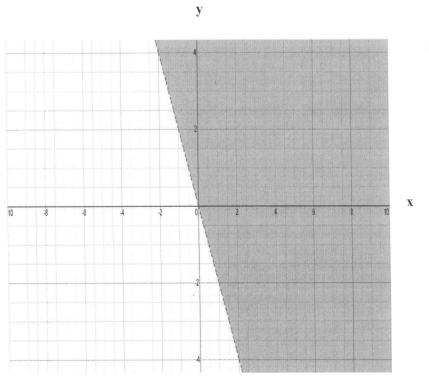

$$y > -2x$$

7)

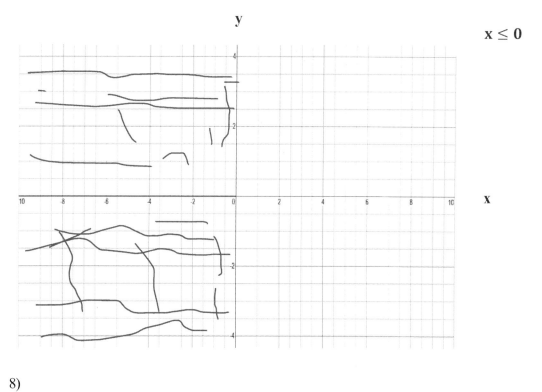

x ≤ 0

8)

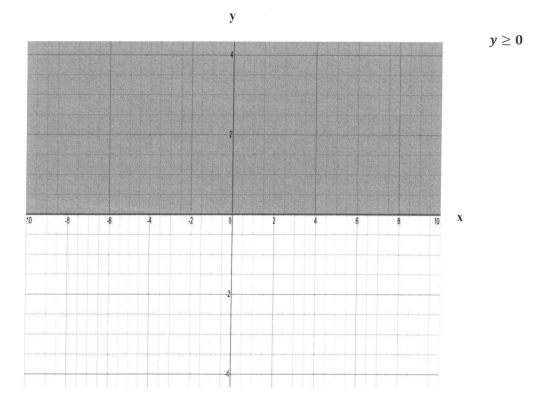

y ≥ 0

B) 1)

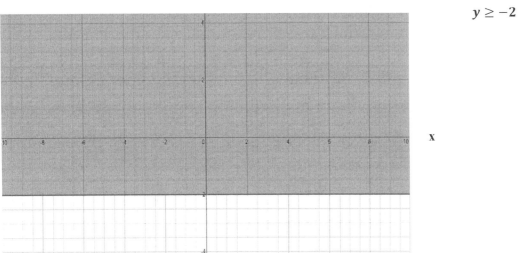

$y \geq -2$

2)

$x > 4$

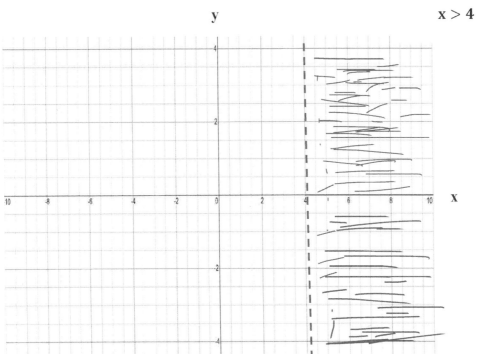

3)

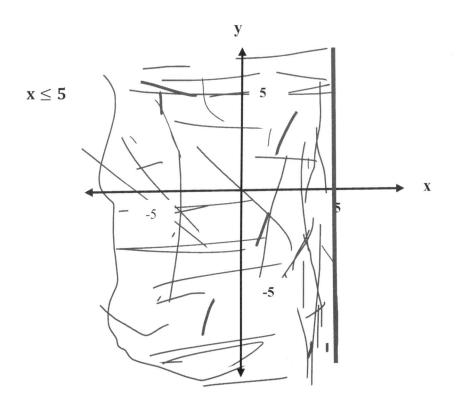

$x \le 5$

4)

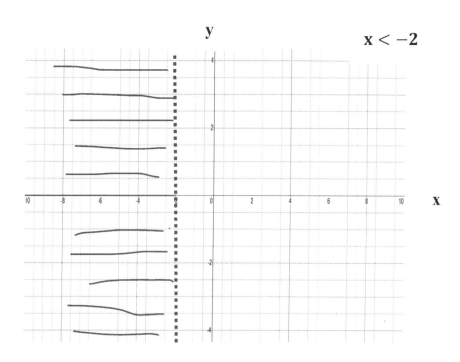

$x < -2$

5)

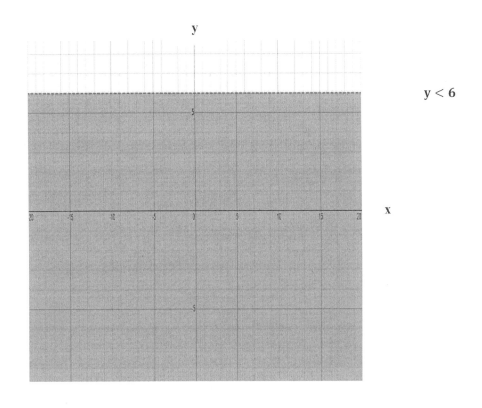

$y < 6$

6)

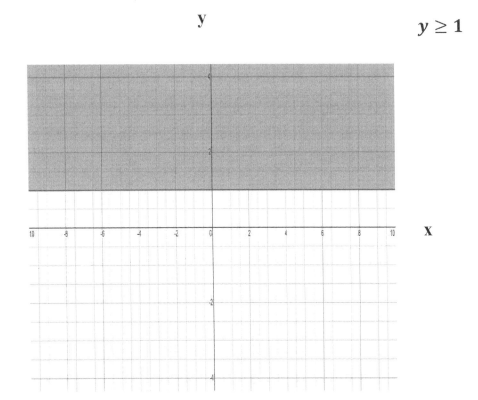

$y \geq 1$

7)

$x > 2$

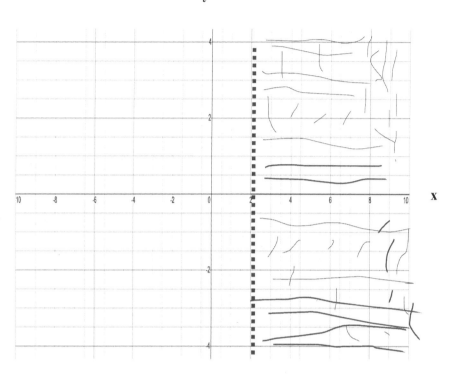

8)

$y < -4$

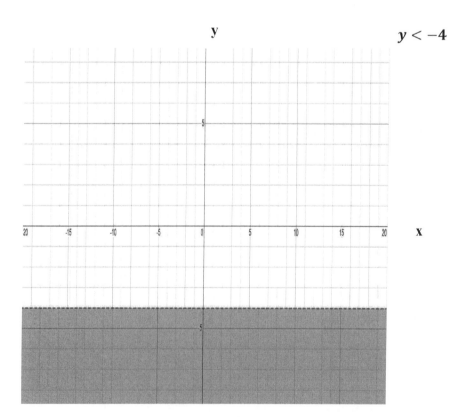

470

CHAPTER-5

Exercise 5.1-5.2.1

1) $(-1, -1)$

2) $\left(\dfrac{5}{7}, \dfrac{-4}{7} \right)$

3) $(2, 4)$

4) $\left(-5, \dfrac{-19}{2} \right)$

5) $(-2, -7)$

Exercise 5.2.2-5.3

a)

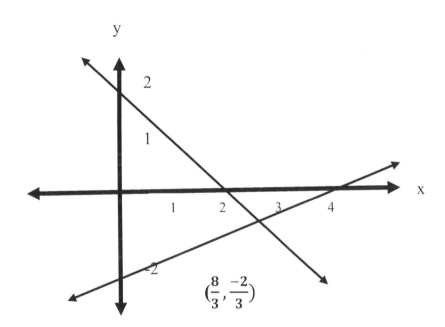

$$x = \dfrac{8}{3},\ y = \dfrac{-2}{3}$$

b)

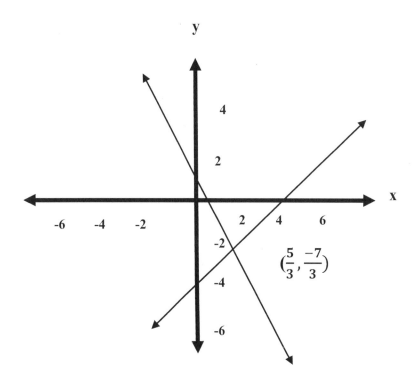

$$x = \frac{5}{3}, \; y = \frac{-7}{3}$$

c)

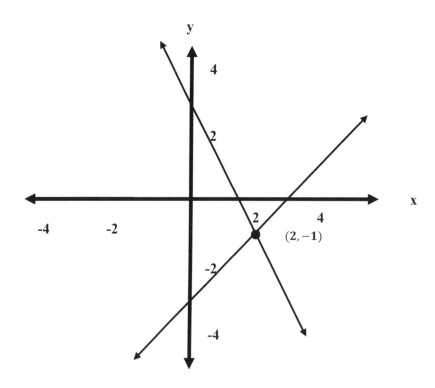

$$x = 2, \; y = -1$$

d)

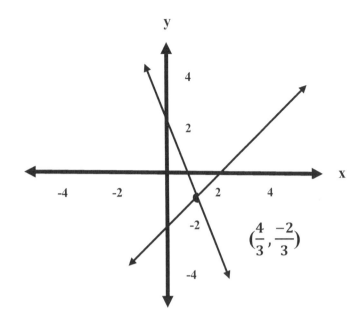

$$x = \frac{4}{3}, y = \frac{-2}{3}$$

e)

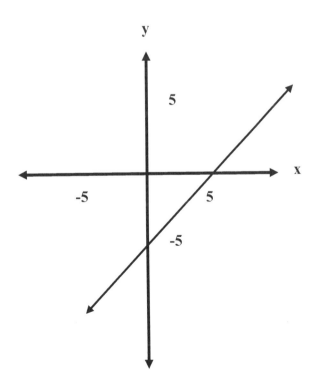

Infinitely many solutions, thus, the two lines are coincident (overlapping). They have the same x and y intercepts.

CHAPTER-6

Exercise 6.2.1

A) 1) $m^{2+5} = m^7$

2) $x^{5+6} = x^{11}$

3) $6^{4+11} = 6^{15}$

4) $\left(\dfrac{1}{2}\right)^{9+4} = \left(\dfrac{1}{2}\right)^{13}$

5) $5^{4+13} = 5^{17}$

6) $y^{16+21} = y^{37}$

7) $8^{4+13} = 8^{17}$

8) $9^{7+6+40} = 9^{53}$

9) $m^{6+4+8} = m^{18}$

10) $\left(\dfrac{1}{3}\right)^{4+6+2} = \left(\dfrac{1}{3}\right)^{12}$

11) $n^{2014+2001} = n^{4015}$

B) 1) $3^{4\times2} = 3^8$

2) $8^{2\times3} = 8^6$

3) $6^{3\times4} = 6^{12}$

4) $a^{5\times4} = a^{20}$

5) $\left(\dfrac{1}{2}\right)^{3\times6} = \left(\dfrac{1}{2}\right)^{18}$

6) $m^{6\times9} = m^{54}$

7) $x^{5\times4} = x^{20}$

8) $\left(\dfrac{1}{3}\right)^{4\times10} = \left(\dfrac{1}{3}\right)^{40}$

9) $6^{\frac{1}{2}(4)} = 6^2$

10) $y^{\frac{1}{8}(8)} = y$

11) $(x)^{\frac{3}{2}(6)} = x^9$

12) $(y)^{\frac{3}{5}\times25} = y^{15}$

C) 1) $(-3)^4 x^4 y^4 = 81 x^4 y^4$

2) $2^3 a^3 b^6 = 8 a^3 b^6$

3) $5^2 m^4 n^6 r^8 = 25 m^4 n^6 r^8$

4) $(-4)^4 x^{20} y^{12} = 256x^{20} y^{12}$

5) $(\frac{1}{3})^5 a^{10} b^{30} = (\frac{1}{243})(a^{10} b^{30})$

6) $2^7 m^{21} n^{28} = 128 m^{21} n^{28}$

7) $x^{\frac{1}{2}} x^{\frac{4}{2}} y^{\frac{4}{2}} z^{\frac{2}{2}} = x^{\frac{1}{2}} x^2 y^2 z$

8) $(\frac{1}{4})^3 x^6 y^3 = \frac{1}{64}(x^6 y^3)$

9) $x^4 y^4 z^4$

10) $a^{15} b^{10} c^{20}$

11) $a^6 b^6 c^6$

Exercise 6.2.2

1) a) $x = 8$
    b) $y = (27)^3$
    c) $x = 2$
    d) $x = 5$
    e) $x = 2$
    f) $x = \dfrac{7}{3}$

    g) $x = 4$
    h) $x = 2$ or -2
    i) $x = -2$
    j) $x = \dfrac{1}{2}$

    k) $x = -8$
    l) $x = 1, y = 0$
    m) $x = \dfrac{12}{13}$

    n) $x = \dfrac{3}{2}$

    o) $x = \dfrac{3}{2}$

    p) $x = \dfrac{4}{3}$

    q) $x = 2$

r) {10}

5 is Extraneous solution

s) {24}

2) a) $\dfrac{256}{81}$

b) $\dfrac{1}{x^3 y^3}$

c) $\left(\dfrac{ab}{c^2}\right)^4$

d) $\dfrac{729}{8}$

e) $\dfrac{1}{n^5 x^{15}}$

f) 27

g) $\dfrac{a^5}{c^5}$

h) $\dfrac{1}{(24)^3 x^6 y^3}$

i) $\dfrac{y^9}{x^6 z^9}$

j) $2x^2 y^2$

k) $12ab|ab|$

l) $\dfrac{1}{xy}$

m) $x^8 y^5 z^6$

n) $\dfrac{z^2}{xy}$

3) a) $\dfrac{x^4 y^2}{z^2}$

b) $\dfrac{1}{4ab}$

c) $m^5 n^7$

d) -1

e) $\dfrac{y^4}{(27)^2 x^2}$

476

f) $\dfrac{3}{4}$

g) $\dfrac{3}{16}$

h) $\dfrac{1}{8}$

i) $8x^3$

4) a) $\dfrac{1}{4}$

b) r^3s

c) $125x^5y^2$

d) $\dfrac{-27}{x^2}$

e) $\dfrac{4x^4}{9y^4}$

5) a) $432x^3y^3$

b) $\dfrac{a^6}{27b^{12}}$

c) 1

d) $\dfrac{n^8}{m^6}$

e) $\dfrac{x^4}{y^8}$

f) $\dfrac{1}{x^3y^3}$

g) $\dfrac{m^2}{n^{8/3}}$

h) $\dfrac{576}{73}$

i) -1

j) $\dfrac{1}{6561}$

Exercise 6.3-6.4

a)

$f(x) = 3^x$	x	−2	−1	0	1	2
	f(x)	$\frac{1}{9}$	$\frac{1}{3}$	1	3	9

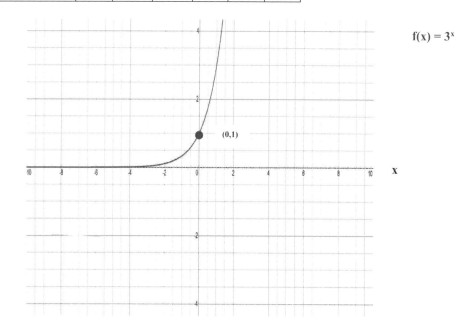

$f(x) = 3^x$

b)

$g(x) = \left(\dfrac{1}{3}\right)^x$	x	−2	−1	0	1	2
	g(x)	9	3	1	$\frac{1}{3}$	$\frac{1}{9}$

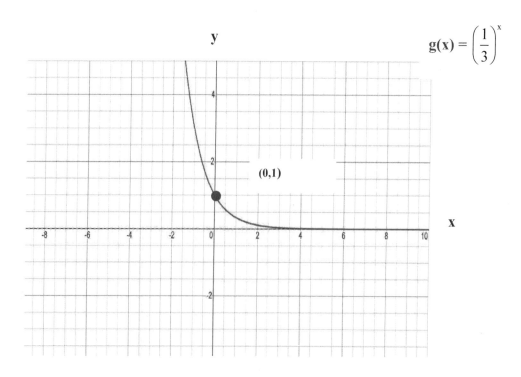

$g(x) = \left(\dfrac{1}{3}\right)^x$

c)

$f(x) = 3^x$	x	−2	−1	0	1	2
	f(x)	$\frac{1}{9}$	$\frac{1}{3}$	1	3	9

$g(x) = \left(\dfrac{1}{3}\right)^x$	x	−2	−1	0	1	2
	g(x)	9	3	1	$\frac{1}{3}$	$\frac{1}{9}$

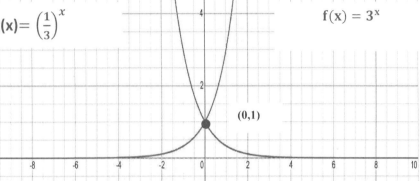

2) a) Domain of $f(x) = (3)^x$
{x: x ∈ |R}

Domain of $g(x) = \left(\dfrac{1}{3}\right)^x$

{x: x ∈ |R}

b) Range of $f(x) = 3^x$
- Set of all positive real numbers.

Range of $g(x) = \left(\dfrac{1}{3}\right)^x$

- Set of all positive real numbers.

c) y - intercept of $f(x) = 3^x$ is 1.
i.e, (0, 1)

y - intercept of $g(x) = \left(\dfrac{1}{3}\right)^x$ is 1.

i.e, (0, 1)

d) f(x) = 3ˣ is an increasing function.

$g(x) = \left(\dfrac{1}{3}\right)^x$ is a decreasing function.

e) Both f(x) = 3ˣ and $g(x) = \left(\dfrac{1}{3}\right)^x$ are one to one function.

f) The x-axis is an asymptote for both f(x) = 3ˣ and $g(x) = \left(\dfrac{1}{3}\right)^x$

Exercise 6.5-6.6

1) a) 4.8×10^{-6}
    b) -3.054×10^{-5}
    c) 8.65403×10^{-1}
    d) 2.05436×10^{1}
    e) -5.63408×10^{-1}
    f) 4.32×10^{-2}
    g) -5.86×10^{0}
    h) 4.6783020×10^{2}
    i) -4.06321×10^{1}

2) a) 6.042×10^{6}
    b) 5.4321×10^{1}
    c) 4.00006×10^{5}
    d) 1×10^{6}
    e) 4.7×10^{7}

3) a)4320000
    b)–0.00043
    c)–0.000005
    d)0.0060083
    e)82.400000
    f)0.0007
    g)0.000565
    h)0.00000602
    i)–0.00000003
    j)–0.0056
    k)603210
    l)9432140

CHAPTER-7

Exercise 7.1-7.17

1) a) {a, e, i, o, u}
   b) {2, 4, 6, 8, 10, 12, 14, 16, 18}
   c) {w, o, r, l, d}
   d) {5, 10, 15, 20, ...}
   e) {0, 1, 2, 3, 4, ...}
   f) {1, 2, 3, 4, 5, 6, 7, 8, 9}
   g) {4, 6, 8, 10, 12}
   h) {4, 8, 12, 16, ...}
   i) {..., –2, –1, 0}
   j) {M, a, r, c, h}

2) a) {x|x is the set of even numbers between 1 and 13}
   b) {x|x is a multiple of 7}
   c) {x|x is the name of the first four days of a week}
   d) {x|x is the set of integers}
   e) {x|x is the set of whole number}

3) a) A ∪ B = {–2, –1, 1, 2, 3, 5, a, b, x, y}
   b) A ∩ B = {–1, 1, 5, b}
   c) A/B = {–2, 2, 3, a}
   d) B/A = {x, y}

4) (i) a) A ∩ B = { }
      b) A ∪ B = {x|x is the set of whole number}
      c) A/B = {x|x is an even number}
      d) B/A = {x|x is an odd number}
   (ii) a) A ∪ B = {x: x is an integer}
      b) A ∩ B {0, 1, 2, 3, ...}
      c) A/B = {..., –3, –2, –1}
      d) B/A ={ }
   (iii) a) A ∪ B = {..., –3, –2, –1, 0, 1, 2, 3, ...}
      b) A ∩ B = {0, 1, 2, 3, ...}
      c) A/B = {..., –3, –2, –1}
      d) B/A = { }
   (iv) a) A ∩ B = {a, 1, 2}
      b) A ∪ B {a, x, y, 1, 2, 4}
      c) A/B = {x, y}
      d) B/A ={4}

5) i)

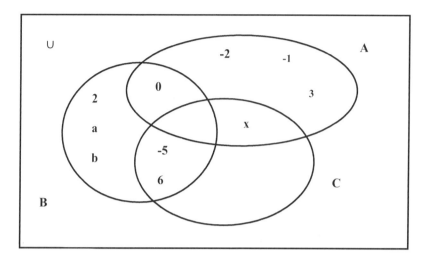

(ii) a)   {0}

b)   {–5, –2, –1, 0, 2, 3, 6, x, a, b}

c)   {x}

d)   {–5, 6}

e)   {–5, –2, –1, 0, 3, 6, x}

f)   {–5, 0, 2, 6, a, b, x}

g)   {   }

h)   {0, x}

i)   {   }

j)   {–5, –2, –1, 2, 3, 6, a, b, x}

k)   {–5, –2, –1, 0, 3}

l)   {–2, –1, 3, x}

m)   {–2, –1, 0, 3}

n)   {0, 2, a, b}

o)   {(–2, 0), (–2, –5), (–2, 6), (–2, 2), (–2, a), (–2, b), (–1, 0), (–1, –5), (–1, 6), (–1, 2), (–1, a), (–1, b), (0, 0), (0, –5), (0, 6), (0, 2), (0, a), (0, b), (3, 0), (3, –5), (3, 6), (3, 2), (3, a), (3, b), (x, 0), (x, –5), (x, 6), (x, 2), (x, a), (x, b)}

p)   {(–2, –5), (–2, 6), (–1, –5), (–1, 6), (0, –5), (0, 6), (3, –5), (3, 6), (x, –5), (x, 6)}

q)   {(–2, –5), (–2, 0), (–2, 2), (–2, 6), (–2, a), (–2, b), (–2, x), (–1, –5), (–1, 0), (–1, 2), (–1, 6), (–1, a), (–1, b), (–1, x), (0, –5), (0, 0), (0, 2), (0, 6), (0, a), (0, b), (0, x), (3, –5), (3, 0), (3, 2), (3, 6), (3, a), (3, b), (3, x), (x, –5), (x, 0), (x, 2), (x, 6), (x, a), (x, b), (x, x)}

r)   {–5, 6, 2, a, b}

s)   {–2, –1, 3, x}

t)   {–2, –1, 0, 3, 2, a, b, x}

(iii) a)   32

b)   64

c)   8

d) 1024

e) 2

f) 2

g) 1024

h) 4

i) 128

j) 4

k) 1

(iv) a) 31

b) 63

c) 7

d) 1

e) 1

f) 3

g) 127

h) 3

v) a) 32

b) 64

c) 8

6) 62 like either of them and 58 like neither of them.

7) a) {c, f, x, y, m, z, n, s, p, t, d, q}

b) {m, z, n, s}

c) {n, s, p, d, q}

d) {s, p, t}

e) {n, s}

f) {s}

g) {s, p}

h) {m, z, n, s, t, p, d, q}

i) {s}

j) {t, p, d, q, c, f, x, y}

k) {m, z, t, c, f, x, y}

l) {m, z, n, d, q}

m) {   }

n) {m, z}

o) {n, d, q}

p) {m, z, n}

q) {m, z, d, p, q}

r) {t, n, d, q}

8) 104

9) d

10) c

11) (i)   {(1, a), (1, b), (1, c), (3, a), (3, b), (3, c), (5, a), (5, b), (5, c)}
    (ii) {(1, x), (1, y), (1, z), (3, x), (3, y), (3, z), (5, x), (5, y), (5, z)}
    (iii) {(1, a), (1, b), (1, c), (3, a), (3, b), (3, c), (5, a), (5, b), (5, c), (1, x), (1, y), (1, z), (3, x), (3, y),
          (3, z), (5, x), (5, y), (5, z)}

12)

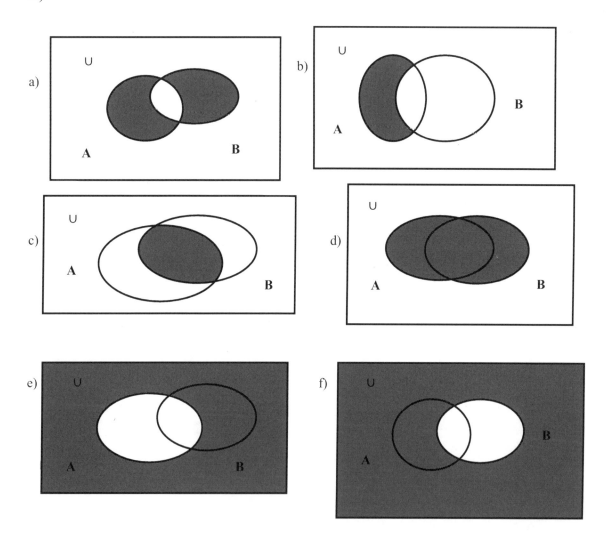

13) 55

14) a)   {b, 5}
    b)   {c, 4, 6, 9}
    c)   {a, 6, 9}

484

d) {a, c, 4, 6, 9}

e) {6, 9}

f) {a, b, c, 4, 5}

CHAPTER-8
Exercise 8.1-8.10

1) a) leading term: $-2x^5$
        leading coefficient: -2
        Degree: 5
        Constant term: -8

    b) leading term: $-4x^6$
        leading coefficient: -4
        Degree: 6
        Constant term: 1

    c) leading term: $3x^4$
        leading coefficient: 3
        Degree: 4
        Constant term: 5

    d) leading term: $4x$
        leading coefficient: 4
        Degree: 1
        Constant term: -1

    e) No leading term.
        No leading coefficient
        Has no degree
        Constant term: 0

    f) leading term: $\sqrt{3}\,x^5$
        leading coefficient: $\sqrt{3}$
        Degree: 5
        Constant term: $\sqrt{8}$

    g) leading term: $\sqrt{5}\,x^2$
        leading coefficient: $\sqrt{5}$
        Degree: 2
        Constant term: 1

    h) leading term: $-8x^{14}$
        leading coefficient: -8
        Degree: 14
        Constant term: 6

    i) leading term: $6x^4$
        leading coefficient: 6
        Degree: 4
        Constant term: 9

2) a) $11x^3 - 2x^2 + 7x + 4$

b) $-9x^3 - 2x^2 - x - 6$

c) $9x^3 + 2x^2 + x + 6$

3) a) $x^2 + x - 7$

b) $x^2 - x + 1$

c) $-x^2 + x - 1$

d) $x^2 + x - 7$

e) $x^3 - 4x^2 - 3x + 12$

4) a) Polynomial

b) Not polynomial

c) Polynomial

d) Polynomial

e) Polynomial

f) Not polynomial

g) Polynomial

h) Not polynomial

i) Polynomial

j) Not polynomial

5) a) $x - 1$

b) $x^2 - 4x + 8 - \dfrac{15}{x+2}$

c) $x^2 + 1 + \dfrac{3x-1}{x^2+1}$

d) $x^2 + 1 - \dfrac{1}{3x+2}$

e) $x + \dfrac{3x+1}{x^2}$ or $x + \dfrac{3}{x} + \dfrac{1}{x^2}$

6) a) 5

b) 3

c) 9

d) 20

e) 27

f) 9

g) 7

h) 17

i) 30

7) 1

8) $\dfrac{-5}{2}$

9) 4

10) $a = \dfrac{-3}{4}$, $b = \dfrac{-1}{4}$

11) 0

12) a) $9x^2 - 4y^2$
    b) $25x^2 - 9y^2$
    c) $9a^2 - 24ab + 16b^2$
    d) $4y^2 - 1$
    e) $x^3 + 6x^2 + 12x + 8$
    f) $x^3 - 6x^2 + 12x - 8$
    g) $9x^4 - 12x^2 + 4$
    h) $64x^3 + 144x^2 + 108x + 27$

13) a) $x = \sqrt{3}$ or $-\sqrt{3}$
    b) $x = 2$ or -2
    c) $y = \sqrt{5}$ or $-\sqrt{5}$
    d) $y = \sqrt{2}$ or $-\sqrt{2}$
    e) $x = 3\sqrt{3}$ or $-3\sqrt{3}$
    f) $x = \dfrac{7\sqrt{2}}{2}$ or $\dfrac{-7\sqrt{2}}{2}$
    g) $x = 3$ or -3

CHAPTER-9

Exercise 9.1-9.5

1) a) $x \neq 1, -1$
   b) All real numbers
   c) $x \neq 3$
   d) All real numbers.
   e) All real numbers
   f) $x \neq -4$
   g) All real numbers
   h) $x \neq 0, 4$
   i) All real numbers
   j) $x \geq 1$
   k) All real numbers

2) a) $x = 4 \pm \sqrt{3}$
   b) $x = \dfrac{-5}{2}$

c) No solution

d) $x = \dfrac{42}{11}$

e) $x = 12$

f) $x = 3$

g) $x = 2$ or -1

h) $x = 1$

3) a) Rational

b) Not rational (Domain is not specified)

c) Rational

d) Not rational

e) Rational

f) Rational

g) Rational

h) Not rational.

4) a) $\dfrac{x + 2}{x}$

b) $x - 5$

c) $\dfrac{3x + 5}{2x + 1}$

d) $\dfrac{9x - 2}{(2x + 1)(x - 1)}$

e) $\dfrac{5x + 2}{3x + 5}$

f) $\dfrac{x - 5}{(x + 2)(x - 2)}$

g) $\dfrac{x + 2}{2}$

h) $\dfrac{2y^3 - 2y}{3(y^2 - 4)}$

i) $\dfrac{2}{x^2}$

5) a) 4

b) $\dfrac{5}{2}$

c) Undefined

d) $\dfrac{-65}{12}$

6) a) $2\sqrt{2}$

   b) 0

   c) $2\sqrt{29}$

   d) 2

7) a) $\dfrac{16x+12}{2x-1}$

   b) $\dfrac{-2x+12}{2x-1}$

   c) $\dfrac{4x+30}{6x-7}$

   d) $\dfrac{-2x+15}{8x-12}$

8) $A = \dfrac{11x-5}{2}$

9) $x = 4 \text{ or } 6$

10) Domain = $\{x: x \neq 0, 4\}$

11) $\dfrac{6x+1}{6x-2}$

12) $\dfrac{1}{x-y}$

13) $\dfrac{r}{r+s}$

14) $\dfrac{a^2b-a^2}{ab^2-b^2}$

CHAPTER-10

Exercise 10.1-10.3

1) a) $2\sqrt{5}$

   b) $3\sqrt{10}$

   c) $2|x|\sqrt{21}$

   d) $|xz|\,y^2\sqrt{xz}$

   e) $3x^2y^3$

   f) $4\,|xy|\sqrt[4]{y^2}$

   g) $x^6y^5\sqrt[3]{2y}$

489

h) $\dfrac{1}{2}\left|x^2y^2z^3\right|$

i) $x^2y^5z^6$

j) $x^2y^3z^5$

k) $|xy|$

l) $|xy^2z^3|$

m) $xyz^2\sqrt[5]{y}$

2) a) $\sqrt[6]{(64)^5} = 2^5 = 32$

   b) $\sqrt{49x^3y^2} = 7|xy|\sqrt{x}$

   c) $\dfrac{1}{\sqrt{x^2y^2z^2}} = \dfrac{1}{|xyz|}$

   d) $\sqrt[3]{x^4y^3z^6} = xyz^2\sqrt[3]{x}$

   e) $\sqrt[4]{a^2b^2c^2}$

   f) $\sqrt[7]{xyz}$

3) a) $m^{3/2}$

   b) $(16a^2b^2c^2)^{1/2}$

   c) $\dfrac{(x^3y^4)^{1/5}}{(a^2b^3)^{1/3}}$

   d) $(32xy^4z^9)^{1/5}$

   e) $\dfrac{(7a^5b^3c^2)^{1/4}}{(27x^3y^4z^5)^{1/5}}$

   f) $(x^3y^2z^6)^{1/3}$

   g) $(2xy)^{1/2}$

   h) $(xyz)^{1/7}$

   i) $(2x^4y^5)^{1/3}$

4) a) -5

   b) $x - 2y$

   c) $2x - y$

   d) $a - b^2$

   e) $b^2 + a + 2b\sqrt{a}$

   f) -1

   g) -23

   h) 26

5) a) $x = \dfrac{5}{2}$

   b) $x = 5$
   c) $x = 4$
   d) $x = 31$

CHAPTER-11

Exercise 11.1-11.4

1) $x = 7$ or 4

2) $x = \dfrac{1}{2}$ or $\dfrac{2}{3}$

3) $x = 5$ or $\dfrac{-1}{2}$

4) $y = 0$ or 1

5) $a = 4$ or $\dfrac{-3}{2}$

6) $x = 8$

7) $x = \dfrac{2}{3}$ or 1

8) $x = \dfrac{7}{2}$ or 6

9) $x = 4$ or 2

10) $x = 7$ or 3

Exercise 11.5

1) $x = 2$ or 1

2) $x = 1$ or $\dfrac{6}{5}$

3) $x = 1$

4) $x = 1$ or 2

5) $x = -5$

6) x = 5 or –4

7) y = –1 or $\dfrac{-5}{2}$

8) x = 9 or –9

9) x = 2

10) No solution set

Exercise 11.6-11.7

A) 1) $x = \dfrac{1+\sqrt{13}}{2}, \dfrac{1-\sqrt{13}}{2}$

2) $x = \dfrac{-2+\sqrt{10}}{3}, \dfrac{-2-\sqrt{10}}{3}$

3) $x = \dfrac{-3+\sqrt{69}}{10}, \dfrac{-3-\sqrt{69}}{10}$

4) x = 2, –4
5) x = 6
6) No solution set
7) No solution set
8) x = 5, –5
9) x = 1, –2
10) x = –2, –3

B) 1) Two solutions
2) One solution
3) No solution
4) Two solutions
5) No solution
6) Two solutions
7) One solution
8) Two solutions
9) Two solutions
10) Two solutions

492

Exercise 11.8

1) $x = 3, -1$

2) $x = 4, -4$

3) $x = 5, -5$

4) $x = 2i, -2i$

5) $x = 5i, -5i$

6) $x = 3\sqrt{2}, -3\sqrt{2}$

7) $x = 4\sqrt{3}, -4\sqrt{3}$

8) $x = 4 + \sqrt{10}, 4 - \sqrt{10}$

9) $x = 12\sqrt{2}, -12\sqrt{2}$

10) $x = 3i\sqrt{5}, -3i\sqrt{5}$

Exercise 11.9-11.9.1

1) a) $x = 3 + 2\sqrt{}, 3 - 2\sqrt{5}$
   b) $x = 5, -1$
   c) $x = 1 + i\sqrt{2}, 1 - i\sqrt{2}$
   d) $x = -3, -5$
   e) $x = \dfrac{-5+\sqrt{17}}{2}, \dfrac{-5-\sqrt{17}}{2}$

   f) $x = \dfrac{-9+i\sqrt{47}}{4}, \dfrac{-9-i\sqrt{47}}{4}$

   g) $x = \dfrac{-1+i\sqrt{26}}{3}, \dfrac{-1-i\sqrt{26}}{3}$

   h) $x = 1 + i, 1 - i$
   i) $x = \dfrac{3+\sqrt{29}}{2}, \dfrac{3-\sqrt{29}}{2}$

   j) $x = \dfrac{1+i\sqrt{7}}{2}, \dfrac{1-i\sqrt{7}}{2}$

2) a) $x = 0, 4$

   b) $x = 2 + 2\sqrt{3}, 2 - 2\sqrt{3}$

   c) $x = \dfrac{1}{8}, \dfrac{-1}{8}$

   d) $x = 0, \sqrt{5}, -\sqrt{5}$

   e) $x = 0, 5, 4$

   f) $a = 3, -3$

Exercise 11.10-11.10.2

1) $k = 3$

2) $k = \dfrac{1}{2}, x = -2$

3) $m = -3, n = 24$

4) $x^2 + x - 6 = 0$

5) $k = \dfrac{41}{32}$

6) $k = -9$

7) $k = 12$

8) $k = 4$

9) $k = 14$

10) $k = -28$ and $k = 26$

Exercise 11.11

1)

x	–3	–2	–1	0	1	2
y	2	0	0	2	6	12

$$y = x^2 + 3x + 2$$

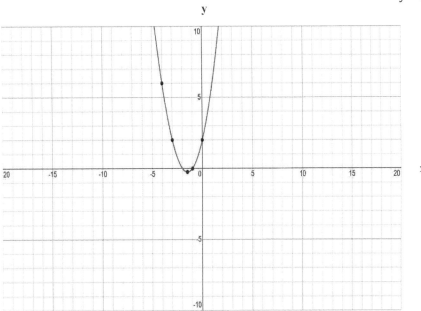

S.S = {–1, –2}

2)

x	–2	–1	0	1	2	3
y	9	3	1	0	1	4

$$y = x^2 - 2x + 1$$

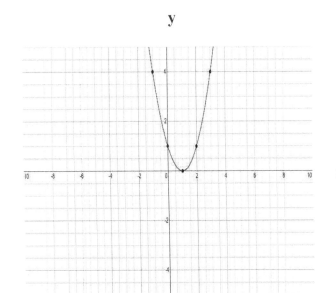

S.S = {1}

3)

x	–2	–1	0	1	2
y	7	4	3	4	7

$$y = x^2 + 3$$

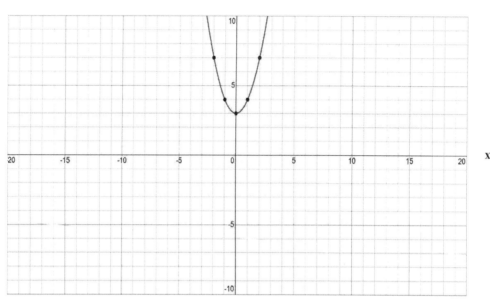

S.S = ∅

4)

x	–3	–2	–1	0	1	2	3	4
y	12	5	0	–3	–4	–3	0	5

$$y = x^2 - 2x - 3$$

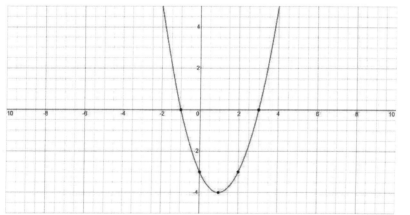

S.S = {–1, 3}

5)

x	−1	0	1	2	3
y	−4	−1	0	−1	−4

$$y = -x^2 + 2x - 1$$

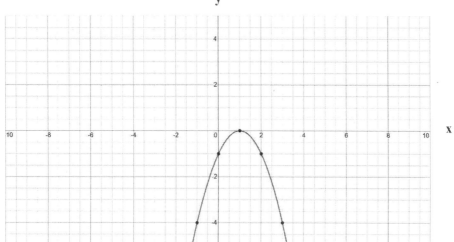

S.S = {1}

Exercise 11.12

1) Vertex: $\left(\dfrac{5}{4}, \dfrac{-17}{8}\right)$

Axis of symmetry: $x = \dfrac{5}{4}$

2) Vertex: $\left(\dfrac{4}{3}, \dfrac{49}{3}\right)$

Axis of symmetry: $x = \dfrac{4}{3}$

3) Vertex: $\left(\dfrac{3}{2}, \dfrac{15}{2}\right)$

Axis of symmetry: $x = \dfrac{3}{2}$

4) Vertex: $(-3, 0)$
Axis of symmetry: $x = -3$

5) Vertex: $(-1, 33)$
Axis of symmetry: $x = -1$

6) Vertex: $(3, -1)$
Axis of symmetry: $x = 3$

Exercise 11.13

1) Maximum value = $\dfrac{4}{3}$

2) Minimum value = 6

3) Maximum value = $\dfrac{-3}{4}$

4) Minimum value = $\dfrac{4}{5}$

5) Minimum value = 0

6) Minimum value = $\dfrac{-105}{16}$

7) Minimum value = 0

8) Maximum value = 10

9) Maximum value = $\dfrac{29}{7}$

10) Minimum value = $\dfrac{26}{7}$

Exercise 11.14

1) a) $t = 4.9$ sec
   b) The object falls about 92.1 m in 3 seconds

2) a) 1,116.25 ft
   b) 3.125 sec

3) a) $h(t) = 20$ ft
   b) $t = 0.3125$ sec

4) a) $x = 150$ meters and the other side is $600 - 2x = 600 - 2(150) = 300$ meters
   b) The maximum enclosed area is (150 m) (300 m) = 45,000 sqm

5) The dimensions of the garden are 20 m and 3 m or 4 m and 15 m.

Exercise 11.15-11.16

1) Set builder notation: {x: x < −3 or x > 3}
   Number line:

Interval notation: **(−∞, −3) ∪ (3, ∞)**

2) Set builder notation: {x: x ≤ −4 or x ≥ −3}
   Number line:

Interval notation: **(−∞, −4] ∪ [−3, ∞)**

3) Set builder notation: {x: 0 < x < 1}
   Number line:

Interval notation: **(0, 1)**

4) Set builder notation: {x: −5 ≤ x ≤ 5}
   Number line:

Interval notation: **[−5, 5]**

5) Set builder notation: {x: x < −7 or x > −4}
   Number line:

Interval notation: **(−∞, −7) ∪ (−4, ∞)**

6) No solution set

7) Set builder notation: {x: x ≤ −7 or x ≥ 7}
   Number line:

   Interval notation: **(−∞, −7] ∪ [7, ∞)**

8) Set builder notation: {x: x ϵ |R}
   Number line:

   Interval notation: **(−∞, ∞)**

9) Set builder notation: {x: 0 ≤ x ≤ 16}
   Number line:

   Interval notation: [0, 16]

10) Set builder notation: {x: x < 4 or x > 4}
    Number line:

    Interval notation: **(−∞, 4) ∪ (4, ∞)**

Exercise 11.17

1)

$$y \leq 2x^2$$

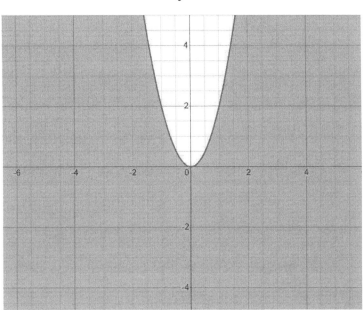

2)

$$y < 2x^2$$

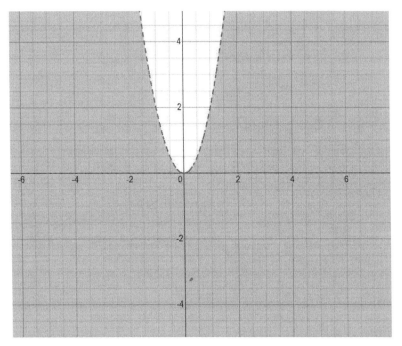

3)

$$y > x^2 - 2x + 1$$

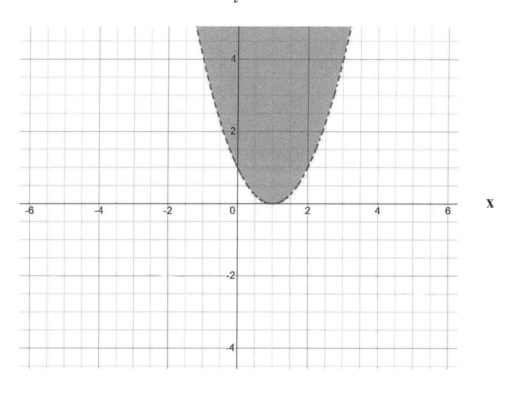

4)

$$y \leq x^2 + 3x + 2$$

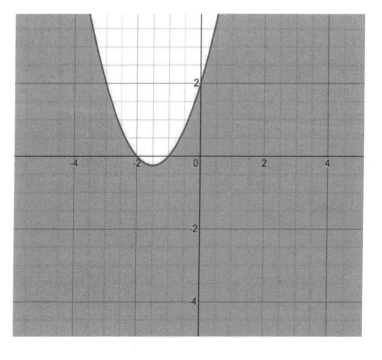

5)

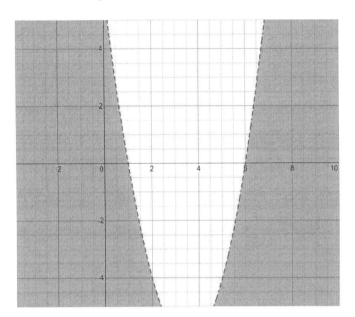

$$y < x^2 - 7x + 6$$

6)

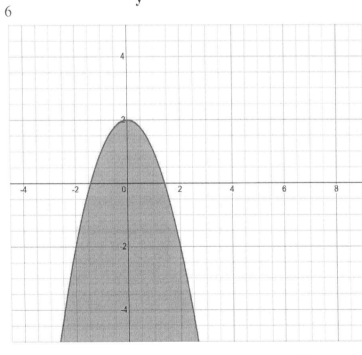

$$y \leq -x^2 + 2$$

CHAPTER-12

Exercise 12.1-12.1.4

1) a) $\dfrac{31}{90}$

   b) $\dfrac{2}{5}$

   c) $\dfrac{1}{4}$

   d) $\dfrac{1}{9}$

   e) $\dfrac{81}{64}$

   f) $\dfrac{1}{2}$

2) a) $x = \dfrac{48}{5}$

   b) $x = \dfrac{8}{3}$

   c) $k = \dfrac{729}{11}$

   d) $x = 36$ or -36

   e) $x = \dfrac{80}{3}$

   f) $x = 2$

3) 54 years

4) a) 200 dollars
   b) 250 dollars

5) a) $\dfrac{1}{2}$

   b) $\dfrac{2}{1}$

   c) $\dfrac{1}{3}$

   d) $\dfrac{2}{3}$

   e) 1

6) The two numbers are -119 and -147

7) The lengths of each piece are 90 cm, 120 cm and 150 cm.

8) The constant of proportionality is $\dfrac{x}{y^2}$.

9) 12 days

10) 9 hours

11) 209 days

12) 90 minutes

13) 30 hours

14) $4.10

15) 9 books for $54.009

16) 7 miles

17) a) $\dfrac{1}{5}$

    b) $\dfrac{6}{25}$

    c) $\dfrac{21}{25}$

    d) $\dfrac{22}{25}$

    e) $\dfrac{9}{20}$

    f) $\dfrac{17}{100}$

    g) $\dfrac{9}{25}$

18) a) 0.17
    b) 0.33
    c) 0.27
    d) 2.34
    e) 0.274
    f) 0.1143

g)   0.01

19) a)   60%
    b)   24%
    c)   125%
    d)   20%
    e)   68%
    f)   50%
    g)   32%
    h)   150%
    i)   18.75%

20) 19.11%

21) 144%

22) 20%

23) 6.6% ≈ 6.7%

24) 60%

25) $4,500

26) $29.52

27) 75%

28) 25%

29) 53.04

30) $200

31) $1000

Exercise 13.1-13.8.2

1)

x	0	1	4	9
y	0	-2	-4	-6

The reflection of the graph of $f(x) = 2\sqrt{x}$ across the x-axis is $f(x) = -2\sqrt{x}$ and its graph is shown below.

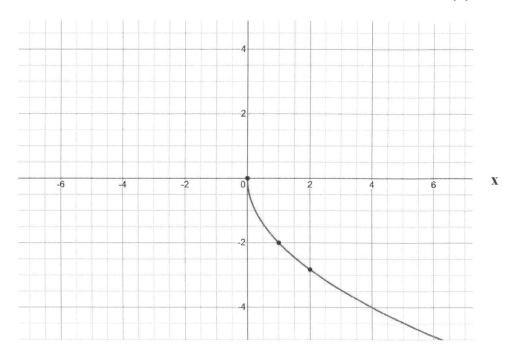

y

$$f(x) = -2\sqrt{x}$$

2) $(-6, 2)$

3) $(4, -1)$

4) $f(x) = -\sqrt{3x}$

5) $f(x) = 4\sqrt{-x}$

6) The reflection of the graph of $f(x) = 2\sqrt{x}$ across the y-axis is $f(x) = 2\sqrt{-x}$ and its graph is shown below.

x	−9	−4	−1	0
y	6	4	2	0

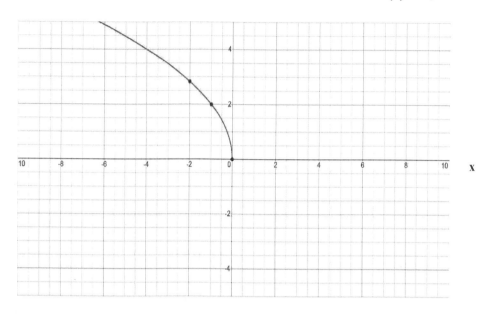

y   **f(x) =2√−x**

7) The reflection of the graph of f(x) = $\sqrt{2x}$ across the x-axis is f(x) = $-\sqrt{2x}$ and its graph is shown below: -

x	0	$\frac{1}{2}$	2	4	8
y	0	–1	–2	$-2\sqrt{2}$	–4

y   f(x) = $-\sqrt{2x}$

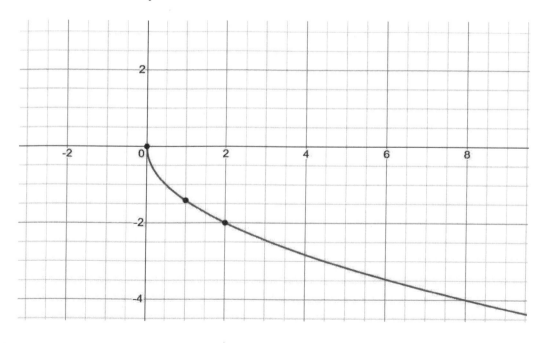

508

8) The reflection of the graph of f(x) = $\sqrt{2x}$ across the y-axis is f(x) = $\sqrt{-2x}$ and its graph is shown below.

x	$\dfrac{-9}{2}$	–2	$\dfrac{-1}{2}$	0
y	3	2	1	0

$f(x) = \sqrt{-2x}$

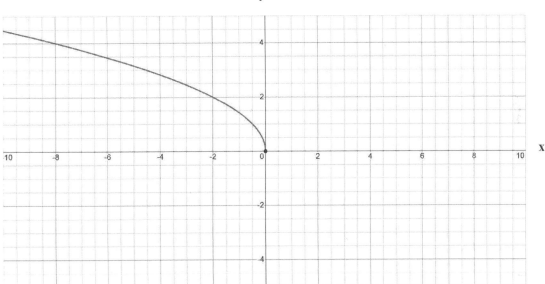

9) (0, 0)

10) (0, 0)

Exercise 13.9-13.9.2

1) a) Domain: $\left\{ x : x \ge \dfrac{-1}{2} \right\}$

  Range: {y: y ≥ 0}
 b) Domain: {x: x ≥ 9}
  Range: {y: y ≥ 0}
 c) Domain: {x: x ≥ –4}
  Range: {y: y ≥ 0}
 d) Domain: {x: x ≤ 6}
  Range: {y: y ≥ 0}
 e) Domain: $\left\{ x : x \le \dfrac{1}{2} \right\}$

  Range: {y: y ≥ 0}

2) a) $(-\infty, 5)$
   b) $(-\infty, -18]$
   c) $(-\infty, 4]$
   d) $(2, 9)$
   e) $(2, \infty)$
   f) $(-2, 0]$
   g) $\left(\dfrac{1}{2}, 2\right)$
   h) $(-1, \mathbf{4})$

3) (i) C   (ii) D   (iii) D   (iv) A   (v) A
   (vi) B   (vii) B   (viii) C   (ix) B   (x) B

CHAPTER-14

Exercise 14.1-14.11.1

1) a) 64
   b) 68
   c) No mode
   d) 66
   e) ≈ 430.857
   f) $\approx \sqrt{430.857}$
   g) 57
   h) 69

2) a) 7
   b) 5
   c) 2
   d) 24
   e) ≈ 43.64
   f) $\approx \sqrt{43.64}$
   g) 45.46%
   h) 18.182%
   i) 36.364%

3) a) 19
   b) 19
   c) 21
   d) 14
   e) ≈ 8.8235
   f) $\sqrt{8.8235}$
   g) $\approx 70.588\%$

h) $\approx 41.18\%$

i) $\approx 23.53\%$

4) 58

5) Mean = 8   Range = 8

    Mode = 11   Variance = **8.3˙**

    Median = 9   Standard D. = $\sqrt{8.3˙}$

6) **87.7˙**

7) a) 1

    b) 0

    c) 0

    d) 11

    e) 14.4

    f) $\sqrt{14.4}$

    g) 40%

    h) 30%

    i) 30%

8)

    a)

v	0	13	15	16	18	19	20
f	1	1	3	4	2	2	3

    b) 16

    c) 16

    d) 16

    e) 21.875

    f) $\sqrt{21.875} \approx 4.677$

9) a) 2

    b) No mode

    c) 6

10) 92

Exercise 14.12-14.20

1) $\dfrac{8}{19}$

2) $\dfrac{1}{36}$

3) a) $\dfrac{1}{6}$

   b) $\dfrac{1}{2}$

   c) $\dfrac{1}{2}$

   d) $\dfrac{5}{6}$

   e) $\dfrac{1}{3}$

   f) $\dfrac{1}{2}$

4) a) The possible outcomes are: HH, HT, TH, TT, thus, the possible outcomes equal to 4.
   b) S = {(HH), (HT), (TH), (TT)}
   c) HH, HT, TH
   d) HT, TH, TT

5) $\dfrac{1}{2}$

6) $\dfrac{5}{6}$

CHAPTER-15

Exercise 15.1

1) 21 and 22

2) David sold 80 oranges in the afternoon and 240 oranges in the morning.

3) 94 and 38

4) 10 and 18

5) 91 and 93

6) 58

7) The present age of Sofia is 12 years and the present age of the father 48 years.

8) The lengths of sides are 34, 34 and 28.

9) Length = 16 m and width = 9 m or length = 9 m and width 16 m.

10) 8, 16 and 20

11) 4 hours

12) 9 hours

13) 196 sqm

14) Distance = 275 km
    Displacement = 125 km East.

15) Distance = 3,200 meters.
    Displacement = 0.

16) a)   4 hours
    b)   At 6 pm

17) 7 meters

18) 18 meters.

the United States
. Taylor Publisher Services